Genesis and Management of Sodic (Alkali) Soils

About the Authors

Dr. S.K. Gupta (Ph.D., FNAE, FAAS, FISAE and FISSRS), (b. 1949) obtained Ph.D. in Civil Engineering and superannuated from the position of Project Coordinator, AICRP on Management of Salt Affected Soils and Use of Saline Water at Central Soil Salinity Research Institute, Karnal. He has been associated in developing irrigation water management technologies, surface and subsurface drainage and eco-friendly strategies for disposal and reuse of saline/waste waters. He has been an international consultant to UNDP through WAPCOS and Ethiopia. He has guided 14 M. Tech. and 4 Ph.D. students. He has contributed 140 papers in referred International and National journals, besides writing 13 books. He is recipient of the prestigious Rafi Ahmad Kidwai Award of the Indian Council of Agricultural Research (ICAR) and is the Chief Editor of the Journal of Water Management.

Dr. I.C. Gupta (b. 1941), was awarded Ph.D. on a subject thesis of environmental soil science by CAZRI/University of Jodhpur in 1968. He worked at Central Soil Salinity Research Institute from 1970 to 1986 (Principal Scientist). Dr. I.C. Gupta served Central Arid Zone Research Institute from 1986 to 2001 where he worked as Head of Regional Research Station, Bikaner and Head, Division of Natural Resources and Environment. A Fellow of Indian Water Works Association, United Writers Association of India & Indian Society of Salinity Research Scientists, Dr. Gupta was elected Vice-Chairman of Arid Zone Research Association of India. He has been working as Secretary of Indian Society of Salinity Research Scientists & Chief Editor Current Agriculture since 1977. Dr. Gupta had been conferred several awards notable being Glaxo Scientific Honours Club, Dr. Rajendra Prasad Puraskar, First Prize, Outstanding Book Award by ISAE, Dr. Gorakh Prasad Puraskar First Prize, S.P. Unvala Memorial Prize by IWWA and Life Achievement Award by IARS. He was nominated as member of the 'Environment Conservation Council of Rajasthan' and of 'Special Committee on Environment Impacts' of INCID, Ministry of Water Resources, Govt. of India. He has about 250 publications including 30 books and two inventions (Jaltripti and Jalshuddhi).

Dr. Gupta was appointed Ph.D. examiner by ten universities. He was also nominated for one of the most INTRIGUING PEOPLE of 2003 by American Biographical Institute, North Carolina, USA, after close scrutiny of biographical archives of over one million individuals.

Genesis and Management of Sodic (Alkali) Soils

S.K. Gupta, Ph.D.
Formerly Principal Scientist and Head,
Division of Irrigation and Drainage Engineering,
Project Coordinator,
INAE Distinguished
Professor,
Central Soil Salinity Research Institute, Karnal

I.C. Gupta, Ph.D.
Formerly Principal Scientist, CSSRI, Karnal
Head, RRS Bikaner
Head, Division of Natural Resources and Environment
Central Arid Zone Research Institute, Jodhpur

SCIENTIFIC
PUBLISHERS

Published by
SCIENTIFIC PUBLISHERS (INDIA)

Jodhpur –
5 A, New Pali Road
P.O. Box 91
Jodhpur - 342 001 INDIA

Delhi –
4806/24, Ansari Road
Daryaganj
New Delhi - 110 002 INDIA

ISBN: 978-93-86102-77-5

Visit the Scientific Publishers (India) website at
http://www.scientificpub.com

Printed in India

PREFACE

Salinity is an age old problem with catastrophic consequences. Civilizations in southern Mesopotamia and Tigris–Euphrates valley were wiped out as a result of widespread problems of water logging and soil salinity. Magnitude of damages being reported in the Aral Sea Basin is the living example of how things go wrong in the absence of proper land and water management practices in irrigated agriculture. Other basins in Australia, China, India, Pakistan and the United States are also grappling with salt related land and water degradation, making these problems global issues that extend to more than 100 countries. As per an assessment of the FAO - Land and Plant Nutrition Management Service, more than 6% of the world's land and 20% of the irrigated lands are afflicted with water logging and/or soil salinity, the latter often characterized as saline and sodic. In India also, current state of the land and water resources has been worrying all the stakeholders considering precarious nature of the food security achieved so far and in view of the recently enacted 'Food Security Bill". To ensure food for all, India needs to produce significantly more food grains to meet the requirements envisaged in the bill. Since twin problems of water logging and soil salinity have emerged as the major abiotic stresses, it has worried the Ministry of Water Resources, Ministry of Agriculture and other organizations connected with natural resources management. It is assessed that if the 'Business As Usual' approach continues, area under water logging and soil salinity by 2050 may treble to 20.0 million ha from the current level of 6.73 million ha, 3.77 million ha sodic and remaining being saline in nature. Amongst the two, soil sodicity has always been considered complex because basic principles of the problem were not understood until recently. It is why, reclamation and management of sodic soils has been considered difficult.

Soil and water salinity including sodic land and water adversely affect farm, regional and national interests that spirals into food security, land and water quality and environmental implications resulting in serious economic and social problems. Considering the looming food security crisis of the 2050, no let-up can be allowed in our efforts to understand the behavior of salt affected lands, salt tolerance mechanisms in crops, develop new technologies and apply the existing ones for prevention, living with salts and/or sustainable land reclamation programs. It calls for an inter-disciplinary approach drawing knowledge and experiences of the experts from soil science, agronomy, engineering, hydrology, other water sciences, plant sciences, environmental sciences, modeling and computer sciences.

CSSRI in its initial years of inception came out with the sodic land reclamation technology in the form of an integrated package and extended the same to the stakeholders. The state governments besides adopting the technical solutions supported the endeavor by framing enabling policies as a result of which significant headway could be made in addressing the sodic problem scene in the country. Current estimate suggests that nearly 2 million ha of sodic lands have been reclaimed by 2015 in various states led by Uttar Pradesh, Punjab and Haryana. In spite of this impressive record, the pace of reclamation so far is much below the expectations as CSSRI estimates show that salt affected areas are likely to increase in future. Therefore, concerted efforts to prevent further degradation and reclamation of already degraded sodic lands need to be strengthened including publications of literature for various stakeholders involved in land reclamation.

To prepare this publication, the authors made a comprehensive review of the current status of sodic land reclamation scene in India and around the globe and extracted relevant information across disciplines. While latest information/technologies are included, references to old information are highlighted since these provide a treasure of scientifically generated information having lot of application potential. Since the book specifically addresses the sodic soils, technologies are segregated to include that specifically deal with this issue although important issues related to soil salinity in general are included. The technologies included are discussed in a self explanatory manner so that field officers and workers can learn about the practical aspects of these technologies. We have also included many issues that are in the course curricula of various universities related to salt stress management. All the chapters have been categorized to deal with a specific issue. Since formation of sodic soils is also governed by the use of sodic water, a special chapter is included that deals with the management of sodic water to get optimum productivity without impacting the soil health. Since the publication gives the readers a holistic picture of reclamation and management of sodic soils, it should cater to the needs of those interested in soil/water sodicity-plant-environment interactions.

The authors believe that this publication should prove to be a valuable reference for all professionals that are associated with soil salinity management research, professional training institutes, technical agricultural centers, irrigation research centers, agricultural extension service and other agencies. Knowledge gained through this publication should help various stakeholders in taking effective and integrated remedial actions as per local needs to eradicate the menace of soil sodicity. Although, most aspects of sodic land management are included, we do not claim that every issue on the subject has been covered in this publication. We would welcome any feedback and suggestions/comments from the readers so that the same can be incorporated in future editions of this publication.

S K Gupta
I C Gupta

CONTENTS

1

ORIGIN, GENESIS AND DISTRIBUTION

While the history of salts in soils is as old as the formation of the soils, its impact on crop productivity has also been known to the mankind since the beginning of agriculture. References to the '*Urvar*' (productive) and the '*Anurvar*' (unproductive) soils in the old Hindu Literature is a testimony to the statement. '*Anurvar*' soils were also referred as '*Usar*' soils, a Sanskrit word derived from '*Ushtra*'. Lands having excess salts such as neutral NaCl or sodic ($NaHCO_3$, Na_2CO_3), were called '*Reh*' by the geologists in mid-nineteenth century to characterize the appearance of salt efflorescence at the surface of lands. The word 'sodic' itself is of Arabic origin meaning ash-like. Since the surface soil looked like calcined wood ash to the Arabs, these soils were named as such. Although salt affected soils are most common to aridic moisture regimes but may be present at any latitude and altitude (Szabolcs, 1998). Secondary salinization due to improper irrigation management was a major cause of land degradation in the Nile Delta, the Mesopotamian Plain, and the valleys of the Yangtze and the Hwang Ho (Dregne *et al.*, 1996). Salinization of the Yellow River irrigated plains was limiting cereal production more than 2000 years ago in China (Dregne *et al.*, 1996). Soil salinization resulting from mismanagement of water and inadequate drainage has demolished flourishing civilizations such as southern Mesopotamia and in several parts of the Tigris–Euphrates valley. In modern times, adverse impacts have come to the fore in the Aral Sea basin (USSR). Currently, salt affected soils either irrigation induced or dry land salinity are naturally present in more than 100 countries of the world. As such, investigations on causes, genesis and nature, kind and degree and reclamation of salt affected soils have become a major issue in natural resource management. This chapter includes nomenclature of salt affected soils besides dwelling upon the origin, genesis and distribution of sodic soils.

TERMINOLOGY FOR SODIC SOILS

Sodic soils in literature have been described by various terms without making a clear distinction amongst the terms. For example; sodicity, alkalinity and basicity have been interchangeably used to describe these soils. The term "basic" for high pH is ambiguous as it can be confused with the term "basis". Therefore, many scientists and authors prefer to speak of alkali instead of basic. The term "alkali soil" is not ambiguous in itself, as the term "alkali" corresponds to "sodic". On the contrary, the term alkaline used to describe a soil with a high pH is a misnomer in the context of sodic soil. According to The Alberta Environmental Farm Plan, (*www.albertaEFP.com* Opened on 10.03.2015), alkaline soils are soils having pH above the neutral value (generally greater than 7.3). Alkaline soils have SAR <13, ESP < 15, EC_e < 4dS m^{-1}. Alkali soil on the other hand have pH_s of 8.2 or greater or a high exchangeable sodium content (15% or more of the exchange capacity) or both, the later being the most common.

Clearly, terms sodic and alkali soils have been and are being used interchangeably in the literature. On the other hand, it may be noted that alkali soils are not always sodic as these can be acidic/saline also. In several cases saline soils have a high sodicity (a high SAR that may be higher than that of even sodic soils), but they are not sodic soils and have usually a good infiltration capacity. Rengasamy (2006) categorized various salt affected soils as acidic–saline soils, neutral saline soils, saline soils, alkaline–saline soils, acidic saline–sodic soils , acidic–sodic soils, non-alkali sodic–acid soils, neutral saline–sodic soils, neutral sodic soils, non-alkaline sodic and strongly sodic neutral soils, alkali saline–sodic soils and alkali–sodic soils. Degraded sodic soils have been reported in deltaic regions of West Bengal which have ESP more than 15 but pH in the range of 5.0-7.0 (Chakravarti and Chakravarti, 1957). It seems that the classifications is quite elaborate and takes care of many local specific situations in Australia and elsewhere. On the other hand for macro-level diagnosis, only two terms alkali and sodic soils are used to describe these soils, the latter being more scientific. It is therefore proposed to use the term sodic although term alkali will be used wherever the authors have used this term in their publications.

Another notable terminology commonly used is white alkali and black alkali. While both these soils have preponderance of sodium, the former has neutral salts such as NaCl and Na_2SO_4 while the latter has Na_2CO_3 and $NaHCO_3$ type of salts. In the former pH and ESP are not high even though SAR may be quite high. These soils are also termed as high SAR saline soils or simply the saline soils. In such soils, it is expected that SAR will reduce during leaching with or without slight increase in pH. The black sodic soils are the real sodic soils, the name being derived from the black spots that are often seen on the surface of these soils. Since the pH of sodic soils is high, organic matter present in the soil profile is dissolved and is transported to the soil surface through capillary forces. Once the water evaporates, salts along with the black coloration are deposited on the soil surface. The nomenclature however has become obsolete but many people continue to use the term.

Besides what is stated, a large number of sub-categories of salt affected soils are also recognized in different parts of the world depending on the dominance of a particular chemical constituent (e.g. calcium chloride rich soils or soils containing excessive quantities of exchangeable magnesium - magnesium solonetz, etc.) or a particular morphological character of the soil profile, e.g. presence of a structural 'B' horizon, etc. Such classification would be indicated wherever required.

CLASSIFICATION OF SALT AFFECTED SOILS

A soil having excess soluble salts and or high exchangeable sodium percentage or both on the soil surface and in the root zone to interfere in crop growth is characterized as a salt affected soil. Many terms as indicated in previous sections have been used to describe salt affected soils. The most commonly used USDA classification of salt affected soils, classifies salt affected soils in 3 groups namely saline, sodic and saline-sodic soils (Richards, 1954).

Saline Soils

In most saline soils, the whole profile is salt laden, strongly saline soils showing a white 1 to 5 cm thick salt layer on the soil surface during the dry period. The salt content in the profile gradually decreases with depth. The excessive salinity in these soils is due to the dominance of sodium chloride followed by sodium sulphate and chlorides and sulphates of calcium and magnesium. The pH_s is less than 8.5 but not less than 7.0, with the exception of some coastal saline soils like those found in Kerala, India, which are distinctly acidic. In some gypsiferous soils, like those found in Gujarat and Rajasthan states of India, calcium sulphate is the dominant salt. The ESP of these soils is less than 15 (Table 1.1). Following three distinct groups of saline soils have been identified.

- Inland saline soils having shallow water table, water quality being saline in nature. These soils suffer from water logging and salinity (Fig. 1.1).
- Normal soils being irrigated over the years with saline and/or sodic ground water. These soils normally have deep water table. However, in certain situations the water table can be shallow. Theses soils generally occur in arid and semiarid regions.
- Coastal saline soils where the problem is due to ingress of sea water resulting in high salinity and temporary flooding. The climate is generally humid to semiarid/arid.

Fig. 1.1 Water logging and soil salinity in Haryana

Due to excessive neutral salts, these soils are usually flocculated and have good physical properties. The structure is however, unstable due to lack of organic matter. As the soils are mostly uncultivated, a hard layer develops below the fluffy salty surface layer (0-10 cm) due to the compaction caused by the hydrostatic pressure of the flood water during rainy season, the prolonged fallowing, and the trempling by farm animals. The conditions like periodic submergence or water logging and the presence of shallow brackish ground water are the severe limitations, as these cause further salinization. Slightly saline soils are moderately cultivated with some salt resistant crops like rice, barley, wheat, sugarcane, oats, *berseem* and *sesbania.* Plants growing on such soils may appear stunted with thickened leaves of dark green colour. On saline lands, salt tolerant grasses appear during the monsoon season but large barren spots are common. Saline soils are potentially productive due to abundant of nutrient reserve and ideal moisture retention and transmission characteristics following reclamation.

Sodic Soils

Sodic soils contain excess of salts like bicarbonates, carbonates and silicate of sodium, capable of alkaline hydrolysis and have sufficient exchangeable sodium to interfere with growth of most crops (Fig. 1.2). The pH of sodic soils is > 8.5, EC_e is variable and ESP >15 (Table 1.1). There is a white encrustation on the surface during dry months, which often shows dark/black colour on wetting. Dispersed organic matter accumulates at the surface of poorly drained areas as water evaporates and imparts a black colour to the surface on wetting, hence the name "black sodic" soils. The clay and organic fractions are dispersed. When wet, the soil becomes sticky and on drying it is cloddy, hard, compact and difficult to work with. Rainwater does not move down and remains standing for a long period. There is a barren look on the land. The excessive sodium carbonate in the soil solution causes the precipitation of calcium of the exchange complex as insoluble calcium carbonate, thus causing a high amount of exchangeable sodium. The ESP values of 80-90 or more are common in the surface layers of these soils. These soils have excessive soluble salts in the upper 0-30 cm soil surface layer. Generally, sodicity and salinity decrease with depth. The impact of sodicity is revealed through surface crusting, reduced seedling emergence, increased run-off and erosion risk, and less ground cover and organic matter resulting in reduced soil aeration and less microbial activity.

Table 1.1 Classification of salt affected soils on the basis of chemical characteristics

Class	EC_e (dS m^{-1})	ESP'	pH_s	SAR[§] (mmol $L^{-1})^{1/2}$	Indian Local Names
Saline soils	> 4*	< 15	< 8.5	< 13	*Thur, Uippu, Lona, Shora, Soula, Pokhali, Khar and Kari*
Saline-sodic soils	> 4	>15	Variable	>13	*Usar, Kallar, Karl, Chopan, Bari, Reh, Choudu and Kshar*
Sodic	< 4	>15	> 8.5	>13	*Usar, Rakkar, Bara, Chopan, Kari*

*Present literature sometimes gives this value as 2.0 dS m^{-1} mainly because many sensitive crops such as gram and vegetables start suffering at this salinity. [§] Sometimes SAR (>13) is used instead of ESP. For Indian conditions, ESP is preferable, for pH_s is measured, the value is taken as 8.2 instead of 8.5.

Fig. 1.2 A view of barren sodic land

Saline- Sodic Soils

These soils have a combination of harmful quantities of salts and a high content of exchangeable sodium, which interferes with the growth of crop plants. The pH_s of these soils is > 8.5, EC_e is > 4 dS m^{-1} and ESP >15 (Table 1.1). These have all the features of a saline soil, and if reclamation procedures are used that do not include gypsum, they become sodic upon leaching.

In the context of ESP and EC_e, the classes of salt affected soils are shown as in Fig. 1.3. Similar figures for pH and EC_e and SAR and EC_e can be drawn.

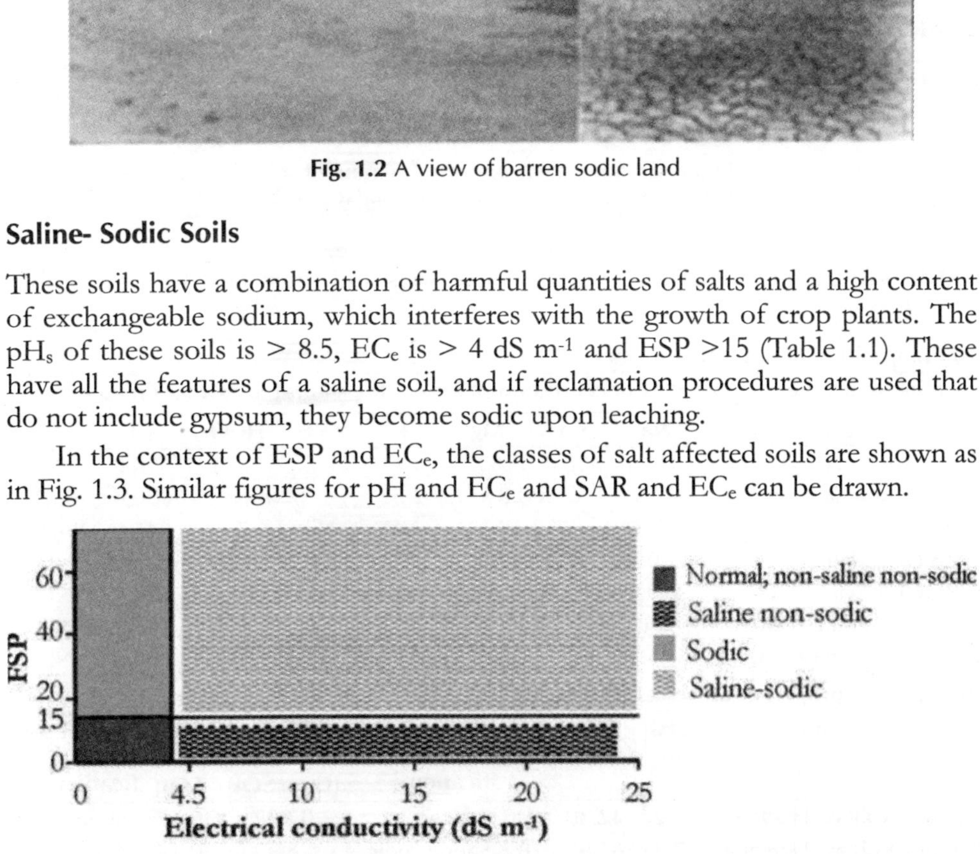

Fig. 1.3 Categorization of salt affected soils based on EC_e and ESP of the soils

In the international and Indian context, several modifications to this classification have appeared over time. For example, in the Russian literature, soils have been classified according to their chemistry, as saline with neutral ($NaCl$, Na_2SO_4, etc.) and sodic with sodic salts (Na_2CO_3, $NaHCO_3$). Besides the characterization of existing salt affected soils, another category is the potential salt affected soils. Soils considered potentially salt affected are those which currently are not or only to a very low degree saline and/or sodic but human intervention, especially irrigation, could result in salinization and/or sodication in future. The term degraded sodic soils is used to describe an advanced stage of soil development resulting from the washing out of salts.

In the Indian context, from management point of view, soils are categorized only in two categories i.e. saline and sodic/alkali. If ESP is used as a classification criterion, the saline sodic soil with ESP > 15 is categorized as sodic. If EC_e and SAR are used to categorize soils, some additional calculations are required to categorize saline-sodic soils as saline or sodic especially from management point of view (Table 1.2). The use of this classification is illustrated through data presented in Table 1.3.

Table 1.2 Classification of saline-sodic soil as saline or sodic

Soils category	Guidelines
Saline	In the saturated paste extracts, the ratio of soluble ions should be $(CO_3 + HCO_3) / (Cl + SO_4) < 1$ $Na/ (Cl + SO_4) < 1$
Sodic	In the saturated paste extracts, the ratio of soluble ions should be $(CO_3 + HCO_3) / (Cl + SO_4) > 1$ $Na/ (Cl + SO_4) > 1$

Source: Chhabra (2003)

Table 1.3 Chemical analysis of saline-sodic soils and their classification (Table 1.2)

(A) Composition of saturation extract

Location of profile	Concentration of ions in the saturated extract (meq L^{-1})							
	Na	Ca	Mg	K	CO_3	HCO_3	Cl	SO_4
Sampla, Rohtak, Haryana	102.5	326.4	163.2	ND	Nil	1.3	590.0	ND
Sampla, Rohtak, Haryana	97.5	500.8		2.53	Nil		680.4	60.7
Gohana, Haryana	975.0	125.7		ND	Nil	7.1	645.0	448.7
CIRB,Hisar, Haryana	700.0	64.2	170.7	ND	Nil	2.4	820.0	ND
CIRB,Hisar, Haryana	440.0	283.9	297.2	ND	Nil	10.0	950.0	ND
Lunkarnsar, Rajasthan	2.1	4.50		0.14	Nil	1.8	4.5	ND
Lunkarnsar, Rajasthan	125.0	117.8		0.52	Nil	1.1	268.5	ND

(B) Chemical properties and classification

Location of profile	pH_s	EC_e, (dS m^{-1})	SAR	Old soil classi-fication	$[CO_3 + HCO_3]/[Cl + SO_4]$	Na / [Cl + SO_4]	Present soil classi-fication
Sampla, Rohtak, Haryana	7.2	42.30	6.6	Saline	0.002	0.17	Saline
Sampla, Rohtak, Haryana	7.3	65.97	18.8	Saline-sodic	Nil	0.14	Saline
Gohana, Haryana	8.1	63.8	123.0	Saline-sodic	0.007	0.89	Saline
CIRB, Hisar, Haryana	8.2	66.20	64.6	Saline-sodic	0.003	0.85	Saline
CIRB, Hisar, Haryana	7.5	69.40	25.8	Saline-sodic	0.010	0.46	Saline
Lunkarnsar, Rajasthan	8.2	0.55	1.4	Saline	0.400	0.47	Saline
Lunkarnsar, Rajasthan	8.0	25.10	16.3	Saline-sodic	0.004	0.47	Saline

Source: Chhabra (2003)

Sodic Soils and Soil pH

Sodic nature of the soil is associated with the presence of sodium carbonates (Na_2CO_3) in the soil, either due to natural weathering of the minerals or brought in by irrigation and/or flood water. The sodium carbonate, when dissolved in water, dissociates into $2Na^+$ (two sodium cations, i.e. ions with a positive electric charge) and $CO_3^=$ (a carbonate anion, i.e. an ion with a double negative electric charge) as per the following reaction:

$$2Na^+ + CO_3^{--} + 2H^+ + 2OH^- \rightarrow 2Na^+ + 2OH^- + H_2CO_3 \quad (1.1)$$

H_2CO_3 being unstable, it further gets dissociated to produce H_2O (water) and CO_2 (carbon dioxide), escaping as a gas. The sodium hydroxide ($Na^+ OH^-$), which is alkali (or rather basic) gives high pH values (pH >9). Each unit increase in pH indicates a tenfold increase of the OH^- concentration.

Not all sodium carbonate follows the above chemical reaction. The remaining sodium carbonate, causes $CaCO_3$ (which is only slightly soluble) to precipitate as solid calcium carbonate (limestone). Hence, the Ca^{++} ions are immobilized as follows:

$$2Na^+ + CO_3^= + Ca^{++} \rightarrow 2Na^+ + CaCO_3 \text{ (solid)} \quad (1.2)$$

Hydrolysis of compounds like $CaCO_3$, and $MgCO_3$ also takes place according to the following reaction:

$$CaCO_3 + 2\,H_2O \leftrightarrow Ca^{2+} + 2\,OH^- + H_2CO_3 \quad (1.3)$$

In this reaction H^+ from water is inactivated as it combines with carbonate to form weakly ionized carbonic acid. Hydroxyl ions are not inactivated through combination with Ca^{2+} resulting in an alkaline solution. The hydrolysis of $CaCO_3$ and of $MgCO_3$ is limited due to their low solubility and therefore, they tend to produce a pH in soils no higher than about 8.0 to 8.2 but in combination with the reactions describe above pH may be as high as 10.0 to 10.5.

Hydrolysis of the exchangeable cations can also take place according to the following reactions:

$$\text{Clay}\begin{matrix}\text{Na}\\ \text{Na}\end{matrix} + H_2O \leftrightarrow \text{Clay}\begin{matrix}\text{Na}\\ \text{H}\end{matrix} + Na^+ + OH^- \quad (1.4)$$

Here, H^+ is inactivated by exchange adsorption in place of Na^+. The displaced Na does not combine with, or inactivate OH^- ions. The increase in the OH^- ion concentration causes increase in the soil pH. The extent to which exchangeable cations hydrolyze depends on their ability to compete with H^+ ions for exchange sites. Na^+ ions are unable to compete as strongly as the more tightly held ions such as Ca^{2+} and Mg^{2+}. For this reason exchangeable Na^{+2} and K^{+2} are hydrolyzed to a much greater extent and produce a higher pH than do exchangeable Ca^{2+} or Mg^{2+}.

ESP Limit for Sodic Soils

As per guidelines given by USDA and others, a soil is considered sodic when the soil ESP exceeds 15. Even today, this limit is used to classify salt affected soils. On the other hand, It may not be out of place to mention that these thresholds for sodic soils (sodium adsorption ratio >13 or exchangeable sodium percentage

>15) are somewhat arbitrary. The change from flocculated (aggregated) to dispersed soil structure occurs gradually with increasing sodium levels. The dispersion of clay particles is strongly dependent on the electrolyte concentration in the soil solution and the relative composition of Ca and Mg on the exchange complex. Clay dispersion is reported to occur at an ESP of five in Ca-Na aggregates, but is observed at an ESP of only 3 in Mg-Na samples (Emerson and Bakker, 1973). An ESP of 6 was proposed by Northcote and Skene (1972) to be the lower limit of soil sodicity. The same value has been reported as the limit to characterize sodicity in Australia (Rengasamy, 2006). The lower limit is related to low concentrations of soluble salts, low content of calcium, and widespread structural deterioration of Australian soils (Rengasamy and Olsson, 1991). Values of 5 (van Beekom *et al.*, 1953) and two (Mitchell, 1976) have been suggested to cause a deleterious effect on soil structure. In the Indian context values ranging between 5 for Vertisols to 10 for heavy textured soils have been reported to seriously impact the soil physical properties and hence the limits to categorize the soils as sodic. A Vermaquept has been classified as sodic if it has an ESP 7 (SAR 6) within 1 m from the soil surface, whereas a Haplocalcid has an ESP of 15 (SAR 13) to be designated as sodic. Several reasons are ascribed to the observations that much smaller amounts appear very injurious in some while even larger amounts cause little problems in the other cases (Table 1.4). Some of the reasons are:

Table 1.4 Percent loss in soil productivity

Exchangeable sodium percentage	Loss of productivity in alluvium derived soils (Fluvisols)	Loss of productivity in black soils (Vertisols)
Up to 5	nil	Up to 10
5–15	<10	10–25
15–40	10–25	25–50
>40	25–50	>50

Source: Sehgal and Abrol (1994)

- The kind of sodic salts or their combinations in the soil water is one of factors that determine the relative degree of toxicity. Hilgard has stated that the limits of tolerance for sodic salts varies as sodium carbonate (0.1% to 0.25% of the soil) > sodium chloride (0.2% to 0.50% of the soil) > sodium sulphate (0.5% to 1.00% of the soil).
- The location and distribution of the sodic in the soil profile.
- The soil texture of the soil i.e. sand, silt, loam, clay, or peat has a profound effect on a given amount of sodicity. Large amount of organic matter contained in a peat soil can greatly reduce the harmful effects of salts on plants growing in the soil.
- The amount and timeliness of the moisture supply in the soil can lower the concentration of sodicity in the soil moisture causing lesser harm due to sodicity.

- Climate and season of the year in relation to the growth period of the crop i.e. more harm may occur in summer than winter season
- Plant characteristics such as rooting behavior viz. salt distribution and tolerance of the plant itself etc.
- Sodicity problems are more pronounced in clay soils than in loamy, silty or sandy soils. The clay soils containing montmorillonite or smectite (swelling clays) are more subject to sodicity problems than illite or kaolinite clay soils. The reason is that the former types of clay have larger specific surface areas and higher cation exchange capacity (CEC).

No deleterious effects on crop productivity have been observed in India at ESPs lower than 15, even when some soil physical properties were adversely impacted. From current standpoint therefore, guideline of ESP greater than 15 should continue to be used to characterize sodic soils until more data are made available that suggest modification in this limit.

ORIGIN OF SALTS IN SOILS

Although salt affected soils occur in all continents and under almost all climatic conditions but their distribution is relatively more extensive in the arid and semiarid regions. Salt affected soils are formed as a result of range of processes varying from very simple to complex in nature and operate at a variety of scales varying from localized to regional scale. Three main indicators to provide insight into areas at risk of salinization at a global scale are climate, drainage and basin lithology. Moisture availability and seasonality are the main factors associated with climate as they exert an influence on the weathering rates, solute release and mobilization. Lithology is important, as weathering is one of the most potent and immediate source of saline material. Topography impacts the formation of salt affected soils as areas of low relief, particularly closed depressions, cause lateral translocation of salts both in surface and ground water and impacts the salt accumulation patterns (Kovda, 1954). As the scale of interest narrows down, the patterns become more complex as other physiographic factors assume greater significance such as proximity to the coast, soil texture, lateral seepage of saline ground water and contribution of airborne salts. Some of the processes that are relevant to the origin, redistribution and accumulation of salts in soils are described herein. These factors may operate either singly or in a combination to cause soil salinization.

Weathering of the primary minerals is the main source of all salts in the soil. In humid regions, soluble salts are carried down through the soil profile by percolating rainwater that ultimately joins the ground water and are transported to sea. On the contrary, in arid and semiarid regions, salts released through weathering are usually deposited at some depth in the soil profile depending upon factors such as the water retention capacity of the soil, seasonal, annual and maximum rainfall, etc. that determines the leaching behavior (Yaalon, 1965). If the salts are deposited beyond the root zone of crops, they may not have any major adverse effect on the crops unless they are redistributed to accumulate in the surface soil layers. This process of formation of salt affected soils as a result of accumulation of salts released during weathering is called primary salinization.

In ancient times large areas over the globe lay submerged under the sea. Gradually the sea disappeared, leaving marine shales with high concentrations of sodium salts. The movement of glaciers mixed the underlying marine shale with surface deposits. In some areas, where glacial deposits are relatively shallow, the total supply of salts is large and the risk of salinity and other salt-related problems is high. Although the salts currently occurring in the ocean have arisen mainly from the weathering processes of the earth crust, the ocean now functions as an important 'source' for redistribution of salts through oceanic derived rain fall, seawater encroachment and ground water containing salts of geologic or marine origin.

Rain water in general qualifies as fresh water resource containing little amount of salts but can have as high as 50 to 200 mg L^{-1} salts near the sea coast. The amount decreases rapidly as rain moves inland. The rain near the seacoast is high in Na, Cl and Mg, whereas inland precipitation is dominated by Ca, Mg and SO_4. Over several thousand years, this atmospheric deposition becomes significant resulting in soil salinization. In the coastal areas, the soils also get enriched with salts from the sea through inundation of surface soil by sea water during high tide and ingress of sea water through rivers, estuaries, ground water inflows, etc.

All irrigation waters contain some salt. The salinity of the river/canal water is generally more than the rainfall (EC = 0.3 to 0.45 dS m^{-1}). The increased salinity in the rivers/canals waters is part of the ongoing geochemical processes. The natural drainage water flowing into the streams through overland flow from watersheds picks up salts from rocks and soils that contain salts. The salt pick up continues as the stream moves down and into the oceans. The pickup of salts is more when the stream flows through soils or rock formations that contain salts. Salts may also be added on its way through seepage and irrigation return flow from irrigated lands, ground water contribution as base flow to the stream and human interference in the form of diverting wastes from industrial/urban development. All these factors together not only determine the quantity of salts present in the water but also govern the nature of the salts present in the water.

If the irrigation water having an EC of 0.5 dS m^{-1} is used to irrigate an area, it would add about 32 kg ha^{-1} of salts. That means, with each irrigation of about 6 cm, one adds 192 or approximately 200 kg of salts per ha. If the EC is 1 dS m^{-1}, the addition is about 400 kg ha^{-1}. When irrigations are continued over an extended period, salts accumulate in the soil and a salinity problem is inevitable to develop. Since river/canal water used in major/medium irrigation projects is of high quality, there is little risk of salt build-up. However, farmers using minor irrigation sources such as tanks and ground water may face salinity problems (Fig. 1.4). In many cases salinity may not be directly associated with irrigation water but could be either traced to the redistribution of salts already present in the soil or in the unirrigated zone towards the tail end as shown in Fig. 1.5.

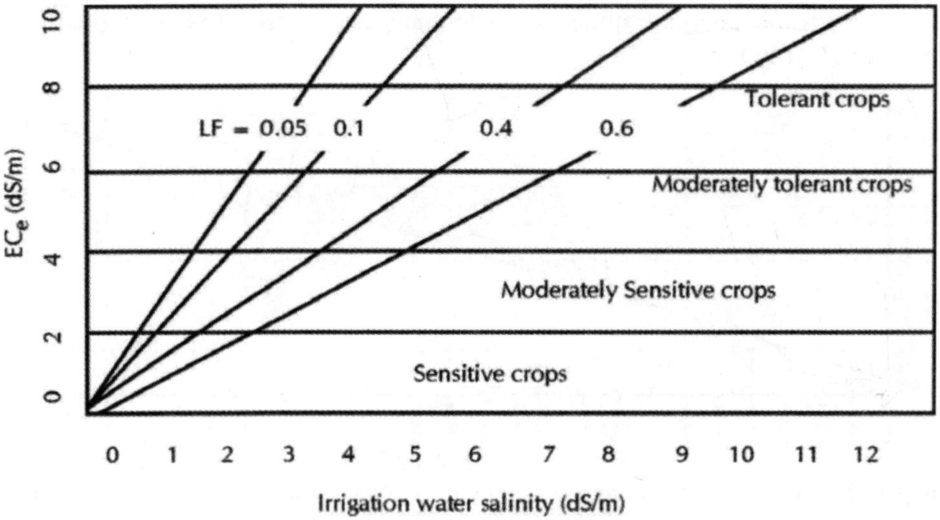

Fig. 1.4 Steady state root zone soil salinity (EC_e) likely to develop with applied irrigation water salinity (EC_i) at various leaching fractions (LF)

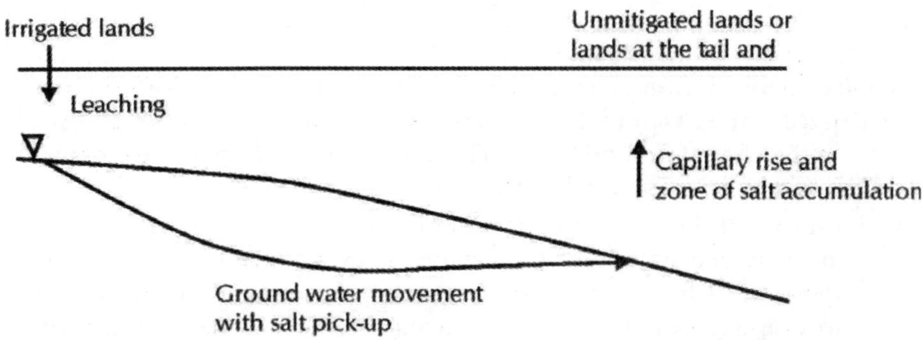

Fig. 1.5 Salt movement in irrigated lands showing areas of salt leaching and accumulation

The rise in water table due to introduction of canal irrigation also increases the risk of soil salinization. Introduction of irrigation results in profound changes in hydrological regime of the area. Prior to irrigation, there exists a balance between rainfall on the one and stream flow, evaporation and transpiration on the other hand. This balance is disturbed because of additional contribution to the ground water in the form of seepage from canals and irrigation channels and percolation of irrigation water from the fields which are often over irrigated. This coupled with unscientific soil, water crop management practices like bringing proportionately more land under high water requirement crops such as rice; presence of poor quality ground water which cannot be used for irrigation to check the water table rise; the poor natural subsurface drainage due to lack of drainage outlets; and other unscientific development activities (construction of roads, railways, irrigation works which nearly completely obstruct natural drainage) results in the gradual rise in ground water table to the critical level (Fig. 1.6). About 6 million ha of the country (covering about 4% of the total cultivable area) are under water logging due to rise of water table in the critical range of 0-3 m.

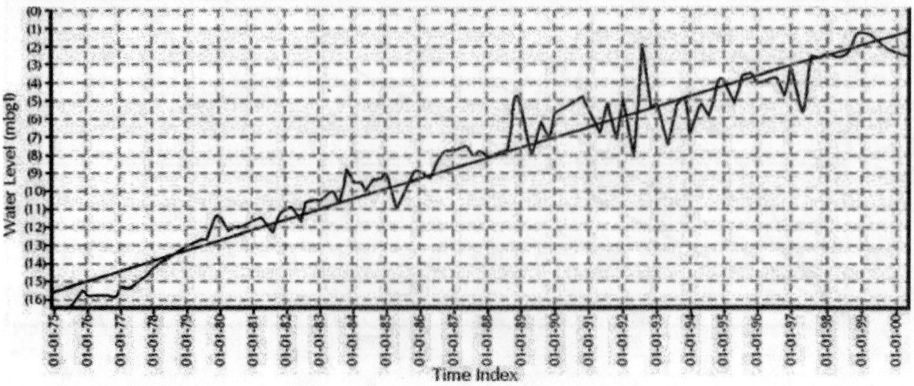

Fig. 1.6 Hydrograph of rising water table at Bhuna in Bhakra irrigation command, Haryana

The other course of rain water is to travel to form ground water and into the oceans. Although the source for both the surface and ground water is rainfall, ground waters are usually more saline than the surface waters. The ground water or drainage water from irrigation system can dissolve appreciable amount of 'fossil salts', derived from either deposits or entrapped solutions in former marine deposits. Most factors that influence salinity of the stream water are also operative in the case of ground water. There are several additional factors which control the salinity of the ground waters.

- Climate of the area including duration and frequency of rain, humidity, intensity and duration of sunshine, wind velocity etc., in as much as these affect the quantity and frequency of recharge and the losses due to evapotranspiration, all influencing the net balance between annual evaporation, precipitation and runoff.
- The rate of local rock weathering with resultant formation of water soluble salts from the decomposed products.
- The rate at which salts are transported in the region by streams, winds, rains from outside sources and deposited in the soil.
- The rate of ground water circulation.
- The interconnection between surface water bodies and the ground water.
- Hydrogeological conditions.

The use of ground water for irrigation may often pose problems related to soil salinity/sodicity depending upon the salt constituents of ground water.

Salt release may occur naturally or may result from human activities. For example; rise of salt bearing ground water through an originally impervious cap that becomes permeable as a result of weathering processes; overlying saline strata illustrates the natural process. Example of the latter are building of canals networks in saline strata resulting in rising water table or use of ground water for irrigation. Irrigation with saline industrial wastes could also add considerable salts and other elements to the soil.

Salt affected soils also occur in regions that receive salts from other areas with water as the primary carrier. Salts contained in flood water may gradually build-up soil salinity especially in undulating terrains where water can accumulate in depressions or in soils with low water permeability as the salts are not easily leached. Eventually the salts contained will be left in the soil after the water has been removed by evaporation and crop use. The presence of sodic incrustations over the soil surface in Indo-Gangetic plains of India is ascribed to this process. Under certain situations, seepage resulting from water inflow in upslope areas can cause severe salinity on the down slope areas, especially when the subsurface water flow takes place through the strata that are rich in salts and/ or marine deposits. Mineral springs owe their salts to such a situation.

Soil salinization is not a static but a dynamic process. While large areas are being reclaimed, more areas are being salinized or identified using new technologies. Current and scenario building studies show that the areas at risk of salinization may increase globally, through expansion of existing areas and the creation of new ones (Schofield and Kirby, 2003).

DEVELOPMENT AND RESEARCHES ON SALT AFFECTED SOILS IN INDIA

Canal Irrigation, Water Logging and Soil Salinization

Water logging and soil salinity has been prevalent even before the canal irrigation and manifests even toady in many rain fed regions. However, introduction of canal irrigation made a huge impact resulting in widespread appearance of water logging and soil salinity in India and elsewhere on the globe. The environmental degradation in the form of water logging and soil salinization appeared within a decade or even lesser number of years after opening of several canal systems. As such, canal irrigation has been considered as one of the most potential factors responsible for the formation of sodic wastelands. Therefore, it is desirable to trace out the development of canal irrigation in India to understand the development of water logging and soil salinity.

The remains of Indus Valley Civilization and Vedic texts bear testimony to the practice of irrigation. The Grand *Anicut* across river Cauvery in Tamil Nadu was constructed in the second century A.D (Fig. 1.7 left). Feroze Shah Tuglaq in fourteenth century AD got the Western Yamuna Canal constructed (Fig. 1.7 right). Irrigation works such as wells all over the country, tanks in south India and inundation canals in north India were constructed during early nineteenth century. The total irrigated area through public works in 1900 stood at 13.3 M ha which increased to 19.3 M ha in 1920-21. At the time of partition, the total irrigated area stood at 28.2 M ha out of which 8.8 M ha went to Pakistan leaving 19.4 M ha with India. Before the plan period started in 1950, the irrigated area had already increased to 22.6 M ha. The gross irrigated area rose to 102.8 M ha in 2006-07 (Table 1.5).

Table 1.5 Irrigation potential created and utilized (M ha)

Plan	Potential created			Potential utilized
	Major & Medium	Minor	Total	Total
At the end of Pre Plan up to 1951	9.7	12.9	22.6	22.6
First (1951- 56)	12.2	14.0	26.2	25.0
Second (1956-61)	14.3	14.7	29.0	27.8
Third (1961-66)	16.6	17.0	33.6	32.2
Annual Plans (1966-69)	18.1	19.0	37.1	35.8
Fourth (1969-74)	20.7	23.4	44.1	41.9
Fifth (1974-78)	24.7	27.3	52.0	48.5
Annual Plans (1978-80)	26.6	30.0	56.6	52.6
Sixth (1980-85)	27.7*	37.5	65.2	58.8
Seventh (1985-90)	29.9	46.6	76.5	68.6
At the end of 1991-92	30.7	50.3	81.0	72.9
Eighth (1992-97)	33.0	53.3	86.3	77.2
Ninth (1997-2002)	37.1	56.9	94.0	81.0
Tenth (2002-07)	42.4	60.4	102.8	87.2
Eleventh (2007-12)	-	-	108.2	

* Reappraised figures

Fig. 1.7 The Grand *Anicut* on river Cauvery in Tamil Nadu (left) and the present day Western Jamuna canal built in 14th century (right)

Development of Sodic Wastelands: Early Investigations

After the opening of western Jamuna canal for regular irrigation in 1840 (Fig. 1.7 right), a first-official notice was lodged in 1855 through a complaint by a grower of village Munak (now in District Karnal) that his lands have been salinized as a result of canal irrigation. Progressive deterioration of lands in the command of the Western Jamuna Canal prompted the Government of Punjab for the first time in 1858 to draw the attention of then Governor-General to the seriousness of the problems of *reh* and *usar* tracts. The chief Commissioner of Punjab reported. "For some time past, it has been known that many villages on the banks of this canal in the Panipat, Delhi and Rohtak districts have been suffering from a destructive saline efflorescence. The accounts of poverty in some of these villages have been quite distressing. At present, about 60 villages or one-tenth of the total number of canal villages are affected. The mischief is increasing yearly. It would soon attain a very considerable proportion and would entail fiscal loss to the Government and suffering to the people. On consultations, Colonel Baird Smith, Superintendent General of Irrigation in NWP, attributed appearance of this efflorescence to the percolation under pressure of the canal water through the soils in various directions (Whitcombe, 1971). It was in the year 1865, an executive engineer of the Punjab Irrigation Department noted that whole of the country in Punjab lying west of Jamuna is impregnated with *reh* (salts). The *Reh* in the form of efflorescence has spread in small and big patches from Lahore to Multan and Shahpoor and onward to the Derajat in the west and foot of the Soolamani mountains in the north (Whitcombe, 1972). Medlicott, the then Professor of Geology at Thomson College, Roorkee, was probably the first scientist who made scientific investigation of the problem of *reh* in 1863. His findings are summarized as follows:

The river water that fed the West and East Jamuna Canals, contained high concentrations of sodium sulphate and sodium chloride, the predominant salt characterizing *reh* and *usar* lands. Soils in which salts accumulated through high rates of evaporation and through capillary action were parched with no natural sub-soil drainage. Because of the change in the rate of movement of sub-soil water through percolation, capillary movement and by the poor quality of irrigation water itself, progressive concentration of soluble and partially soluble salts, originally distributed more widely throughout the profile, takes place in the upper strata of the soil to a degree toxic to plant growth. The canal water was an exacerbating agent but not a prime cause and the natural salt lands were made increasingly so by inefficient drainage and insufficient application of water.

The correspondence on *reh* published in 1864 represents the first collected documentation of the problem and its scientific analysis. Copies were sent to the Secretary of State in England in 1865, with specimens of *reh* soil collected from village Munak and Asan-Kalan, both in the present district of Karnal, on the Western Jamuna Canal. Samples of canal water were also sent to W.J. Ward in the metallurgical laboratory of the Royal School of Mines for the chemical analysis. Ward observed that canal waters contained notably low amounts of chlorides and sulphates. He also concluded that *reh* was a mixture of highly soluble salts and that water was the vehicle by which it was brought into prominence in the soil. The cure is optimum combination of sufficient irrigation with adequate drainage. A further set of independent chemical analysis of *usar* samples made by T.E.B. Brown, Chemical Examiner of Punjab, in 1863 corroborated that the one of kind of local *usar*, known as '*kallar*' in Punjab, was *Chikna Kallar*' having excess of sodium chloride, sulphate and carbonate. Meanwhile, samples of *reh* were also sent for analysis to Dr. Thomas Anderson, Professor of Chemistry at Glasgow. He corroborated in general the reports of his predecessors that *reh* may be described as mixture of common salt (sodium chloride), sulphates, soda, potash, lime and small quantity of alumina. The appearance of *reh* was ascribed to the presence in the soil of some alkali mineral, the decomposition of which is promoted by the irrigation water. As a result, large quantities of these substances are converted into soluble form that gradually accumulate until they become so abundant as to become noxious to plants. Drainage was suggested as the cure of the problem.

The Eastern Jamuna Canal was opened in 1830 and latter Gang canal and Agra canal were constructed to irrigate large areas in Uttar Pradesh. By 1864, *usar* was well known in Faizabad, Behraich, Lucknow, Sitapur and Hardoi and reports were pouring in with the Government of India. *Usar* was in general a combination of *reh*, a mixture of highly soluble salts and *kankar*, a residual insoluble precipitation or hardpan comprising of impure calcium carbonate. Amongst soluble salts, carbonates rather than sulphates and chlorides predominated. Occurrence of *usar* was ascribed to the capillary action. A good account of degraded soils is available from the diary of Mr. Sleeman who wrote that large areas are lying barren in the United Province near Agra and surrounding areas. He also mentioned some of the techniques used by the farmers to reclaim such lands near Raebareilly. The Lieutenant Governor, NWP, recommended by way of immediate action that experiment on leaching and drainage should be initiated in the central portion of the Eastern Jamuna Canal. It emerged that the practices have little practical value as the returns on investment were not immediately forthcoming. For example, canal silt enabled crops to be grown on *usar*, but the operation was not remunerative. By 1875, the most practical operation seemed to raise non-agriculture plantations.

In 1876, another alarm was raised about the disastrous spread of *reh* in parts of Aligarh, Meerut and throughout the Kali Nadi Valley of Uttar Pradesh. Through a petition, David Robarts of *pargana* Sikandra Rao, Aligarh district, requested the Board of Revenue for relief due to deterioration of his lands and appearance of saline efflorescence as a result of the introduction of canal irrigation. The condition had become far more obvious, sufficient to cause serious concern on the part of Government. This led to the formation of the historical Indian *Reh* Committee in 1877 (Agarwal and Gupta, 1968). Dr. J. A. Voelcker was first Agricultural Scientist appointed by Government to report on Indian agrarian conditions. He recorded that it is hardly surprising that enormous tracts especially in the plains of Northern India are affected by *reh*. In the NWP alone, it was estimated to cover between 10,000 and 12,000 km^2 (Voelcker, 1897).

Center (1880) studied sodic soils and saline well water of Punjab and ascribed occurrence of *reh* in soils to the decomposition of the elements of rocks and soils under the action of air and water. He also discussed nature of *reh* and its varieties, formation of *kankar*, causes of salt accumulation and methods of cure by sluicing and irrigation, manure and cultivation, arboricultural (Plantation of trees and grasses), chemical manure and nitrate of lime. Uses of *reh* for the recovery of Na_2SO_4 etc. have also been mentioned. The report of the *Reh* Committee was made available in 1886, which focused attention on many important points related to canal irrigation, drainage and spread of sodic soil in the state. The inquiry was far from adequate. No account, for instance, was given of the extent of *usar* tracts in the provinces except saying that it covered 'immense areas'. The committee recommended that indiscriminate use of water through the flush system of irrigation (flood irrigation) should be stopped. Provision of deep drainage, straightening and opening of natural drainage ways and preparation of ground water maps were also suggested. It was also recommended that experiments in reclaiming *usar* tracts which began in 1874 under the supervision of the newly created Department of Agriculture should be continued. The results had consistently shown that *usar* could be brought back into cultivation only by careful watering accompanied by intensive manuring. Nothing, however, was done on any scale to increase the local supply of manure near these tracts in order to keep pace with the increase in irrigation from the canal. Obviously the outcome of the modest series of reclamation experiments was not very fruitful. Whilst it was stated, for example, in 1888 that Government of India attached great importance to the experiments, Colonel J.G. Forbes, Irrigation Secretary to Govt., NWP and Oudh, admitted that so far all attempts at reclamation had proved unsuccessful.

SCIENTIFIC INVESTIGATIONS

UNDP Investigations in Haryana

Introduction of canal irrigation in 1.2 M ha in Haryana has proceeded with the rise of water table and consequential development of water logging and soil salinity (Fig. 1.6). It has been estimated that State of Haryana is suffering annually a loss of 420 million due to twin problems of water logging and salinity. Further, this loss is likely to increase to Rs. 1000 million by 2000 AD, if the problem is not contained. Out of the total geographical area of 4.4 million ha in Haryana, around 2.8 million ha is underlain with saline ground water. The cost of reclamation of the entire area in Haryana for potential agricultural production through drainage and other measures has been estimated to be Rs. 7200 million (HSMITC, 1984). Recently a Master Plan for drainage of agricultural lands has been prepared. It is estimated to cost Rs. 23000 million (Table 1.6). Clearly, in the absence of timely actions, the cost of land reclamation would increase tremendously.

Table 1.6 Salient recommendations on drainage in the drainage master plan of Haryana

Project	Quantity/area to be treated	Cost (Rs. million)
Surface drainage	(1,400 km^2) 14 lakh ha, 1,432 km length	12,310
Horizontal subsurface drainage	47,000 ha area	2,570
Vertical tube wells drainage	1,200 Nos. (24,000 ha)	430
Conjunctive use tube wells	1 lakh Nos. (2 lakh ha)	2,500
Bio-drainage	2 lakh ha area	3,850
Repair of canal water courses, water supply tanks	–	440
Road raising & cross drainage	–	680
Total Cost		22,780

Source: Anonymous (2000)

Gujarat

NEDCO (1968), a consultancy organization of the Netherlands, operated a Pilot Polder Project (1,500 ha) in Gujarat in order to suggest measure for the reclamation of the *Bhal* area, representing a former tidal flat abandoned by the sea. It comprised of Vertisols in the highest part and Halaquepts in the lower part, lying as wasteland or in use as poor pasture. The results obtained in the project had been poor due to various reasons including the presence of a layer of very low permeability in the upper meter of the soil profile. Swelling nature of the soil, low amount of fresh water carried by the rivers during the monsoon, irregular and short duration floods and absence of adequate sources of fresh ground water were mentioned as other causes. Discharge of water from the pilot polder could not be disposed off due to severe siltation of the outlet channel. Desalinization of the soil did not respond to improved drainage. A major recommendation emerged from the studies was not to proceed with the plans for reclaiming the entire area of saline soils in *Bhal* unless after extensive research more was known about desalinization. Other pre-requisites identified in the project were: whether or not in combination with flooding and drainage desalinization could occur, possibilities for storage of water (small dams) for irrigation in the area west of the Bhavnagar-Ahemdabad road, availability of fresh ground water in the area and possibilities of water storage in the Mithapur Depression. Measures to check erosion especially in the valley of the Khalkhalia and encroachment of the Sea in the Sonrai creek area by low levels around the threatened arable lands were also suggested for investigations.

Maharashtra (Deccan)

In the Deccan tract, the Nira valley irrigation canal was opened in 1884 to irrigate for the first time on a large scale the deep black cotton soil area and soon after, the land began to be affected by salts and water logging. Reports of water logging and soil salinity appeared in 1886 and by 1905 an area of 1600 ha was taken out of cultivation (out of an irrigated area of 14,800 ha) increasing to 5570 ha in 1947. In 1886, Dr. Ozanne, reported that damage was due to seepage and percolation from canals. In 1893, Leather visited the tract and sampled some of the affected soils. The problem, in this tract, became so serious that in 1903, the Government of Bombay appointed a committee to consider the measure to be adopted to check the spread of the damage. The committee recommended that all *nalas* (natural drainage lines) should be opened without delay, salt affected lands should be surveyed and reclamation trials carried out. Mann and Tamhane (1910) estimated that 6-7% of the area was being damaged annually, and even those lands on which sugarcane was being raised had accumulated soluble salts, whose adverse effect was being suppressed by the amounts of manure and irrigation water being used. Drainage trials laid out remained unsuccessful. A symposium on the 'Sodic soils of the Deccan' was held as early as in 1940 to take stock of the available information on this subject.

By 1947, for Deccan black cotton soil tract, where altogether different geomorphology existed, effective drainage and land improvement techniques were developed (Inglis and Gokhale, 1928), genetic soil survey was carried out and soil classification was made (Basu and Tagare, 1943). Talati (1941) reported that Deccan had four types of damaged soils excluding waterlogged areas: (i) mixed saline soils, (ii) saline soils, (iii) sodic soils and (iv) strongly sodic soils. These required different treatments for their improvement. The first type of soil improved merely by leaching while the saline soils became sodic either under natural rainfall or artificial conditions of leaching and required fertilizer for their improvement. Gypsum and Sulphur in combination with FYM were found the best. Sugarcane variety CO 290 was observed to be sodicity resistant and was tried after *dhaincha* (*Sesbania aculeata*) green manuring with a basal dose of gypsum and FYM. Talati (1947) also conducted field experiments on the reclamation of waterlogged salt affected lands in Baramati of the Bombay Deccan. The water table was lowered to a depth of more than 1.2 m by drainage followed by ploughing and levelling. Rainfall of 400 cm reduced salt content from 3 to less than 1% and reduced pH in top 1.5 m but increased in the lower horizon. For reclaimed soils under sugarcane, tilth was good up to pH 8.6 and yield decreased if the pH was beyond this value. Addition of $CaSO_4$ with FYM or sheep manure gave best results.

Punjab

Most of the canals in Punjab were constructed in the last quarter of nineteenth century. Bernee in 1908 conducted the survey of Lower Chenab canal area and field experiments on land reclamation began at Narwala in district Layallpur in 1914. By 1918 the problem of water logging and salinity was of so much concern that Government set up a Drainage Board. After the abnormal excessive rains in 1925, a Water Logging Inquiry Committee was constituted to carry out statistical investigations and experimental work pertaining to water logging and *thur* (saline soil). The Chakanwali Reclamation Farm in Gujranwala district (Now in Pakistan) was also opened with the simultaneous establishment of the Punjab Irrigation Research Laboratory at Lahore in 1924 to investigate all the problems connected with irrigation, drainage and salinity. This research laboratory rose to the status of a full-fledged Research Institute by 1931-32. Scientists like Mackenzie, Taylor, Puri, Ashghar and Mehta devoted themselves to research on the problems of soil salinity and water use. Standards of soil and water quality were established and the means of reclamation of deteriorated soils found. The implementation of these research findings in the field started in the late thirties. Since the problem continued to increase beyond all proportions, 'Water Logging Inquiry Committee' was expanded and converted into 'Water Logging Board'. In 1929, Punjab Government agreed to construct main and branch drains but the responsibility for the construction of field drains was left with the landowners. The position was reviewed annually in water logging conference presided over by the Governor.

Commencing from 1927 to 1941 was a period of fruitful research in Punjab on the subject of saline-sodic soils. Puri and Associates (1933, 1937), Singh and Nijhawan (1932), Hoon and Associates (1937, 1949), Mehta (1940); Taylor and Associates (1933, 1934, 1941) in the undivided Punjab made valuable contributions to the subject of sodic soils, their characterization, diagnosis and reclamation. Singh and Nijhawan (1932) reported that calcium salts followed by farmyard manure can improve the sodic soils to the status of normal soil. Puri (1933) reported the relationship between exchangeable sodium and crop yield in Punjab soils. He observed that treatment of a Montgomery soil with $CaCl_2$ and leaching reduced base saturation from 67 to 20% and increased yields from 32 to 100 kg ha^{-1}. During the same period, some of the factors determining the hydrogen ion concentration in suspension of Punjab soils were also examined. Taylor *et al.* (1933) statistically analyzed the water table data for the last several decades and showed that a regular rise of water table has taken place in canal irrigated tracts. They attributed this to the combined effect of seepage from the eastern channels, heavy percolation from irrigated fields, pronounced effect of rainfall on the water table on account of percolation through ploughed cultivated fields, blockade of natural drains and above all lack of drainage.

Taylor *et al.* (1934) pointed out that the crop yields declined above pH 8.5. Application of gypsum was suggested to prevent base exchange reaction. Mehta (1940) observed that salt could be removed from *reh* soil by one or two crops of rice and the reclaimed land could produce excellent crops of cotton, wheat, sugarcane etc for eight years without deterioration. The inclusion of rice in the crop rotation was recommended. Taylor and Mehta (1951) reported that taking a rice crop at the initial stage of irrigation could prevent land deterioration due to excess salts. Many leaching, drainage and reclamation trials were conducted at the Chakanwali Reclamation Farm and effective methods suited to Punjab soils were developed. Water rating, the problems of irrigation and rise of water table were also simultaneously studied at length to check the deterioration of land against indiscriminate use of water.

In 1940, 'Land Reclamation Board' distinct from 'Water logging Board' was constituted and in 1945 a separate 'Land Reclamation Directorate' was created.

Valuable work on the movement of water and the salts in the soil was initiated before 1947. Unfortunately, the partition of the country in 1947 interrupted this work. Most of the reclamation and drainage work was confined to areas that now constitute a part of Pakistan.

The problem created by the abnormal floods of 1947, 1950, 1954, 1955 and 1956 aggravated the water logging and *thur* problems in Punjab and drainage became a quite serious issue. The Punjab Government set up an Action Committee in 1957 in order to tackle this problem effectively on a top priority basis. The Action Committee recommended the following remedial measures:

- Minimization, if not total prevention of flooding of the affected area from river, big streams and creeks through improvements in surface drainage.
- Installation of an adequate number of tube wells and revival of percolation wells fallen into disuse to stabilize the sub-soil water at the proper level.
- Lining of watercourses, main canals and distributaries, if economically possible.
- Increase in the water allowance in *thur* affected areas so as to attain an intensity of 200%.
- Construction of seepage drains parallel to the existing channels.

Hoon and Associates (1950, 1953, 1955), Dhawan and Associates (1950, 1953, 1955, 1958), Asghar and Associates (1947, 1948), Kanwar and Associates (1960, 1961, 1962) contributed greatly in Punjab. Central Board of Irrigation and Power published three research monographs on Land Reclamation viz. Mehta (1951), Hoon (1955) and Dhawan (1964). In first monograph, formation of *thur* lands, movement of moisture and salts in the soil profile, rise of water table, water logging, reclamation in deep and shallow water table areas etc had been discussed. Hoon (1955) pointed out that due to ability to withstand salt in the soil and a fair range of variation in irrigation intensity, sugarcane offered immense possibilities for soil amelioration, when the top of the soil profile is somewhat free from excessive salt content. Dhawan (1964) observed the main method of reclamation consisted of land levelling followed by planting of *dhaincha*. The water was allowed to stand 8 to 10 cm deep for about two and half months. *Dhaincha* was buried when its age was about 6 weeks. After these operations, the fields were dried, ploughed and rice seedlings were transplanted preferably in rows with two or more seedlings at each spot. After harvesting rice in October, gram or *berseem* was grown in the winter depending upon the degree of reclamation. Construction of drains was recommended.

The problems of water logging and soil salinity are still prevalent in southwest Punjab and have assumed serious dimension. As a result Planning Commission (2013) constituted a high power committee to study the problems and suggest remedial measures. The report lists a number of reclamation and management options including lining of canals and subsurface drainage. A comprehensive review paper on water logging and soil salinization in India highlighting the package of reclamation and its application potential in southwest Punjab has been published (Gupta, 2015). By March 2015, 0.55 M ha of sodic lands in Punjab have been reclaimed.

Uttar Pradesh

Dr. J.W. Leather was appointed as first Agricultural Chemist in Uttar Pradesh and deserves to be called the first scientific worker on soil salinity in the country. He made a thorough systematic and scientific multi-pronged attack on *reh* (sodic) problem. Beginning his studies in 1893, Leather published several research findings on the 'reclamation of *reh* or *usar* lands' in Agricultural Ledger during 1893 and 1897. The first and the foremost scientific treatise on the subject 'Investigations on *usar* Land in the United Provinces' by Leather, was published, in the form of a report, by Government Press, Allahabad in 1914 (Leather, 1914). Some of the salient conclusions which emerged from his studies in the districts of Etah and Mainpuri, in the vicinity of Lower Ganges canal, are as follows:

- *Usar* soils contained practically sodium carbonate and bicarbonate, which amounted to 0.06 to 0.90% respectively in the surface 60 cm depth. Total soluble salt content seldom exceeded 2%. A good arable soil contained less than 0.1% soluble salts.
- Infertile (*usar*) patches were impervious, the water penetration being 1/10 to 1/1000 of normal soil and always less than 0.1cm h^{-1} and associated with water logging at about 2.8 to 3.6m from the soil surface. Sub-soil water was considered to rise and fall freely below the impervious patches of *usar* soil, being 2.8 to 3.2m thick.
- In respect of frequency, infertility, nature and amount of salts or physical condition, there was substantially no difference between *usar* patch irrigated with canal water or unirrigated. The sodicity in the soil was presumed to be a product of the soil itself. Canal water would possibly not create such patches, but there could be possible extension of *reh* due to water logging upon introduction of canal irrigation.
- Lowering of sub-soil water or opening of drains was though necessary to reclaim these soils but experiments showed that surface and subsurface drainage were ineffective.
- Gypsum and canal silt were found to reclaim *usar* land but the cost was prohibitive.
- Deep cultivation coupled with heavy manuring reclaimed only surface 30 cm of the soil. Scrapping of salts was useless.
- Allowing 60 cm of water to percolate through the soil for a prolonged period in well maintained dykes (*bunds*) coupled with small amounts of gypsum of fine size was also supposed to remove salts to a lower stratum.
- Reclamation by tree (*babul* or *shisham*) planting or by growing of grasses after enclosing the land to avoid grazing for some period was found encouraging.

- Pot culture experiments revealed that sodium carbonate up to 0.4% and chloride and sulphate up to 1.0% were not generally fatal to maize, wheat, barley, cotton etc. but germination was severely affected at 0.7% sodium carbonate or sodium sulphate. The germination of leguminous plants, gram, peas and pigeon pea was interfered with 0.2-0.4% sodium carbonate or sodium chloride and 0.7% sodium sulphate.
- Investigation on the quality of waters of Mathura district showed that well waters were considerably saline as compared to canal waters, which were the purest form of irrigation water available. Some of the well waters contained 1,250-10,000 ppm salts. The upper limit of TSS of waters suitable for irrigation was fixed as 1000 ppm (EC 1.5 dS m^{-1}). This limit was considered safe only when the irrigation water was used in excess of what was required for maturing the crops.
- The origin of the salts was ascribed to four possible ways: (a) sub-soil bed of salts brought at least in part by the rivers when the alluvium of the plains was deposited; (b) those portions of the salt which have been rendered soluble by water, air and plants and rejected by the latter, which are not drained away and (c) brought in part by the canal water used for irrigation. Imperfect drainage, surface washing, wind drift and canal were considered as causes for the accumulation of *reh*.

Duthie (1896) compared the *usar* reserves in Chherat and Gursikran of Aligarh district in 1893-94 with that in 1888. The grass *bat (Diplachne fusca)* was observed to be more useful than *usar* grass (*Sporobolus pallidus*) on the low lying regions, where water was liable to accumulate for any length of time. Planting of *Dhak* (*Butea frondosa*) tree was recommended. The *frash* tree (*Tamarix artculata*) also thrived on saline soil. Although, Leather recognized as early as 1897 that calcium salts coagulated *usar* soil and made them permeable, he calculated gypsum requirement of the soil as simply the quantity which was equivalent to soluble carbonates present. The formation, characterization and reclamation of sodic soils, in fact, remained obscure until the implication of the principles of Base Exchange were fully known.

Dhar and Mukherji (1936) and Dhar (1939) tried a number of reclaiming agents particularly molasses and press mud for the amelioration and reclamation of saline-sodic soils. But the gravity of the problem went on increasing as a consequence of which an *Usar* Land Reclamation Committee was constituted in 1938 by the State Government of Uttar Pradesh to advise on all matters related to the characterization, reclamation and prevention of spread of *usar* land. The committee reviewed the work done on such soils, and examined the problem in great detail and reported that the sodic conditions were mainly due to sodium carbonate. The profiles of these soils are generally associated with *kankar* or definitely a hard pan, which restricted the upward movement of water by capillary action and downward flow of water by percolation under force of gravity. The setting-up of a Wasteland Reclamation Board to deal with all the problems of wastelands was suggested.

The 15 years of post-independence period (1947-1962) saw a great progress in research on saline-sodic soils in UP (India). Agarwal and Associates (1956, 1957, 1961) and Khan (1950) studied several aspects of soil salinity problem in Uttar Pradesh. Currently, CSSRI, Regional Research Station, Lucknow and UP *Bhumi Sudhar Nigam*, Lucknow are operating in the state to develop new technologies to reclaim sodic lands and extend the land reclamation programs to the farming community in various districts of the state. It is heartening to note that the efforts have brought fruits resulting in reclamation of 0.70 M ha of sodic lands in the state by March 2015. The current pace of land reclamation needs to be enhanced.

Rajasthan

Rajasthan is located in the northwest part of India and is the second largest state in the country. With only 1% of India's total water supply in the state, the canal systems supplying much of the water used for irrigation purposes depend upon water sources located outside the state. Some of the identified problems include declining water table except in canal commands, increasing salinity of ground water and large pockets of waterlogged salt affected lands in the irrigation commands. As an illustration by 1972, approximately 10 years after irrigation development began, it was estimated that approximately 20,000 ha area had a water table within 1.5 m of ground surface. The affected area at that time was increasing at the rate of 1%. Even after 40 years the problem is expanding as revealed through satellite imageries.

The situation is no different in the Indira Gandhi Canal Project, one of the biggest irrigation projects of the world implemented in the state. The project cover CCA of 19.64 lakh ha and irrigate 15.17 lakh ha in the districts of Sriganagnagar, Hanumangarh, Churu, Bikanar, Jaisalmer, Jodhpur and Barmer. Without doubt the project has transformed the destiny of this desert land and its inhabitants. In phase I, crops like wheat, rice, sugarcane, pulses, mustard, cotton and groundnut are being grown raising the economic standard of the people living in the area. Loonkaransar terrain comprising of a sub-soil hard pan, traditionally known as the no-man's-land (*Kalapani* of Bikaner) has been transformed into a prosperous land of plenty with the arrival of water by lift irrigation. This one time forbidden land now produces groundnut worth Rs. 100 million annually. While, these are issues of pride, environmental degradation resulting from lack of drainage has become a matter of concern both for the project and for the nation, which has invested heavily in the project. In less than a decade, the scarcity-prosperity-scarcity cycle seems to have run a full circle. Introduction of canal irrigation in the desert of western Rajasthan has completed journey from one wasteland (water starved) to another wasteland (water logging). In the year 1997-98, it was estimated that water logging has affected an area of 43,428 ha (area inundated with water and water table within 1.5 m of the ground surface) in stage I and 3,978 ha in stage II. The most serious aspect of this problem is that affected area is increasing over the years. Sub-soil conditions resulting in water logging, a rising water table (43 to 83 cm a year) and the inevitable salinity have already rendered major part of the land useless for agriculture (Fig. 1.8). It is anticipated that 20 to 60% of the canal-irrigated area may develop drainage and salinity problems within 30 to 100 years.

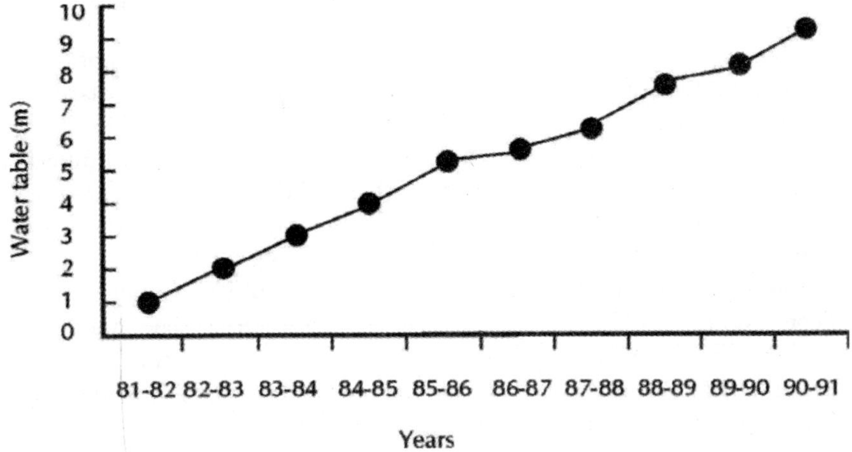

Fig. 1.8 The rise in water table in Indira Gandhi *Nahar pariyojana,* Phase I (Anonymous, 2002)

It is a matter of satisfaction that some respite in the problems has been observed in recent years as the vertical drainage is being extended to several areas in the command. The conjunctive use of ground water and canal water is gaining momentum.

Other States

Studies were conducted throughout the length and breadth on India where problems of water logging and soil salinity existed i.e. in West Bengal (Banerjee, 1959); Gujarat (Shah and Associates 1954, 1958), Madras (Satyanarayana, 1953), Andhra Pradesh (Krishna *et al.*, 1960); Maharashtra (Basu, 1950); Delhi (Raychaudhary and Associates, 1952, 1957). The Indian Council of Agricultural Research organized a well-planned Seminar on the Recent Advances in Agronomy and Soil Science at Shimla in 1959. It included a session on saline and sodic soils. The Council desired that Dr. R.R. Agarwal and Dr. R.N. Gupta of Uttar Pradesh collect related information from the available sources and write a bulletin on the subject. In consequence, a bulletin entitled 'Saline-Alkali Soils in India' was prepared which covered research work on the subject and was published by ICAR in 1968.

A seminar on 'Salinity and Sodic Soil Problems', held at Indian Agricultural Research Institute, New Delhi in 1962, recommended reclamation of saline and sodic soils on a war footing at the national level. It also recommended that establishment of a suitably manned and equipped Central Salinity and Sodic Laboratory, already approved in the Third Five Year Plan should be given the highest priority.

A number of schemes to conduct research on fundamental and applied aspects of the salinity problem were launched by the Indian Council of Agricultural Research. The scheme operated in Punjab from 1958 to 1961, Rajasthan from 1960 to 1963 and in Andhra Pradesh from 1955 to 1960. A scheme for research on possibilities of using charcoal in improvement and reclamation of soils was also sanctioned by ICAR (Puri, 1960). A scheme on treatment of saline waters was also sanctioned in Punjab during 1960-61. Another scheme was sanctioned for the study of composition of plants growing on *usar* lands under reclamation to Lucknow University during 1957-60. During 1960-70 research work on the problem of soil salinity continued to be strengthened at various places such as Central Arid Zone Research Institute, Jodhpur; University of Udaipur, Udaipur and State Deptt. of Agriculture, Rajasthan; University School of Sciences, Gujarat University, Ahmedabad; Irrigation Research Laboratory, Baroda; Govt. Agriculture College, Kanpur; B. R. College, Agra; B.H.U., Varanasi; Dept. of Chemistry, Allahabad University, Allahabad; Botany Deptt. Lucknow University, Lucknow; College of Agriculture, P.A.U., Ludhiana; Irrigation Research Institute, Amritsar in Punjab and Indian Agricultural Research Institute, New Delhi.

While more or less systematic research work continued in some parts of the country, increasing need was being felt seriously for the intensification of the studies on the problem soils. Planning Commission Committee on Natural Resources (1963) prepared a study report on wastelands, including saline sodic and waterlogged lands and their reclamation measures. The report recommended setting up of three Regional Salinity Laboratories in the country. This recommendation was not accepted immediately and concerted efforts to solve the problem on national basis remained lacking, though the salinity continued to induce creeping death to the soil.

USAID-INDIAN TEAM

In 1967, Indian Council of Agriculture Research appointed a Consultant Team to suggest the plan for research on water use and soil management towards meeting India's food shortages. The team made a thorough study of the problems. The team recommended that to increase research effectiveness on immediate food production goals, researches on applied or adaptive nature should begin immediately. ICAR should coordinate on water use and soil management research and training activities. In states, Agricultural Universities should carry out researches that have regional significance. To provide essential technological information for the water and soil management component of the 'Package of Practice', several research projects were recommended. Great emphasis was laid on pilot projects. In addition, the team recommended for early initiation of the following measures:

- Establishment of a coordinated drainage-salinity pilot research project at one national center and regional centers (Black soils-Maharashtra or Madhya Pradesh; mixed Red and Black soils-Karnataka or Odisha;

Coastal alluvial soils- Odisha or West Bengal) to evaluate the application of principles of drainage and management of salinity and sodicity in cultivated soils of India.

- In states other than those selected for the centers and where salt problems are acute, appointment of an Indian Scientist having in depth knowledge to aid cultivators.
- A research program on basic properties in relation to soil and water management practices and production of crops should be initiated at four centers in India.
- Establish a Central Water Management Research Center at the Indian Agricultural Research Institute to study water management problems of national significance.
- Research on problems of well design, construction and maintenance, including materials and equipment and evaluation of pumping costs.
- Research on tanks at catchment level and storage structures for water supplies.
- Studies on the problems of recharge of underground aquifers.
- Examination of existing approaches to research on rainfed lands and a substantial strengthening of programs at the Soil Conservation Research, Demonstration and Training Centers throughout India.

As a consequence of these recommendations, ICAR launched in 1968 an All India Coordinated Scheme for 'Studies on Soil Salinity and Water Management' at different locations and established Central Soil Salinity Research Institute (CSSRI) at Karnal and Water Technology Center at New Delhi (Indian Agricultural Research Institute) in 1969. It is just by coincidence that CSSRI happens to be located on the banks of the Western Jamuna Canal and the village Munak where first reports of salinity appeared, is just 30 km away from its campus (Fig. 1.7 left). Another Coordinated scheme on 'Use of Saline Water in Agriculture' came in operation in 1972 at 5 centers in the country in the states of Uttar Pradesh, Rajasthan, Karnataka, Andhra Pradesh and Maharashtra. Since then the scheme has been expanded and has 14 centers including the coordinating unit at Karanl with its mandate widened to cover new challenging problems. With the establishment of two Research Institute/Centers and two Coordinated Schemes, and with sufficient budget grants, the lacunae on the part of the Government had almost been removed. The outcome of the findings on the problem and its solution was expected to pave the path for sustainable irrigated farming in future. A brief list of important activities that have happened in India are listed in Table 1.7.

Table 1.7 List of important activities related to soil salinity/drainage in India

Decade	Important Activities
19th Century	
8th	Whitcomb Report published, Ist subsurface drainage experiment established, Water logging conference organized at Aligarh
9th	*Reh* Committee Report released
10th	Dr. J.A. Voelcker report made available, Leather report documenting research on drainage and other activities in land reclamation published
20th Century	
1st	Drainage project in Amritsar Punjab and Coimbatore in Tamil Nadu laid
2nd	Mole drainage experiment laid out in Lyallpur (now in Pakistan), Experiments conducted at Baramati (Maharashtra)
3rd	Report on land drainage published by Inglish and Gokhle, Chakanwali Reclamation Farm established (now in Pakistan) and drainage experiments initiated
4th	Report on land drainage published by Evershed, Seminar on sodic soils of Deccan organized, Land Reclamation Board constituted in Punjab, Usar Land Reclamation Committee constituted in UP
5th	Land Reclamation Directorate in Punjab established
6th	Land reclamation report brought out by CBIP, New Delhi; Experiment on SSD laid out in Hansi (Haryana), Punjab action committee on drainage constituted, A special session on land reclamation organized at Simla conference of ICAR
7th	Several subsurface drainage experiments notably in Punjab, Rajasthan, Uttar Pradesh, Odisha established, ICAR scheme on Soil Salinity and Water Management launched, Technical reports by CBIP and ICAR published Central Soil Salinity Research Institute established, Planning commission Report on Wastelands released
8th	Drainage experiment laid out at Karnal and Sampla, Central Board of Irrigation and Power organized Seminar on Water logging and Remedial Measures, Technical report by ICAR revised
9th	Experiments at Sampla extended under Indo-Dutch collaboration, Several experiments established under ICAR Drainage Scheme
10th	Indo-Dutch Network Project launched, RAJAD project came into being, Haryana Operational Pilot Project (HOPP) began, Ministry of Water Resources and Ministry of Rural Development funded drainage projects launched, Drainage Master Plans of several states finalized, National Conference on Salinity Management organized, National Drainage Water Policy under preparation, ICID Drainage Workshop organized, drain laying machines imported under HOPP
21st century	
1st	ICID International Drainage Workshop organized, HOPP imported two more drain laying machines; Water Resources Department Punjab also imported two macines, Rex Polyextrusion Ltd. Joined hands with Water Resources Department in Maharashtra, Imported the drain laying machine under Public-Private Partnership mode, CSSRI expanded its activities at national and international level, Funding under *Rashtriya Krishi Vikas Yajana* (RKVY) initiated in several states
2nd	Manual on Land Drainage submitted to Education Division of the Indian Council of Agricultural Research, CBIP symposium on Water Logging and Soil salinity in Irrigated Agriculture

EXTENT AND DISTRIBUTION OF SALT AFFECTED SOILS

The extent of sodic soils is not precisely known. A first attempt to compile information on the extent of salt-affected soils on a worldwide basis was made by Massoud based on the FAO/UNESCO Soil Map of the World (Annexure 1). Approximately 1000 M ha is assessed to be salt affected (Dudal and Purnell 1986). Szabolcs (1974) also reported the salt affected soils of Europe and presented maps showing the distribution of salt affected soils in most continents. Szabolcs (1979) gave a distribution of saline/sodic soils (Szabolcs, 1974; Table 1.8). Estimates given by Dregne *et al.* (1991), Oldeman *et al.*, 1991 and FAO (1997) suggest that around 831 M ha area is salt affected out of which 397 M ha is saline and 434 M ha is sodic. More than 77 M ha of land worldwide has become salt affected due to human causes (secondary salinity); of this 45.4 M ha are irrigated land (in the world's semi-arid and arid regions) and the remaining rain fed (Table 1.9). In India, several estimates ranging from 7.0 to 26.0 M ha were reported by various agencies. For example, MOA (1994) reported 6.3 M ha, NCA (1976) 7.16 M ha, SPWD 7.17 M ha, Singh (1992) 8.11 M ha, NRSA (1996) 3.90 M ha, NBSS&LUP 6.20 M ha, Massoud (1974) 23.2 M ha, Abrol and Bhumbla (1971) 7.0 M ha. The data in Annexure 1 reveal that India has about 23.8 M ha of salt affected soils with 23.22 M ha of saline and 0.58 M ha of sodic soils. The current data compiled by CSSRI, Karnal do not support this as only 6.73 M ha area is salt affected with 3.77 M ha sodic and 2.95 M ha saline including coastal saline (Table 1.10). Salinity problem is acute in four major tracts of India:

- The arid tract of Rajasthan and Gujarat
- Semi-arid alluvial tracts of Punjab, Haryana and Uttar Pradesh
- The arid and semi-arid tracts of Southern States, and
- The coastal alluvium.

As per the future projections made on an all India basis, an area of about 11 M ha is likely to be affected by the problems of water logging and soil salinity in the irrigation commands of India by 2025 and 20 M ha by 2050 (CSSRI, 2013). It calls for a fresh look on the global distribution of salt affected soils ending up with a new compilation. The distribution of sodic soils in India is given in Fig. 1.9 and Table 1.11 showing that several states are affected by the problem.

Table 1.8 Global distribution of saline and sodic soils

Continent	Area (million ha)		
	Saline	Sodic	Total
North America	6.2	9.6	15.8
Central America	2.0	–	2.0
South America	69.4	59.6	129.0
Africa	53.5	27.0	80.5
South Asia	83.3	1.8	85.1
North and Central Asia	91.6	120.1	211.7
Southeast Asia	20.0	–	20.0
Europe	7.8	22.9	30.7
Australasia	17.4	340.0	357.4
Total	351.5	581.0	932.2

Source: Szabolcs (1989a)

Table 1.9 Global estimate of secondary salinization in the world's irrigated lands

Country	Total land area cropped (M ha)	Area irrigated		Area of irrigated land that is salt-affected	
		M ha	%	M ha	%
China	97	45	46	6.7	15
India	169	42	25	7.0	17
Soviet Union	233	21	9	3.7	18
United States	190	18	10	4.2	23
Pakistan	21	16	78	4.2	26
Iran	15	6	39	1.7	30
Thailand	20	4	20	0.4	10
Egypt	3	3	100	0.9	33
Australia	47	2	4	0.2	9
Argentina	36	2	5	0.6	34
South Africa	13	1	9	0.1	9
Subtotal	843	159	19	29.6	20
World	1,474	227	15	45.4	20

Source: Ghassemi *et al.* (1995) compiled from FAO data for 1987

Table 1.10 Extent of salt affected soils in India (ha)

State	Saline	Sodic	Coastal saline*	Total
Andhra Pradesh	0	196609	77598	274207
Andaman & Nicobar Island	0	0	77000	77000
Bihar	47301	105852	0	153153
Gujarat	1218255	541430	462315	2222000
Haryana	49157	183399	0	232556
Karnataka	1307	148136	586	150029
Kerala	0	0	20000	20000
Madhya Pradesh	0	139720	0	139720
Maharashtra	177093	422670	6996	606759
Odisha	0	0	147138	147138
Punjab	0	151717	0	151717
Rajasthan	195571	179371	0	374942
Tamil Nadu	0	354784	13231	368015
Uttar Pradesh	21989	1346971	0	1368960
West Bengal	0	0	441272	441272
Total (ha)	1710673	3770659	1246136	6727468
Total (M ha)	1.71	3.77	1.25	6.73

Source: Adapted from Mondal *et al.* (2011). There is mismatch between the figures quoted since the extent as high as 3.1 M ha for coastal saline soils are being reported.

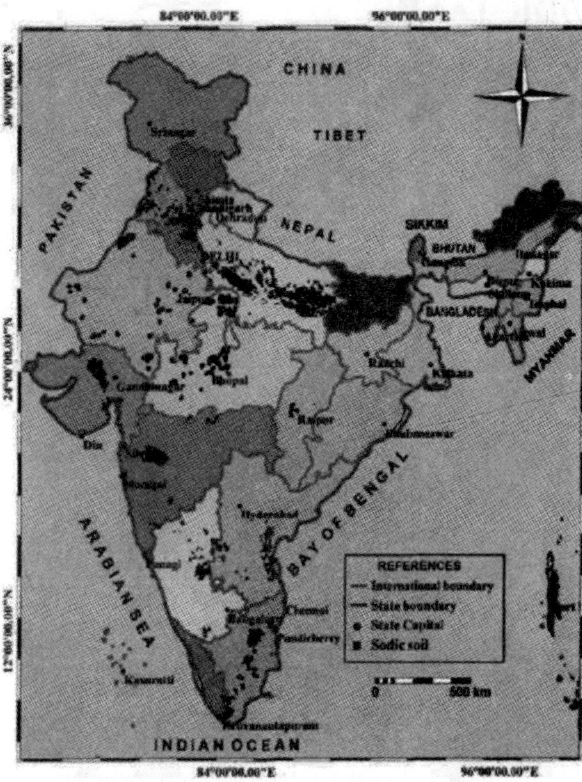

Fig. 1.9 Distribution of sodic soils in India

Table 1.11 Distribution of sodic soils in India

State	Districts affected by the problem
Andhra Pradesh	Nizamabad, Karim Nagar, Kurnool, Guntur, Anantapur, Cuddapah
Bihar	Muzaffarpur, Shahabad
Delhi	Delhi
Gujarat	Navsari, Bharuch, Surat, Bhavnagar, Kaira
Haryana	Karnal, Kurukshetra, Jind and Parts of Rohtak, Sonepat and Gurgaon
Karnataka	Bellary, Bijapur, Dharwar, Chital, Durg, Gulbarga
Madhya Pradesh	Gwalior division, pockets in West and East Nimar districts and along Narmada in black soil area, Bhind, Morena
Maharashtra	Sholapur, Nasik, Jalgaon, Parbhani, Ahmednagar, Satara
Punjab	Amritsar, Kapurthala, Sangrur, Patiala, Gurdaspur and Ropar
Rajasthan	Parts of Jaipur, Kota, Tonk, Bharatpur, Sawai Madhopur and Bhilwara
Tamil Nadu	South Arcot, Salem, Thanjavur, North Arcot, Coimbatore, Madurai and Ramanathapuram
West Bengal	24 Parganas, Midnapur
Uttar Pradesh	Aligarh, Etah, Etawah, Mainpuri, Hardoi, Kanpur, Unnao, Lucknow, Fatehpur, Rai Bareli, Sultanpur, Azamgarh, Ghazipur, Partapgarh and Meerut

GENESIS OF SODIC LANDS

A distinction is made between two types of sodicity- primary and secondary also referred as natural or man-made (anthropogenic). Primary sodicity is a consequence of the parent material producing sodium carbonate (Na_2CO_3) upon weathering or other geological processes. Secondary sodicity is a consequence of changes in land use or other anthropogenic interventions. It may be due to introduction of surface irrigation or removal of vegetation causing rising water tables, which brings soluble salts to the soil surface. Besides it may also occur because of accumulation of salts applied with irrigation water, sea water inundation in coastal regions and deposition of windblown salts. In most cases poor drainage is a major factor in the genesis of sodic soils.

The investigations on the origin of sources of the sodic salts in the soils led to quite a number of theories. As the investigations progressed, some early theories were discarded while many new theories and obscure facts came into limelight. In spite of considerable difference of opinion, we have a better perspective and much clearer idea of the sources of the soluble salts in the soil today than before. Nonetheless, there is a need to generate more data before it is possible to state definitely the causes and genesis of sodic soils under different agro-climatic conditions. The following paragraphs discuss some of the most common theories postulated to trace the formation of sodic soils.

During the process of chemical weathering which involves hydrolysis, hydration, solution, oxidation, carbonation and other processes, the salt constituents are gradually released and are transported away from their source of origin through surface or ground water. According to Kovda *et al.* (1973), mixed salts while migrating fractionate the constituents in a manner that the least soluble salt are retained close to the source, whereas the most soluble and mobile salts move to accumulate at the farthest end. Similarly, Hardy and Eugster (1970) suggested that in an evaporating water containing mixed salts, first to precipitate is calcite followed by gypsum or sepiolite (magnesium silicate) depending upon the ionic environment, the residual solution being made up of more soluble salts. These two theories are not very different to conclude that salt precipitation depends on the relative solubility of the ionic species, most soluble ions staying in solution at high concentrations. Ions with higher solubility move closer to the soil surface than less soluble ions, when the upward movement of water in the soil prevails. Saline soils thus, show stratified deposition of different salts in different layers. The predominant ions near the site of weathering in the presence of carbon dioxide will be carbonates and bicarbonates of calcium, magnesium, potassium and sodium albeit with low concentrations. As the water moves from the more humid to the arid regions, the salts gets concentrated. During evaporation, sodium salts separate from calcium salts because of differential solubility. The soluble sodium salts concentrate high in the soil profile. As the soil solution becomes more concentrated with sodium, the percentage of sodium on the exchange complex also increases. Eventually there is enough exchangeable sodium to cause dispersion of clay and organic matter.

The first scientifically supported theory on the origin of alkali/sodic soils was presented by E. W. Hilgard, who pointed out the difference between the littoral and the terrestrial salty soils, latter being designated as sodic soils. According to this theory, these soils are the result of insufficient rainfall to leach out the salts that are formed progressively by the weathering of soils minerals; therefore, the more arid the climate, the more chance there is for the accumulation of sodium in the soil. Hilgard treated the sodic soils as more or less impregnated with different sodium salts and emphasized largely the obnoxious effects of the latter on crops and fruit trees (As reported by De'sigmond, 1925 and translated by Waksman). Considering many critical comments, De'sigmond (1925) concluded that sodic soils may originate under arid and semiarid conditions, where in the past there has been a temporary abundance of humidity in the soil intersected by dry periods. In the case of the Hungarian sodic lands it is historically indicated by the large inundational territories of earlier times, resulting partly in swampy and peaty lands and partly in sodic soils. Another combination of favorable conditions for the origin of sodic soils was the collection of seepage water in the lower basins and water rivulets in the large sandy territory between the two main rivers, the Danube and the Tisza (Theiss), the bottom of these lowlands being always impervious clay or calcareous hardpan. In both cases there was an opportunity for the sodium salts to accumulate on the surface, and the more or less concentrated sodium salt solutions reacted with the zeolitehumate- complex in the soil replacing in some degree the calcium, magnesium, and potassium by the sodium cation (De'Sigmond, 1924). Harris (1920) also presented an exhaustive treatise on origin and genesis of sodic soils.

Most sodic soils seem to have passed through four stages, namely, salinization, sodication, desalinisation and degradation. Thus, one can say that the formation of sodic soils is primarily traced to soil salinization. Whether or not an sodic soil will pass beyond the salinization stage will depend on the composition of the soluble salts that accumulate at the soil salinization stage. If sodium salts predominate, sodication of the exchange complex is bound to take place. Desalinisation of sodicated soil is usually accompanied by the formation of Na_2CO_3 and by pronounced deflocculation. If $CaCO_3$ is present, the formation of Na_2CO_3 is accompanied by the replacement of absorbed sodium by calcium. Under these conditions the degradation stage will not be attained. If $CaCO_3$ is absent, base-unsaturated constituents arise under leaching conditions and more or less of the inorganic-exchange complex becomes decomposed into simple oxides (Kelley, 1934).

De'Sigmond (1924, 1925, 1938) considered degraded sodic soils to be an advanced stage of soil development resulting from the washing out of salts. The details of the type of soil developed as the leaching proceeds depend on local conditions, particularly soil texture and type of clay present. As a result of the leaching processes there is a tendency for the dispersed clay and organic matter to move down the profile resulting in the formation of a dark, extremely compact layer having a sharply defined upper surface and merging gradually into the subsoil with increasing depth. As a result, while a few inches of the surface soil may be relatively coarse in texture and friable; but below, where the clay accumulates, the soil may develop a dense layer of low permeability that may have a columnar or prismatic structure. A non-saline-sodic soil develop characteristic morphological features only after sufficient time has elapsed.

The term sodication was first used by the Soil Science Society of America for pedogenic processes that increase the exchangeable Na^+ content of the soil. Sodication may also be a natural process in the presence of Na^+-rich sedimentary materials (e.g. Na-evaporites or Na-rich glacial till) or rocks that can weather releasing Na^+. On the other hand, the mechanisms responsible for the formation of sodium carbonate in soils which characterize sodic soils have been discussed in several standard works (Kelley, 1951). Sodic soils of Egypt in Wadi Tumilat, Ferhash and Wadi-El-Natroun are reported to have been formed by desalinization in the absence of enough divalent cations in some parts of the Nile Delta, by high carbonate and bicarbonate water in Wadi Tumilat and by denitrification and sulphate reduction under anaerobic conditions in Wadi-El-Natroun (Elgabaly, 1971). Reduction of sulphate ions under anaerobic conditions and in the presence of organic matter has been reported to result in the formation of sodium carbonate as per the following reaction (Whitting and Janitzky, 1963):

$$Na_2SO_4 \rightarrow Na_2S \rightarrow Na_2CO_3 + H_2S \quad (1.5)$$

The sulfide gas leaves the system.

Salt pans in many soils can be attributed to aeolian addition originating from the nearby sea through wind flows towards the inland. Many soils have a surface accumulation of secondary lime, which may also be aeolian. The salinization of irrigated soils by ground water has become a major process of soil formation in many areas of the Sultanate, particularly in coastal areas. These sodic soils are not saline but their soil solution is more concentrated than in brown steppe soils. The abrupt boundary between the brown steppe soils and the sodic soils is related with morphological and geochemical changes, resulting from calcite and fluorite precipitation in the soil solution. The soil distribution (brown steppe soils and sodic soils) probably inherited from former mesomorphology which includes a pond, arise from two hydrological regimes in contrast. From a geochemical point of view, sodication sets on by concentration of the soil solution, which evolves in a carbonate alkaline way, with precipitation of calcite, fluorite and probably magnesium silicates (Barbiéro and Van Vliet-Lanoe, 1998).

Sodication may occur when saline ground water moves toward the soil surface by capillary rise from a shallow water table by lateral flow seeping out or by artesian flow. Desalinization in absence of enough divalent cations is the major cause of secondary solonization and solodization that occur in human-degraded sodic soils (FAO-AGL, 2000). In arid and semiarid regions, where annual precipitation is insufficient to meet the evaporation needs of plants, salts do not get leached and start accumulating in the root zone. Even if downward movement of water is not a limiting factor, the salt will move downward to the wetting front. On the other hand, if plants take up most of water, then salt will accumulate near the surface. In such cases, under steady sate conditions, salt in the soil tends to match that of irrigation water although leaching fraction of water, which is the portion of water passing through the root zone, controls the salt profile. Beek and van Breemen (1973) pointed out that highly sodic soils can develop in a closed basin with an excess of evaporation over precipitation if the inflowing water has a positive residual sodicity. In Pakistan 2.3 M ha area is sodic as a result of irrigation with sodic water (Qureshi and Barret-Lennard, 1998). Water from shallow ground water also moves up by capillary action to the soil surface, where it evaporates and leaves behind the salts, accumulation being quite high in the top 15 cm of soil (Fig. 1.10). The ground water containing residual sodicity would result in the formation of sodic soils.

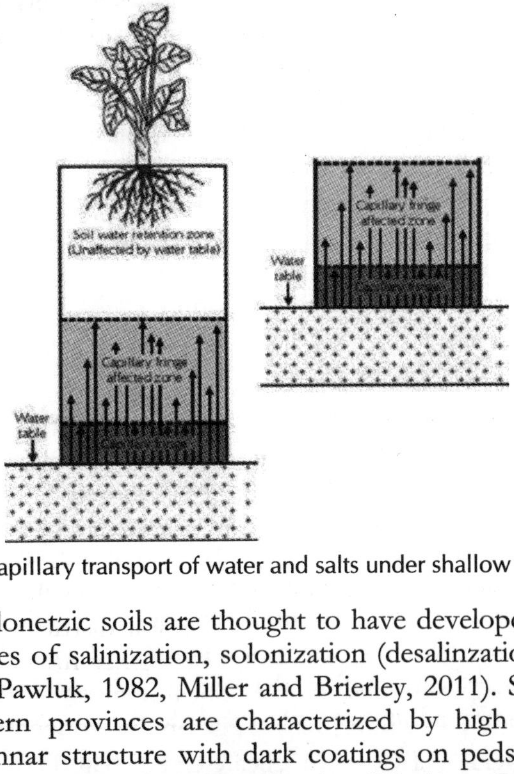

Fig. 1.10 Capillary transport of water and salts under shallow water table

In Canada, Solonetzic soils are thought to have developed via the stepwise pedogenic processes of salinization, solonization (desalinzation and sodication), and solodization (Pawluk, 1982, Miller and Brierley, 2011). Solonetzic or sodic soils in the western provinces are characterized by high exchangeable Na, prismatic or columnar structure with dark coatings on peds, and hard to very hard consistence when dry (Soil Classification Working Group 1998). These soils have a Solonetzic B horizon designated as a Bn or Bnt horizon. Soluble salts are brought into the soil pedon by capillary movement which gets concentrated through evaporation and freezing of water. Solonization proceeds when desalinization lowers the total salt content and sodication is initiated by high exchangeable Na. Solodization occurs when anisotropic flow conditions or a change in vertical hydraulic gradients prevent capillary rise and replenishment of soluble Na in the Bn horizon. Two common Solonetzic catenas are found in the prairies. In the first sequence, Gleyed Solonetz or Solonetz occur in the depressions of the landscape, and soils then grade through Solodized Solonetz,

Solods, and in some cases, Chernozems or normal zonal soils at higher elevations. In the second sequence, Solods are found in the lowest topographic position, while Solodized Solonetz, Solonetz and Chernozems are found at progressively higher slope positions.

Large areas in western Canada, Australia, USA and other countries have profile morphology typical of solonetz/solod soils although sodium forms only a minor proportion of the exchangeable ions. It is surmised that these soils originally might have enough exchangeable sodium for the solonetz-solod morphology to develop in the profile but most of this sodium has now been lost through leaching. As the leaching of these desalinized soils proceeds, the upper horizons deepen and often become slightly acidic in reaction and the amorphous silica content increases. For example; non-saline-sodic soils in some areas of western United States have exchangeable-sodium-percentages considerably above 15, and yet the pH reading, especially in the surface soil, may be as low as 6. They occur only in the absence of lime and the low pH reading is the result of exchangeable hydrogen. The physical properties, however, are dominated by the exchangeable sodium and are typically those of a non-saline-sodic soil.

In a study on geochemical origin of sodic soil along the Phugwane using Zr/Ti ratios, it was observed that basalt is the most likely parent material for the sodic soils. The parent material for the sodic soil is terrace soil regardless of what occurs upslope, granite or basalt soil (Khomo *et al.*, 2006).

India

Origin and accumulation of salts in Indo-Gangetic plains of India has been a subject of immense interest. Agarwal *et al.* (1979) made a general assessment on the subject based on various hypotheses available in the world literature. They agreed with earlier workers who discounted marine formations or salt beds underneath as the source of salts but agreed that aridity, weathering and shallowness of the mineralized ground water are well within the realm of possibility. Bhargava *et al.* (1980) attributed alkaline hydrolysis of feldspar in the uplands plains of Haryana for the origin of sodium and weathering of alumino-silicate minerals through carbonation that yields solutions of carbonates and bicarbonates of sodium in addition to silica and alumina.

$$\text{Na-primary mineral} + H_2O \rightarrow \text{H-silicate clay} + NaOH \quad (1.6)$$

$$\text{H-silicate clay} + H_2O \rightarrow Al_2O_3.H_2O + H_2SiO_3 \quad (1.7)$$

$$NaOH + CO_2\,(\text{air}) \rightarrow NaHCO_3 \quad (1.8)$$

$$NaHCO_3 \rightarrow Na_2CO_3 + CO_2 + H_2O \quad (1.9)$$

These salts migrated with the subterranean and surface waters and accumulated in undrained area being one of the chief contributing factors in the formation of sodic soils in indo-Gangetic plains and many other regions. The relative differences in relief between parts of the plains and Siwaliks and within the plains facilitate the accumulation of salt laden run-off to accumulate in the depressions. Relief and climate together facilitated seasonal flooding and evaporation of the accumulated run-off resulting in formation of sodic soils in the form of localized stretches within the micro-depressions. Sodic soils, dominated by sodium carbonate-type salts, in the Indo-Gangetic pains of India (Haryana, Punjab, Uttar Pradesh, Bihar and Madhya Pradesh) have been formed through the direct process of sodification. The geographic setting of sodic soils, their historic background, and their physicochemical, hydrological and hydro-chemical characteristics establish the origin of these soils through the direct process of sodification commencing initially at the surface. Hence, they differ from the sodic soils (solonets) occurring in the parts of Europe, former USSR and USA, where soils have evolved through the sequential processes of salinization, sodification and desalinization.

The model proposed by Bhargava *et al.* (1980) fails to explain the presence of sodic soils in hilly areas with high rainfall, udic moisture regime and steep slopes since the soils with sodic character have been reported to occur in several parts of Punjab, Haryana and Himachal Pradesh such as Nurpur Bedi (Dist. Hoshiarpur); catchment of Sukhana lake, Sukhomajri, Nada and Bunga in the union territory of Chandigarh; Panchkula in Haryana; and in Nahan, Nalagarh, Bilaspur areas of Himachal Pradesh (Kumar, 1992). Recently, Szabolcs (1989a) also reported that salt affected soils (sodic) in mountainous regions at high and medium heights are more frequent than was previously realized. These sodic soils also have lithological discontinuities. Such morphological features do not find support from the earlier proposed mechanism of development of sodicity from weathering of feldspars in these Entisols with sodic character. Earlier, research based on thermodynamic principles on weathering of soil minerals by Kittrick (1971), Lindsay (1979) and Kumar (1992) strongly indicate that feldspars including albites become highly stable in high pH soil environment and therefore, feldspar weathering is not possible at such high pH. Further, the alumino-silicate weathering as the source of salts cannot explain the presence of appreciable amount of chlorides and sulphates in these soils along with sodium carbonates. Therefore, Kumar (1992) suggested that the salts might have come from further up i.e. the Himalayas and specifically from the marine environment of Tethyan sea during Miocene to Eocene periods (25-40 million yr.). After many geological upheavals these materials might have been transported by ancient river system in alte Pleistocent to Recent period (1 million yr.) and deposited to form the present day alluvial plains (Kumar *et al.,* 1993, 1994). Recent finding of salt, silt, sodium and smectite impregnated alluvium underneath normal soils in south-west of Punjab, which is similar to the sodic parent material occurring in Siwalik Hills, confirms the transported nature of salinity. This salt impregnated strata is source of nitrate, chloride, sulphate, fluoride, carbonate, potassium, calcium, magnesium, sodium, arsenic, molybdenum, boron, chromium, nickel and rare earth elements such as praseodymium, uranium and erbium in traces.

According to Dhir (1998), perfect fractionation as postulated by Kovda *et al.* (1973) rarely occurs in nature. According to him several processes together might have resulted in the formation of salt affected soils. A model that seems to explain the origin and accumulation of salts in the Indo-Gangetic plains is that sodium chloride in the soils and ground water has come from the Himalayan catchments (from both geological sources and rainfall) and partly from rainfall occurring in the plains. Rainfall may contain as high as 50 to 200 mg L^{-1} of salts near the sea coast. This amount decreases rapidly, however as rain moves inland. The rain near the sea coast is high in Na, Cl and Mg, whereas inland precipitation is dominated by Ca, Mg and SO_4. Over several thousand years, this atmospheric deposition becomes significant. The dominant source for sulphate seems to be the rock formations of the hills. Alkaline earth carbonate has originated mainly from the weathering of plagioclase and scarce ferromagnesian minerals in rocks and in soils and sediments. The sodium carbonate, a dominant constituent of sodic soil is mainly from the weathering of minerals in soils and sediments of the plains. Distant sources have played a major role in the formation of these soils. He postulated that accumulation of salts in semiarid tracts can be ascribed to a variety of sources such as rain water, aeolian nature of salts and release of salts from silicate weathering. In arid regions salts released from rocks and minerals of the earth's crust tend to accumulate because of limited leaching.

Efflorescent crusts consisting predominantly of the sodium carbonate minerals trona (Na_2CO_3. $NaHCO_3$. $2H_2O$) with minor thermonatrite (Na_2CO_3. H_2O) are commonly developed on less vegetated portions of the Indo- Gangetic plains of the state of Uttar Pradesh in India. On the basis of this observation, Datta *et al.* (2002) observed that trona alone or possibly a mixture of trona and thermonatrite, which crystallize as a surface efflorescence on flood plains of the river Ganges, represent the end product of evaporative concentration of seasonal flood waters mixed with the local ground water drawn up within the soils by capillary action. In the process of evaporation, calcite precipitation occurs below the surface, whereas trona and thermonatrite form on the surface. Crystallization of salts raises the pH of near-surface soil to 10.5. It seems that occasional flooding followed by fast evaporation in this extensive flood plain is a possible cause of the formation of this mineral (Datta *et al.*, 2002).

Vertisols contain both pedogenic calcium carbonate (PC) and non-pedogenic calcium carbonate (NPC) irrespective of the ecosystems to which they belong. Icromorphological studies indicate that the NPCs are part of the parent material of Vertisols. Dissolution of NPCs and recrystallization of dissolved Ca^{2+} ions are responsible for the formation of PC. Vertisols of arid and semiarid climates contain more PC in their soil control section (SCS) than those of sub-humid climates (Pal *et al.*, 2000). Formation of PC is the prime chemical reaction responsible for the increase in pH, the decrease in the Ca/Mg ratio of exchange site with depth and in the development of subsoil sodicity (Srivastava *et al.*, 2002). The dissolution of the NPC to release Ca^{2+} ions and their recrystallization as $CaCO_3$ (PC) is the prime chemical reaction of the pedogenesis of these sodic Vertisols. This reaction modifies pH, exchangeable Ca and Mg, and induces soil sodicity and hydroxyl interlayering in the smectite interlayers. The observed petrographic features of $CaCO_3$ and lack of alteration in plagioclase and micas validate the hypothesis that the formation of Vertisols reflects a positive entropy change (Smeck *et al.*, 1983).

BENCH-MARK SOIL SERIES

Soil series are abstract names sometimes taken from some local geographic feature near the site where the series is first established. Examples include the Zarifa Viran Series of sodic soils first identified near the village Zarifa Viran in Karnal district. Each soil series is classified with *Soil Taxonomy* (Soil Survey Staff, 1975) nomenclature providing Order, Suborder, Great group, Subgroup, family and series in that order. Each soil series represents a specific soil class with a unique set of physical, chemical and mineralogical characteristics and is the most homogenous category in taxonomy. As such, these constitute an essential vehicle for transferring soil information and knowledge from one place (where it was obtained) to another (where there are similar soils). National Bureau of Soil Survey and Land Use Planning, Nagpur (NBSS&LUP, 1986) have identified broadly 64 soil series out of which several pertain to sodic soils. A brief discussion of the three series namely (1) Zarifa Viran (Established) series, (2) Hirapur series (Tentative) and (3) Sakit series (Tentative) are as follows.

Digging a Soil Profile

A soil profile is a vertical section of the soil from the surface through all its horizons including C-horizon. The horizons are roughly parallel to the soil surface. They can be differentiated on the basis of colour, texture, structure, organic matter content, stoniness or other soil characteristics. The horizons are designated as O, A, E, B, and C. These are used to designate the master horizons either singly or in a combination.

A pit of about 2 m length, 1.5 to 2 m depth and 1 m width is dug to study the soil profile. In shallow water table areas, depth is restricted to the depth to water table. A vertical face is prepared which reveals the layered pattern as discussed in the previous paragraph. Bhargava (2003) prepared a training manual that outlines the procedure to study the genesis of sodic soils.

Zarifa Viran Series

The Zarifa Viran series is a member of the fine-loamy, mixed, hyperthermic family of Typic Natrustalfs. Zarifa Viran soils have pale yellow to yellowish brown very strongly alkaline loam to clay loam A horizons, olive brown to light olive brown very strongly alkaline loamy C horizons (Table 1.12 a, b). They have developed in old alluvium in micro depressions on nearly level to flat lands at an elevation of 200 to 250 m above mean sea level (MSL) in Karnal district, Haryana and in parts of Uttar Pradesh. The climate is semiarid sub-tropical with mean annual air temperature of 25°C and mean annual rainfall of 600 to 700 mm (Fig. 1.11).

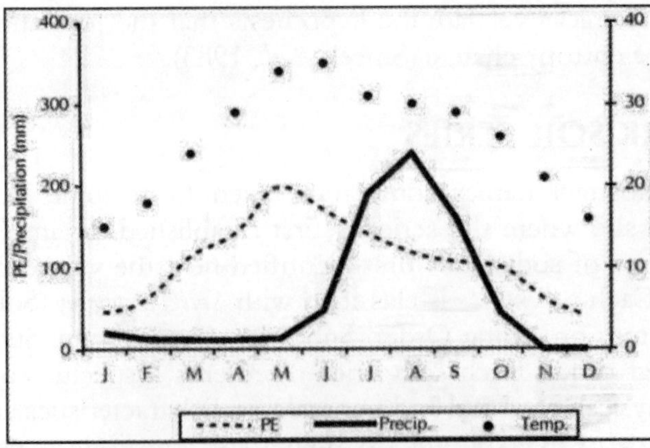

Fig. 1.11 Precipitation, temperature and water balance at Karnal

Table 1.12a Physico-chemical characteristics of Zarifa viran series (Typic Natrustaf)

Hori-zon	Depth (cm)	Size class			Coarse fragments >2 mm, % of whole soil	Organic carbon (%)	Carbo-nate as $CaCO_3$ <2mm (%)	pH (1:2.5 soil water suspen-sion)	Bulk density (g cc^{-1})	EC (dS m^{-1})	CEC/ Clay ratio
		Sand (2-0.05 mm)	Silt (0.05-0.002 mm)	Clay (<0.002 mm)							
A11	0-5	43.4	34.6	22.0	-	0.30	0.5	10.3	-	8.2	0.46
A12	5-24	33.4	37.6	29.0	-	0.30	0.9	10.3	1.48	8.0	0.44
B21t	24-56	30.6	36.2	33.2	-	0.20	1.4	9.8	-	1.9	0.45
B22t	56-85	26.0	42.6	31.0	4	0.20	3.3	9.8	1.63	1.4	0.46
B3ca	85-118	33.2	40.0	26.8	10	0.10	12.4	9.6	1.47	1.0	0.41
Cca	118-140	45.0	32.0	23.0	-	0.10	20.5	9.2	-	0.9	0.41

Table 1.12b Physico-chemical characteristics of Zarifa viran series (Typic Natrustaf)

Depth (cm)	Exchangeable bases (c mol (p^+) kg^{-1}					Water at satur-ation (%)	Water extract from saturation paste (m mol L^{-1})							DTPA extractable micronutrients (mg L^{-1})			
	Ca+ Mg	Na	K	Excha-ngeable Na (%)	CEC NH_4O AC		Ca + Mg	Na	K	CO_3	HCO_3	Cl	SO_4	Zn	Cu	Mn	Fe
0-5	0.2	9.9	0.1	97	10.2	33	0.4	85.3	0.2	30.0	36.5	12.2	5.5	0.36	2.01	12	9
5-24	0.3	12.1	0.4	94	12.8	39	0.4	83.9	0.1	27.0	40.5	14.5	5.5	0.36	2.32	9	16
24-56	0.8	13.4	0.6	90	14.8	32	0.6	18.0	0.1	2.5	8.5	5.5	3.0	0.34	1.38	11	6
56-85	1.8	12.4	0.4	85	14.6	40	0.5	13.5	0.1	2.2	7.6	2.5	2.0	0.21	0.80	8	5
85-118	3.1	7.6	0.5	68	11.2	40	0.8	10.0	0.1	2.2	4.5	2.5	1.0	0.21	0.50	6	4
118-140	5.6	3.8	0.4	39	9.8	44	1.0	9.2	0.1	1.5	5.0	2.5	1.0	-	-	-	-

Clay fraction mineralogy is MI4 KK3 M3; MI=mica, KK=Kaolinite and SM= Smectite (4 means half to one third, 3 means one third to one fifth and 2 means one fifth to one twentieth

Typifying Pedon : Zarifa Viran Loam – Saline Waste

Al l 0-5 cm --- Pale yellow (2.5 Y 7/4 D) loam, brown to dark brown (10 YR 4/3 M); moderate fine to medium platy structure; slightly plastic; strongly effervescent; few fine roots; many very fine discontinuous vertical and oblique tubular imped pores; pH 10.3; abrupt smooth boundary.

A12 5-24 cm --- Yellowish brown (10 YR 5/4 M) clay loam; moderate medium sub-angular blocky structure; firm, sticky and plastic; strongly effervescent; few very fine roots; many very fine discontinuous vertical and oblique tubular imped pores; pH 10.3; clear smooth boundary.

B21t 24-56 cm --- Olive brown (2.5 Y 4/4 M) clay; strong medium to coarse angular blocky structure; firm, very sticky and plastic; strongly effervescent; 2 to 3 mm size iron-manganese concretions 5% by volume, few very fine roots; patchy thin cutans; many very fine discontinuous vertical and oblique tubular imped pores; pH 9.8; clear smooth boundary.

B22t 56-85 cm --- olive brown (2.5 Y 4/4 M) clay loam; strong coarse angular blocky structure; very firm very sticky and plastic; strongly effervescent; 2 to 3 mm size iron-manganese concentrations 4 to 7% by volume; 5 to 15 mm size lime concretions; patchy thin cutans; many very fine discontinuous vertical and oblique tubular imped pores; pH 9.8; clear wavy boundary.

B3ca 85-118 cm --- Light olive brown (2.5 Y 5/4 M) loam; many fine distinct light yellowish brown (10 YR 5/6) mottles; strong coarse angular blocky structure; very firm, very sticky and plastic; strongly effervescent; 50 mm size lime concretions 30% by volume; 5 mm size iron-manganese concretions; patchy thin cutans; many very fine discontinuous vertical and oblique tubular imped pores; pH 9.6; clear smooth boundary.

Cca 118-140 + cm --- Light olive brown (2.5 Y 5/4 M) loam; many medium distinct light yellowish brown (10 YR 5/6) mottles; moderate medium to coarse angular blocky structure; very firm, sticky and plastic; violently effervescent; 1 to 3 mm size iron manganese concretions 3 to 5% by volume; many very fine discontinuous vertical and oblique tubular imped pores; pH 9.2.

Micromorphology: The high sodium saturation causes peptization of the soil material when wet. Clay is released with drainage resulting in sorting of grain sizes. Over the depth of 85 cm studied few to common cutans and common complete infillings of voids composed of clay (eg. Argillans) and/or clay sized material and/or silt-sized material occur. Often, the cutans and complete infillings are laminated. Carbonates are accumulating over the studied depth. In the A horizon they occur to a minor extent, mainly as neocalcans or calcans. In the B horizon, the quantity increases rapidly and nodules of increasing diameters are found. Micas are under alteration in this soil. Some sesquioxides have been released and accumulate to a minor extent in the ground mass.

Range in characteristics: The thickness of the solum is more than 100 cm. The estimated MAST is 26°C. MSST is 30.7°C and MWST is 14.0°C. The moisture regime is ustic.

Drainage and permeability: Imperfectly drained with moderately slow permeability.

Use and vegetation : Mostly barren with salt-loving grasses.

Distribution and extent : Extensive in alluvial plains in Karnal and adjoining districts of Haryana and parts of Uttar Pradesh.

Type location : Village Gudha, Tehsil and District Karnal, Haryana; CSSRI Farm, about 200 m south-west of tube well No. 1. 29°25'N 76°55' E.

Series proposed : Central Soil Salinity Research Institute, Karnal.

Interpretation: Zarifa Viran soils are saline and highly sodic with moderately slow permeability. Due to their location in depressions, water stagnates during rainy season. The soils need to be reclaimed to bring it under cultivation. The soils of this series are grouped as:

- Land capability sub-class: IIIs
- Irrigability sub-class: 3d
- Productivity potential: Low

Hirapur Series

The Hirapur series is a member of the fine-loamy, mixed, hyperthermic family of Aeric Halaquepts. Hirapur soils have light olive brown to olive brown very strongly alkaline very fine sandy loam A horizons, olive brown very strongly alkaline sandy clay loam B horizons and olive brown to light olive brown very strongly alkaline sandy clay loam C horizons (Table 1.13 a,b). They have developed in alluvium in Aligarh district of Uttar Pradesh on nearly level to gently sloping lands of 1 to 3% slopes at an elevation of 150 to 200 m above MSL. The climate is semiarid sub-tropical with mean annual air temperature of 26.7°C and mean annual rainfall of 775 mm (Fig. 1.12). A fluffy layer of salts, about 2.5 cm thick, is observed on the surface during dry months. Water table fluctuates between 0.2 and 2 m depth.

Typifying pedon: Hirapur very fine sandy loam – saline waste.

A1 0-10 cm --- Light olive brown (2.5 Y 5/4 M) very fine sandy loam; moderate medium platy structure; friable and slightly plastic; strongly effervescent; salt cutans present; many fine and medium fibrous roots; pH 10.4; abrupt wavy boundary.

A3 10-42 cm --- Olive brown (2.5 Y 4/4 M) sandy clay loam; moderate fine to medium sub-angular blocky structure; friable and slightly plastic; strongly effervescent; salt cutans present; 0.5 mm size iron-manganese concretions 1 to 2% by volume; irregularly scattered sand pockets, stratified; common fine and medium fibrous roots; many very fine to fine discontinuous vertical tubular imped pores; pH 10.3; clear smooth boundary.

B2 42-71 cm --- Variegated colours – olive brown and very dark greyish brown (2.5 Y 4/4, 3/2 M) sandy clay loam; moderate fine to medium sub-angular blocky structure; friable and slightly plastic; strongly effervescent; 35 mm size iron-manganese concretions about 7 to 10% by volume; irregularly scattered sand pockets, stratified; common fine and medium fibrous roots; many very fine to fine continuous vertical tubular imped pores; pH 10.2; diffuse smooth boundary.

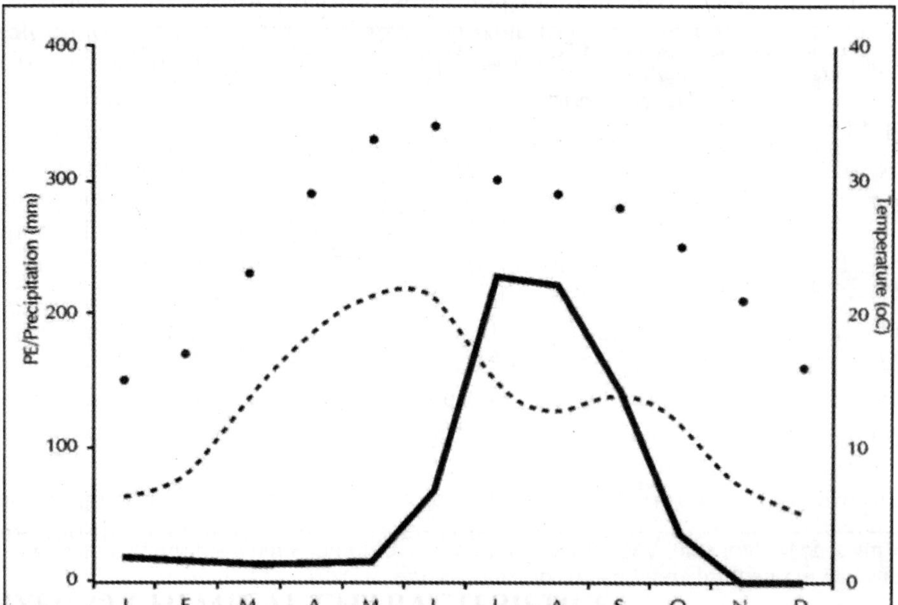

Fig. 1.12 Precipitation, temperature and water balance at Hirapur

Table 1.13a Physico-chemical characteristics of Hirapur series (Aeric Halaquept)

Horizon	Depth (cm)	Size class			Coarse fragments >2 mm, % of whole soil	Organic carbon (%)	Carbonate as $CaCO_3$ <2mm (%)	pH (1:2.5 soil water suspension)	DTPA extractable micronutrients (mg L^{-1})			
		Sand (2-0.05 mm)	Silt (0.05-0.002 mm)	Clay (<0.002 mm)					Zn	Cu	Mn	Fe
A1	0-10	65.4	19.8	14.8	-	0.30	2	10.4	0.27	0.94	4	10
A3	10-42	56.9	20.1	23.0	-	0.20	2	10.3	0.15	1.42	9	20
B2	42-71	54.0	20.8	25.2	-	0.20	2	10.2	0.07	0.58	10	7
B3	71-106	50.2	22.6	27.2	-	0.10	3	9.8	0.07	0.27	6	4
C1ca	106-122	53.0	20.6	26.4	-	0.10	8	9.7	-	-	-	-
C2ca	122-150	57.0	22.6	20.4	-	0.10	7	9.3	-	-	-	-

Table 1.13b Physico-chemical characteristics of Hirapur series (Aeric Halaquept)

Depth (cm)	Exchangeable bases (c mol (p $^+$) kg $^{-1}$)					Water at saturation (%)	Water extract from saturation paste (m mol L^{-1})							EC (dS m^{-1})	Clay fraction mineralogy
	Ca+Mg	Na	K	Exchangeable Na (%)	CEC NH_4 OAC		Ca+Mg	Na	K	CO_3	HCO_3	Cl	SO_4		
0-11	0.1	7.9	0.3	95	8.3	47	0.4	225.2	0.5	101.0	114.0	15.0	14.0	24.0	MI5 CH2
11-38	0.4	11.8	0.4	94	12.6	46	0.5	115.3	0.4	38.5	51.5	9.5	13.5	11.1	
38-48	0.9	12.8	0.4	89	14.1	62	0.6	95.0	0.2	30.5	50.5	6.5	8.0	8.7	MI4 SM2 VM 2
48-69	1.4	13.2	0.6	89	15.2	56	0.5	74.2	0.4	12.5	45.2	6.5	9.5	6.5	VM2
69-94	2.2	10.2	0.4	80	12.8	53	0.8	34.7	0.2	5.5	20.5	4.5	4.2	3.2	
94-121	3.9	4.5	0.6	50	9.0	47	1.2	19.1	0.2	4.2	9.5	5.0	3.5	1.8	
121-142	5.8	2.5	0.5	28	8.8	57	1.2	13.2	0.2	2.2	6.6	4.0	2.2	1.3	

MI=mica, SM= Smectite, VM= Vermiculite, CH= Chlotite (5 means more than half, 4 means half to one third and 2 means one fifth to one

B3 71-106 cm --- Olive brown (2.5 Y 4/4 M) sandy clay loam; massive tending to form weak fine sub-angular blocky structure; firm and slightly plastic; strongly effervescent; 2 mm size iron-manganese concretions about 1% by volume; few fine and medium fibrous roots; many very fine to fine continuous vertical tubular imped pores; pH 9.8; abrupt smooth boundary.

Clca 106-122cm --- Variegated colours-olive brown and dark greyish brown (2.5 Y 4/4, 4/2 M) sandy clay loam; massive; friable; strongly effervescent; 2 to 7 cm size calcium carbonate concretions about 5 to 6% by volume; few fine fibrous roots; pH 9.7; abrupt smooth boundary.

C2ca 122-150+cm --- Light olive brown (2.5 Y 5/4 M) sandy clay loam; massive; friable; strongly effervescent; 3 to 4 cm size calcium carbonate concretions about 5 to 6% by volume; pH 9.3.

Micromorphology: The fauna have played an important role in producing most of the voids and causing homogenization of the soil material. The soil material is very unstable when wet. Micas are altering and part of the micas has been completely altered to clay minerals. High sodium saturation causes peptization of clay. This material flows into voids where sorting of grain sizes occurs whereby few fine clay cutans and common infillings are observed. The quantity of cutans and laminae composed of clay is relatively high and is probably due to the addition of clays derived from altered mica. No change in occurrence of cutans and infilling is observed with depth. Carbonates, accumulated in thin calcans and neocalcans, increase with depth. Similarly, sesquioxides are accumulating in nodules and increase with depth.

Range in characteristics : The thickness of the solum is 100 to 110 cm. The estimated MAST is 26.6°C. The MSST is 29.9°C and MWST is 16.0°C. Moisture regime is aquic, with water table within 0.2 m depth during some part of the year.

The A horizon is 30 to 45 cm thick. The colour is in hue 2.5 Y, value 4 to 5 and chroma 3 to 4. The texture is fine sandy loam to loam. Thick salt fluffy encrustation is noticed on the surface in dry season. The B horizon is 60 to 70 cm thick. Its colour is in hue 2.5 Y, value 3 to 4 and chroma 2 to 4. The texture is sandy clay loam to clay loam. Few fine mottles are occasionally present but iron-manganese concretions are invariably present.

Drainage and permeability: Imperfectly drained with moderately slow permeability.

Use and vegetation: Mostly barren with sparse thorny vegetation and grasses.

Distribution and extent: Extensive in Aligarh and adjoining districts of Uttar Pradesh.

Type location: Village Hirapur, tehsil Sikandararao, District Aligarh, U.P.; about 1.5 km south – west of Hirapur and 7 km from Akbarabad on the left of Aligarh-Etah road, 27°45 N 78°20 E.

Series proposed: Central Soil Salinity Research Institute, Karnal

Interpretation: Hirapur soils are imperfectly drained and are saline-sodic. Water table fluctuates between 0.2 and 2 m depth indicating the need for surface drainage. These soils can be grouped as:

- Land capability sub-class: IIIs
- Irrigability sub-class: 3d
- Productivity potential: Low

Sakit Series

The sakit series is a member of the fine-loamy, mixed, hyperthermic family of Typic Natrustalfs. Sakit soils have light brownish grey to light olive brown very strongly alkaline fine sandy loam to sandy clay loam A horizons, variegated coloured dark brown to dark yellowish brown very strongly alkaline clay loam Bt horizons, yellowish brown to dark greyish brown mottled very strongly alkaline gravelly sandy clay loam to clay loam C horizons (Table 1.14 a,b). They have developed in old alluvium on flat terraces in Etah and adjoining districts of Uttar Pradesh at an elevation of 150 to 200 m above MSL. The climate is semi-arid sub-tropical with mean annual air temperature of 26.0°C and mean annual rainfall of 675 mm (Fig. 1.13). Principal associated soils are Hasanpura series, which is a Halaquept, and Kazi Nagla series a Haplustalf.

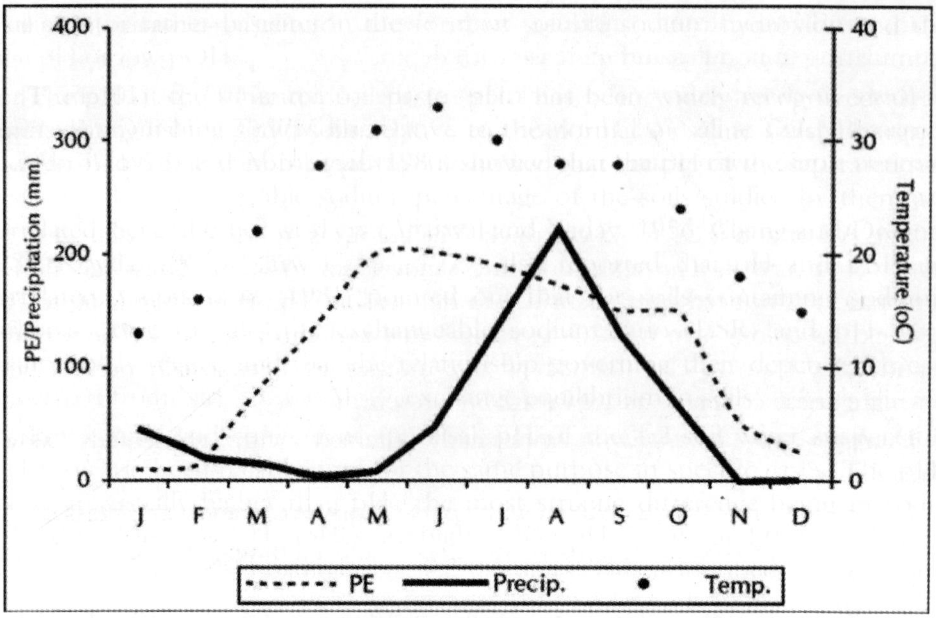

Fig. 1.13 Precipitation, temperature and water balance at Sakit

Typifying pedon: Sakit very fine sandy loam – saline waste.

A11 0-11 cm --- Light brownish grey (2.5 Y 6/2 D) very fine sandy loam, greyish brown (2.5 Y 5/2 M); moderate fine to medium platy structure; hard, friable; strongly effervescent; salt cutans; common fine roots; pH 10.5; abrupt wavy boundary.

A12 11-38 cm --- Light olive (2.5 Y 5/4 M) sandy clay loam; weak to moderate medium sub-angular blocky structure; friable, slightly sticky and slightly plastic; strongly effervescent; few 1 to 2 mm size iron-manganese concretions 1 to 2% by volume; salt cutans; few fine fibrous roots; common very fine discontinuous vertical tubular imped pores; pH 10.4; clear wavy boundary.

B21t 38-48 cm --- Variegated colours – dark brown (10 YR 3/3 M) and dark greyish brown (2.5 Y 4/2 M) clay loam; moderate medium columnar structure; firm, slightly sticky and slightly plastic; strongly effervescent; 1 to 2 mm size iron-manganese concretions 1 to 2% by volume; salt cutans; few fine fibrous roots; many very fine discontinuous vertical tubular imped pores; pH 10.2; abrupt wavy boundary.

B22t 48-69 cm --- Variegated colours – dark yellowish brown (10 YR 4/4 M) and dark greyish brown (2.5 Y 4/2 M) clay loam; strong medium columnar structure; firm, sticky and plastic; very strongly effervescent; 2 to 5 mm size calcium carbonate concretions about 5% by volume; common fine distinct strong brown (7.5 YR 5/6) mottles; very few thin clay cutans in pores and on ped faces; few fine fibrous roots; many very fine discontinuous vertical tubular imped pores; pH 10.1; clear wavy boundary.

B23t 69-94 cm --- Variegated colours – mainly dark brown (10 YR 3/3 M) and dark yellowish brown (10 YR 4/4 M) clay loam; moderate coarse columnar structure; firm, sticky and plastic, strongly effervescent; 2 to 5 mm size calcium carbonate concretions 1 to 3% by volume; common fine distinct strong brown (7.5 YR 5/8) mottles; patchy thin clay cutans in pores and on ped faces; few fine fibrous roots; many very fine discontinuous vertical tubular imped pores; pH 10.0; abrupt wavy boundary.

C1ca 94-121 cm --- Variegated colours – mainly light yellowish brown (10 YR 6/4 D), brownish yellow (10 YR 6/6 M) and light brownish grey (10 YR 6/2 M) gravelly sandy clay loam; moderate coarse angular blocky structure tending to be massive; very firm, slightly sticky and plastic; violently effervescent; calcium carbonate concretions about 50% by volume; pH 9.6; abrupt wavy boundary.

IIC2ca 121-142 + cm --- Variegated colours – mainly dark grayish brown (2.5 Y 4/2 M) and light olive brown angular blocky structure tending to be massive; friable to firm, sticky and very plastic; violently effervescent; 5 to 20 mm size calcium carbonate concretions 1 to 2% by volume; few fine brown fibrous roots; pH 9.0.

Micromorphology: The fauna are responsible for most of the recent and former voids. The soil material is unstable when wet. High sodium saturation causes peptization releasing clay which moves down, fills the voids and forms cutans. The composition of the cutans and infillings become finer grained with increasing depth. In the B21t horizon the former fine to medium prismatic structure elements are embedded in infilled, often coarser grained zones about 10 mm wide. Carbonates are accumulating throughout the B horizon. Carbonate and sesquioxide nodules also occur.

Range in characteristics: The thickness of the solum is 85 to 100 cm. The estimated MAST is 27°C, MSST is 30.7°C and MWST is 17.5°C. The moisture regime is ustic.

The A horizon is 30 to 40 cm thick. Its colour is in hue 2.5 Y, value 5 to 6 and chroma 2 to 4. The texture varies from sandy loam to clay loam. Salt incrustations at the surface and salt cutans in lower part of the A horizon are common. The Bt horizon is 50 to 60 cm thick. It has variegated colours in hues 10 YR and 2.5 Y, value 3 to 4 and chroma 2 to 4. Texture is dominantly clay loam. Clay cutans and distinct mottles are present. The Cca horizon, which is more than 50 cm thick is of variegated colours, and mottled.

Drainage and permeability: Imperfectly drained with slow permeability.

Use and vegetation: Mostly barren with salt-loving grasses.

Distribution and extent: Extensive in Etah and adjoining districts of Uttar Pradesh.

Typical location: Village Ramgarhi – Hasanpur, Tehsil Jalesar, District Etah, Uttar Pradesh 27°29' N 78°18' E.

Series proposed: All India Soil and Land Use Survey, Regional Centre, Delhi.

Interpretation: Sakit soils are highly saline and sodic. These have drainage problems and are subject to water stagnation during monsoon. Ground water rises to within 2 m from the surface. These soils need to be reclaimed to bring under cultivation. The soils can be grouped as:

- Land capability sub-class: IIIs
- Irrigability sub-class: 3d
- Productivity potential: Low

Table 1.14a Physico-chemical characteristics of Sakit series (Typic Natrustalf)

Horizon	Depth (cm)	Size class			Coarse frag-ments >2 mm, % of whole soil	Organic carbon (%)	Carbo-nate as $CaCO_3$ <2mm (%)	pH (1:2.5 soil water suspe-nsion)	DTPA extractable micronutrients (mg L^{-1})			
		Sand (2-0.05 mm)	Silt (0.05-0.002 mm)	Clay (<0.002 mm)					Zn	Cu	Mn	Fe
A11	0-11	64.6	17.2	18.2	-	0.40	0.8	10.5	0.58	0.81	7	30
A12	11-38	52.4	22.4	25.2	-	0.30	1.0	10.4	0.31	2.56	8	24
B21t	38-48	41.0	25.4	33.6	-	0.30	0.9	10.2	0.30	1.47	5	15
B22t	48-69	41.2	27.4	31.6	-	0.20	0.9	10.1	0.12	0.59	4	11
B23t	69-94	42.8	29.4	27.8	-	0.20	3.0	10.0	0.15	0.30	3	7
C1Ca	94-121	47.6	25.0	27.4	-	0.10	20.1	9.6	-	-	-	-
IIc2Ca	121-142	50.2	25.2	24.6		0.10	21.6	9.0				

Table 1.14b Physico-chemical characteristics of Sakit series (Typic Natrustalf)

Depth (cm)	Exchangeable bases (c mol (p^+) kg^{-1})					Water at satur-ation (%)	Water extract from saturation paste (m mol L^{-1})							EC (dS m^{-1})	Clay fraction mineralogy
	Ca+Mg	Na	K	Exch-angea-ble Na (%)	CEC NH_4OAC		Ca+Mg	Na	K	CO_3	HCO_3	Cl	SO_4		
0-10	0.1	6.5	0.2	96	6.8	39	0.4	460.1	0.2	226.5	200.5	22.0	12.0	44.4	MI5 VM2
10-42	0.4	9.1	0.3	93	9.8	38	0.4	147.5	0.2	48.5	72.5	17.5	11.5	13.9	
42-71	0.9	9.0	0.5	86	10.4	35	0.2	29.6	0.2	7.0	14.0	6.0	5.0	3.3	
71-106	1.6	10.0	0.6	82	12.2	35	0.4	20.1	0.1	3.0	12.0	3.0	3.0	2.1	MI4CH2IT2
106-122	2.8	7.2	0.5	68	10.5	40	0.5	19.2	0.2	2.5	7.5	6.0	3.7	1.6	
122-150	04.3	3.1	0.4	39	7.8	37	0.6	12.5	0.2	2.0	4.5	3.8	1.9	1.0	

MI=mica, VM= Vermiculite, CH= Chlotite and It = Interstratified (5 means more than half, 4 means half to one third and 2 means one fifth to one twentieth

Besides these three classical soil series, characteristics of a black sodic soil (Vertisols) in Madhya Pradesh is illustrated in Table 1.15.

Table 1.15 Physico-chemical characteristics of typical Vertisol (Sahjapur, Madhya Pradesh)

Depth (cm)	Size class			Carbonate as $CaCO_3$ <2mm (%)	pHs	CEC (Meq $100g^{-1}$)	EC_e (dS m^{-1})	ESP	Water extract from saturation paste (m mol L^{-1})				
	Sand (2-0.05 mm)	Silt (0.05-0.002 mm)	Clay (<0.002 mm)						Ca	Mg	Na	HCO_3	Cl
0-13	15.5	50.5	34.5	9.8	8.3	12.0	3.5	60	5.1	5.1	26	6.3	23.0
13-31	30.0	45.5	29.5	10.4	8.4	12.8	1.7	65	4.2	1.0	8	9.2	9.0
31-62	3.0	56.0	42.0	12.2	8.2	27.0	2.8	73	2.1	4.8	14	14.2	8.8
62-94	4.0	52.5	43.5	10.5	9.4	19.0	1.6	51	2.2	4.9	8	10.2	8.1
94-126	7.0	46.0	47.4	21.4	9.1	25.0	6.5	51	6.1	8.8	59	87.0	21.0
126-158	3.0	54.0	48.0	18.7	8.7	17.7	-2.5	56	2.9	5.1	18	8.8	10.8

❑❑❑

2

NATURE AND PROPERTIES

Nature and properties of sodic soils are quite different than the normal, acid or saline soils. As a result, these soils pose insurmountable problems during cultivation because of poor physical state and excessive chemical constituents. To understand the nature and properties of a sodic soil, it is of utmost importance to have chemical and physical tests performed on representative soil samples. The kind and number of tests required to assess the kind and degree of problem vary from soil to soil and one kind of assessment to another. The need to test these soils also arises because there is a gradual improvement is soil's physical, chemical and biological properties over the years. The dynamics of these changes depends upon many factors and is of paramount importance because it tells the current status of the soil to decide appropriate management options and cropping patterns. This chapter describes the soil sampling procedure, chemical/physical tests relevant to sodic soils and discusses the chemical and physical properties of these soils that have a bearing on the crop productivity. Changes in the soil properties likely to occur over the years are also discussed in continuity with the initial soil properties.

SOIL SAMPLING

The most critical aspect of soil testing is to draw a soil sample that is representative of the real field situation. If there is large sampling error and the collected sample does not represent the field, it is impossible to provide a reliable assessment of the degree and kind of problem. In several cases sampling errors far exceed the errors in the laboratory analysis. Since most soils are heterogeneous with wide variability, even in the so called uniform fields intensive soil sampling is the most effective way to get a representative sample. On the contrary, funds and resources available determine the intensity with which samples can be drawn.

An area that can be represented by a composite soil sample is defined as a unit area. In the assessment of kind and degree of salinity problems, unit area is usually an acre (0.4 ha, Annexure II). The soil samples are collected over the unit area dividing the fields into zones or on a grid basis (Fig. 2.1). Within these zones/grids, soils samples are drawn randomly or from near the intersections. If large variability is noticed, it may be appropriate to determine this variability by categorizing the areas as strongly, moderately and slightly affected. To prepare a composite sample in such a case, the number of samples from each kind of zone must be drawn in proportion to the affected area. The number of samples for combining into each representative composite sample varies from 5 to 20 samples. Irrespective of what is stated, the unit area may be reduced if there are variations in appearance, slope, drainage, soil types and past treatment etc. Some areas/spots can also be sampled separately. Samples of salt crusts are usually taken separately and are not mixed with the soil samples. Usually a composite sample of one kg of soil is taken from a field.

Soils should be sampled when the analysis is capable to determine the effect of salinity/sodicity on plant growth. A freshly leached soil (immediately after a rain event) may not give the true picture of salt affected soil and its nature. Normally, for arable cropping in salt affected soils, sampling soils up to 30 cm depth will suffice. For deep-rooted crops, it may be extended to 60 cm while for tree crops, the depth may extend to 1-1.8 m. The samples are usually drawn in depths range of 0-15, 15-30, 30-60, 60-90, 90-120, 120-180 cm. A uniform sample is taken from the surface to the desired depths. The same procedure is repeated and composite samples of various depths are prepared. Mixing of samples from various depths is not desirable.

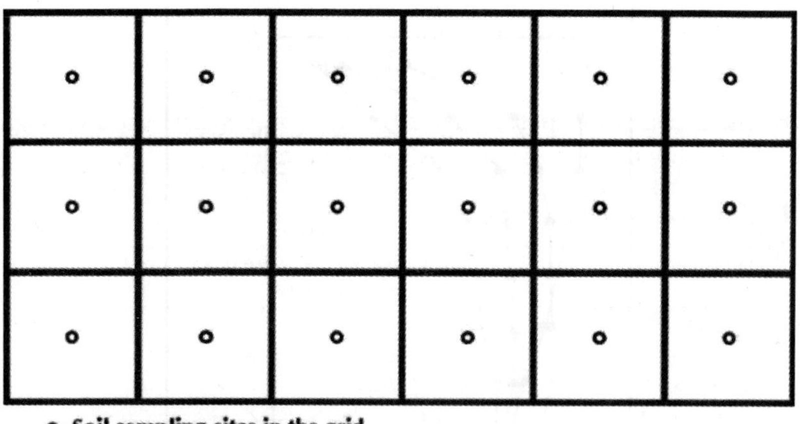

Fig. 2.1 Soil sampling plan in a regular grid

Preparation of the Composite Soil Sample

The preparation of composite samples in the laboratory involves drying, grinding, sieving, mixing and partitioning. After uniform mixing of the samples, 'Quartering Technique' is used to reduce the size of the sample. This technique consists of placing the soil sample in the center of the mixing sheet. The cone of the sample is flattened and divided through the center. It is further halved to make the four quarters. Two diagonally 'opposite quarters' are discarded and the other two are mixed thoroughly. The process is repeated, until 250-500 g composite soil material is left. The sample may be weighed, if necessary and stored in an air tight container.

Preparation of Saturated Soil Paste

Weigh 300 g of soil and place it into a plastic beaker. Add distilled water to the sample while stirring with a spatula. Add distilled water until the soil is nearly saturated. At saturation, the soil glistens as it reflects light, flows slightly when the container is tipped and the paste slides freely and cleanly off the spatula for all soils but for soils with a high clay content. Allow the paste to stand for few hours so that soil is fully wet. Add more water, if necessary, to form a saturated paste. Cover the beaker and leave it this way overnight. Some more distilled water or weighed amount of dry soil is added, if necessary. Mix the sample. Neither the free water should collect on the soil surface nor should the paste stiffen markedly. It should also not lose its glistening appearance on standing. The saturated soil paste is used to measure the EC and pH of the soil sample. The paste is also used to determine the saturation per cent defined as mass of water in the saturated paste expressed as a percentage of oven dry weight of soil. It is measured as follows:

Transfer a portion of the saturated soil paste to a tared soil can with lid. Weigh, dry to constant weight at 105°C and weigh again. The saturation percentage (SP) is then given by:

$$SP = \frac{\text{Loss in weight upon drying}}{\text{Weight of oven dry soil}} \times 100 \tag{2.1}$$

Approximate saturation percentage of some soil textural groups is given in Table 2.1 which provides a first guess of the likely value.

Table 2.1 Saturation percentage of some soil textural groups

Texture	Saturation percentage (g H_2O per 100 g oven dry soil)
Coarse	32* (16-43)
Medium	43 (26-60)
Fine	60 (42-79)
Organic	142 (81-255)

Source: (Richards, 1954); * Average followed by range in parenthesis

Preparation of Saturation Extract

Apparatus

Pressure extractor, compressor pump, filter paper and test tubes/culture tubes/ plastic bottles of 50 ml capacity

Procedure

Place the saturated soil paste in a Buchner funnel lined with a Whatman 50 filter paper. Apply the vacuum and collect the extract in a bottle or test tube as per requirement. The filtration is finished when air starts to pass through the paper. Add sodium hexametaphosphate solution to the filtrate (approximate one drop for each 25 ml of filtrate) to prevent calcium carbonate to precipitate. The saturation extracts so collected are used to determine electrical conductivity and chemical analysis of soluble cations such as calcium (Ca), magnesium (Mg), sodium (Na) and potassium (K) and anions such as carbonate (CO_3), bicarbonate (HCO_3), chloride (Cl) and sulphate (SO_4).

SOIL TESTS

Soil testing is an important tool to manage salt affected soils in general and sodic soils in particular. The commonly determined chemical, physical and biological properties of interest are given in Table 2.2. The units and interpretation of few chemical investigations are given in Table 2.3. In this chapter, analysis procedures for only few properties required to diagnose the sodicity problem i.e. EC, pH, CEC and exchangeable cations are included. Procedures to determine the quantity of gypsum required to reclaim the sodic lands are discussed in chapter 3. For other investigations readers may refer Richards (1954) or Gupta *et al.* (2012) or any other standard book on laboratory investigations.

Table 2.2 Chemical, physical and biological parameters for monitoring soil health

Chemical	Physical	Biological
EC	Bulk density	Biological worm population
pH	Organic matter	Bio-diversity
ESP	Soil water holding capacity	Crop yield
SAR	Soil porosity	
Soil nutrients	Hydraulic conductivity	
	Soil crusting	
	Infiltration rate	
	Soil cracking	
	Depth to ground water	
	Surface stagnation	

Electrical Conductivity (EC)

EC is a measurement of the dissolved material in an aqueous solution, which also relates to the ability of the material to conduct electrical current through it. As such, it is used to measure soil salinity or the salinity of irrigation water. Commonly, an Ohm meter or electrical conductivity bridge is used for measurement. The principle used is "for constant geometry and constant distance of separation and a constant electrical potential (voltage), the amount of current produced is a linear function of the concentration of total dissolved salt (or electrolytes) in the solution". At a constant potential, the current is inversely proportional to the solution's resistance or directly proportional to the solution's conductance. The unit of conductance is 1/Ohm or mho (Ω^{-1}). Since, the probe consists of two metal electrodes each of 1 cm^2 spaced 1 cm apart (thus the unit of measurement is microSeimens or milliSeimens per centimeter). The basic unit of EC is mho cm^{-1}, which is too large for most natural waters. Therefore, mΩ^{-1} cm^{-1} is often used. Sometimes, $\mu\Omega^{-1}$cm^{-1} is also used. In SI units, the unit of conductivity is Siemens meter^{-1} or decisiemens meter^{-1} abbreviated as dS m^{-1}. The effects of electrode geometry are incorporated into a term called cell constant, which is commonly obtained by calibration with standard solutions of known concentration. For a quick visualization, the EC of some of the waters are given in Table 2.4.

Table 2.3 Soil chemical analysis for assessment of salt affected soils

Analysis	Units	What it measures?	Interpretation of test data
Electrical conductivity (EC)	dS m^{-1}	Ability of soil solution to conduct electricity	• The higher the soil EC, more the dissolved ions it contains. • An indicator of potential salinity of the soil
pH	Scale of 0–14	Acidity or alkalinity	• Assesses whether pH is favorable for target crop. • Indicates potential solubility of soil minerals and nutrients. • An indicator of potential sodic condition.
Exchangeable cations	meq 100 g^{-1}	Sodium (Na) Calcium (Ca) Magnesium (Mg) Potassium (K)	• Concentration of cations adsorbed on negatively charged surfaces in soil. • Estimates sodium hazard as exchangeable sodium percentage (ESP). • Estimates gypsum requirement for sodic soils/waters
Exchangeable sodium percentage (ESP)	Percentage of the total cation exchange sites occupied by sodium	Sodium hazard increases with increasing ESP	ESP is used to determine gypsum requirement for treatment of sodic soils.
Cation exchange capacity (CEC)	meq 100 g^{-1}	Capacity of soil to hold positively charged ions (cations).	More gypsum is needed to adjust SAR and ESP of soils having higher CEC.
Calcium carbonate equivalent	Per cent	Percentage of soil by weight that is undissolved limestone	If calcium carbonate is present, usually it is not economical to adjust pH below 8. In this case, acidifying soil amendments such as elemental sulfur or sulfuric acid can be used as soil reclaiming agent.

Table 2.4. Guide to typical salinity limits for waters

Product	EC ($dS\ m^{-1}$)	TDS ($mg\ L^{-1}$ or ppm)
Distilled water	0.001	0.67
Typical deionized water	0.1	64
Rainfall	0.03	20
Sewage effluent	0.84	565
Freshwater	0–1.5	0–1000
Good quality ground water	0.7–1.0	470–670
Brackish water	1.5–15	1000–10050
Upper limit recommended for drinking	1.6#	1070
Salt water swimming pool	5.9–8.9	4000–6000
Seawater	55	36 850
Dead Sea	110	73 700

If there is no other source, limit in India can be relaxed to 3.0 dSm^{-1}
Note: To convert from $dS\ m^{-1}$ to $mg\ L^{-1}$, multiply by 0.64-0.73 to get approximate value.

Measurement of EC

The conductivity of a water sample can be determined theoretically by the following relation:

$$EC = \sum_{i=1}^{n} Ci * Fi \tag{2.2}$$

Such that EC is the EC of water in $\mu S\ cm^{-1}$, Ci is the concentration of the ionic specie in solution, $mg\ L^{-1}$ and Fi is the conductivity factor for ionic specie and n is the number of ionic specie present in the water. Commonly found ions and their Fi values are given as follows:

Ion	Fi ($\mu S\ cm^{-1}$ per $mg\ L^{-1}$)
Ca^{2+}	2.60
Mg^{2+}	3.82
K^{+}	1.84
Na^{+}	2.13
HCO_3^{-}	0.72
Cl^{-}	2.14
SO_4^{2-}	1.54
NO_3^{-}	1.15

The use of eq. (2.2) is illustrated in Table 2.5. While the concentration of various ions in the water is given in column 2, the EC in $\mu S\ cm^{-1}$ is given in column 4. Adding these values, the EC of the solution is given as 1258.8 $\mu S\ cm^{-1}$ (Table 2.5).

Table 2.5 Ion specific conductivity as per eq. (2.2)

Ion	Concentration (mg L^{-1})	Factor (μS cm^{-1} per mg L^{-1})	Conductivity (μS cm^{-1})
Ca^{2+}	85.0	2.60	221.0
Mg^{2+}	43.0	3.82	164.3
K^{+}	2.9	1.84	5.3
Na^{+}	92.0	2.13	196.0
HCO_3	362.0	0.716	258.8
Cl	131.0	2.14	280.3
$SO_4{}^{2-}$	89.0	1.54	137.1
NO_3	20.0	1.15	23.0
	Total		1285.8 μS cm^{-1}

Besides, there are several empirical relationships for converting one type of water quality measurement to another. For solutions in the EC range from 0.1 to 5 dS m^{-1}:

Sum of cations (mmol$_c$ L^{-1}) = EC (dS m^{-1}) x 10 (2.3)

Ionic strength (I) = 0.013 x EC (dS m^{-1}) (2.4)

All such relations to determine EC are either approximate or difficult to use. Therefore, EC is commonly measured using an EC meter (Fig. 2.2). To measure EC, turn on the EC meter and calibrate the probe using a standard solution of known conductivity. Choose a standard close to the likely EC of the samples. Since the ionic activity in an aqueous solution increases with increasing temperature, it plays an important role in the conductivity measurements. It is customary to reference conductivity measurements to 25°C. The coefficient used to correct for changes in temperature, β is expressed as a percentage per degree Celsius. EC meter with built-in temperature compensation are available. The calibration procedures are devised for both the situations. To calibrate a meter with temperature correction, rinse conductivity cell with KCl, 0.01M (EC = 1.412 dS m^{-1}) and immerse the electrode in the standard KCl solution. Adjust temperature compensation dial to 0.0191/°C (Hydrology Project, 1999). Read the actual EC from the meter. The cell constant is given as:

$$Kc = (1.412/EC_m) \qquad (2.5)$$

Here Kc is the cell constant and EC_m is the measured EC, dS m^{-1}. The design of the plates in the conductivity cell, size, shape, position and condition determines the conductivity measured and is reflected in the so-called cell constant. The cell constant is subject to slow changes in time, even under ideal conditions. Thus, the cell constant must be checked on a regular basis.

If EC meter is without built-in temperature compensation the electrode is rinsed in KCl. The electrode is then immersed in the standard solution and its EC and temperature are noted to calculate the cell constant as follows:

$$Kc = (1.412/ EC_m)[0.0191(t-25) +1] \qquad (2.6)$$

Here t is the observed temperature of standard KCl solution, °C. All other notations have the same meaning as for the previous equation.

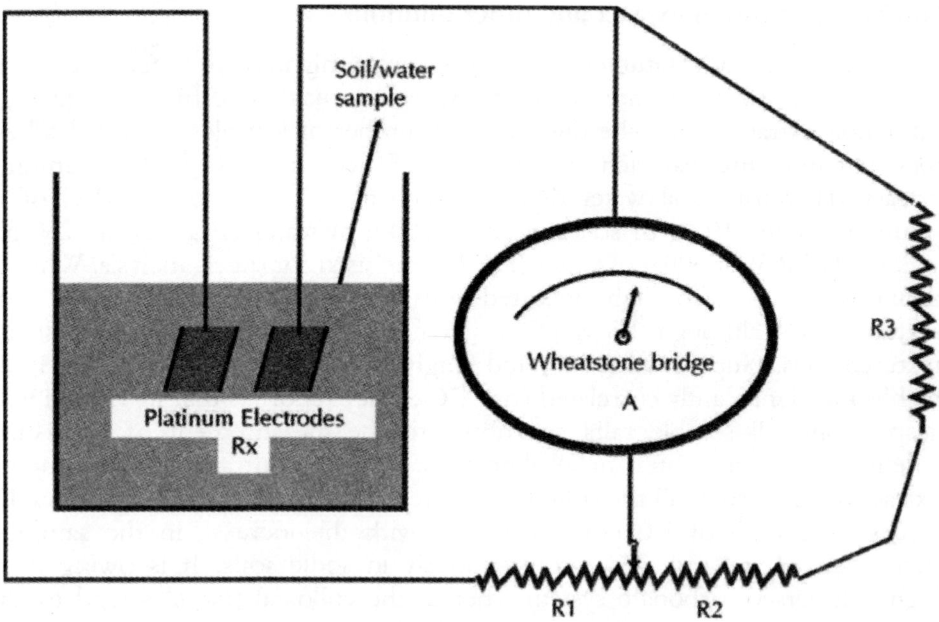

Fig. 2.2 Schematic diagram of an EC meter

To determine the EC of the sample, rinse cell with portions of sample and immerse the cell in remaining portion of the sample. Read and note its conductivity. Measure temperature and calculate EC at 25°C.

$$\text{Electrical Conductivity (dS m}^{-1}) = (EC_m * Kc) / [0.0191(t-25) + 1] \quad (2.7)$$

Such that EC_m is the measured EC of the sample, dS m^{-1}.

Portable and laboratory instruments are available normally having similar features. Since the calibration procedures vary from one instrument to another, it is recommended that the manufacturer's instructions should be followed (Hydrology Project, 1999). For continuous measurement systems, a controller and sensors are used. The units of measurements and their conversion to other known units are given in Table 2.6.

Table 2.6 Conversion factors for converting various salinity units to dS m^{-1}

Unit	Conversion factor
mS cm^{-1} to dS m^{-1}	1.0
mmho cm^{-1} to dS m^{-1}	1.0
mS m^{-1} to dS m^{-1}	0.01
µS cm^{-1} (EC units) to dS m^{-1}	0.001
mmho cm^{-1} (EC units) to dS m^{-1}	0.001
meq L^{-1} (NaCl) to dS m^{-1}	0.0912
ppm or mg L^{-1} (TSS) to dS m^{-1} (for mixed salts)	0.0015625
ppm or mg L^{-1} (TSS) to dS m^{-1} (for NaCl)	0.002
bar or kPa to dS m^{-1}	-2.78

TSS = Total soluble salts

Soil EC : saturation extract and other dilutions

More often than not, saturation extract is commonly used to determine the EC of the soils. But there are situations where it may be difficult to take out saturation extract. It may be due to large number of samples or samples being repetitive in nature that each time it may be difficult to resort to draw saturation extract. Therefore, soil water dilutions such as 1:2, 1:2.5 or 1:5 (EC of the solutions having 100 g of soil and either 200 g of water (EC 1:2) or 250 g of water (EC 1:2.5) or 500 g of water (EC 1:5) are used for quick analysis. While 1:2 dilutions are used for light textured soils, 1:5 dilutions are used for heavy textured soils although 1:5 dilutions have also been used even for all light textured soils. Gupta *et al.* (1969) and Singhla *et al.* (1974) observed that EC_2 is highly and significantly correlated with EC_e. They reported that in 1:2 soil water suspensions, all soluble salts contribute to the measurement of EC equally whether these are neutral or alkaline salts. On the contrary, in the saturation extract of sodic soils, all the soluble salts are not recovered, resulting in low EC_e values. The ratio of EC_e/EC_2 decreases with the increase in the saturation percentage of the soil, more conspicuously in sodic soils. It is owing to the highly dispersed carbonate system wherein the colloidal particles tend to align themselves and form an impervious layer, obstructing the dissolution of a part of the total soluble salt in the soil solution.

Once the EC of the soil water dilutions are known, the EC of the saturation extract is determined using the conversion factors reported in Table 2.7. In practice however, the conversion factors are determined either theoretically or graphically as per the following procedures. Theoretically,

$$\text{Multiplier factor} = \text{WC in the diluted sample/WC at saturation} \tag{2.8}$$

Such that WC is the water content by weight.

Under field conditions where one anticipates repeated sampling, soil samples at the initial stage are analyzed both in terms of saturation extract and the desired dilution. A graph is plotted between the desired dilution and saturation extract values. The slope of the line provides the multiplier factor. Equation that describes the relationship between the two values can also be used to convert EC of the given dilution to the saturation extract value.

Table 2.7 Conversion factors to estimate EC_e using $EC_{1:5}$ values for various soil textures

Texture	EC 1:5 to EC_e
Sand	× 12.5
Sandy loam	× 10.0
Loam	× 5.0
Clay loam	× 8.0
Light clay	× 7.0
Heavy clay	× 6.0

It may be noted that EC of any dilution would be less than the saturation extract and the plants would be exposed to even higher salinity values as the salinity of the soil water at field capacity (EC_{fc}) is higher than the saturation extract. It is common to assume field capacity moisture content as half of the moisture content at saturation so that the EC_e is about ½ of the EC_{fc}. On the basis of EC, the soils are evaluated as normal, critical for germination, critical for the growth of salt-sensitive crops and severely injurious to crops if EC_2 is <1, 1-2, 2-3 and > 3 dS m^{-1} (Muhr *et al.,* 1963). The categorization however varies widely in the literature as revealed by the data in Table 2.8. Qualitatively, the soils having EC_e values < 1.5, 1.5-3.0, 3.0-5.0 and 5-10.0 dS m^{-1} may be called as low, medium, high and very high saline soils respectively.

Table 2.8 Soil and water salinity criteria based on plant salt tolerance groupings[a]

Plant salt tolerance groupings	Water or soil salinity rating	Average root zone salinity, EC_{se} (dS m^{-1})
Sensitive crops	Very low	< 0.95
Moderately sensitive crops	Low	0.95 – 1.9
Moderately tolerant crops	Medium Low	1.9 – 4.5
Tolerant crops	High	4.5 – 7.7
Very tolerant crops	Very high	7.7 – 12.2
Generally too saline	Extreme	>12.2

[a] Adapted from DNR (1997a, 1997b)

Precautions

In an EC meter, probe is the key component that determines the accuracy of measured EC. Therefore, probe must be maintained properly. Electrodes normally get coated with impurities especially if left to dry out. The degree to which this occurs will depend upon the nature of the samples being tested. It will also depend upon whether or not the electrode is rinsed with distilled water and properly stored after each use. It is also not desirable to clean the plates in a manner that might cause scratches on the plates. Always rinse the electrodes in distilled water and allow to air dry before replacing the cap. Absorbent paper can be used to dry out the electrode before storing. If the electrodes are being used frequently, store the electrode in distilled water.

pH

The water molecules in an aqueous solution gets dissociated into H^+ and OH^- ions. The pH of the aqueous solution is log of the inverse of hydrogen ion activity and pOH the log of the inverse of hydroxide ions activity as follows:

$$pH = -\log [H^+] \qquad (2.9)$$

$$pOH = -\log [OH^-] \qquad (2.10)$$

Kw, the ion-product constant for water is given as $Kw = [H^+][OH^-] = 1 \times 10^{-14}$ and may be written as:

$$pH + pOH = -\log Kw = 14 \qquad (2.11)$$

The pH values on a scale of 0 to 14 characterize the aqueous solutions as acidic, alkaline or neutral. A solution which is neither acidic nor alkaline is neutral and corresponds to a value of 7 on the pH scale. Acidity indicates a higher activity of hydrogen ions showing a pH value below 7. Alkaline solutions characterized by a low hydrogen ion activity show a value above 7. At 0 pH, hydrogen ion activity (mol L^{-1}) is 1.0 being 0.0 at 7.0 and 0.0000001 at pH 14.0. On the other hand, OH^- activity would be 0.00000000000001 at pH 0.0, 0.0 at 7.0 and 1.0 at pH 14.0. Thus, a difference of one pH measurement unit represents a tenfold increase or reduction of hydrogen ion activity in the solution. Based on pH, solutions are characterized as ultra acids to very strongly alkaline (Table 2.9). While milk is nearly neutral, orange juice and colas are acidic in nature. Sea water shows a pH above 7.0 (Table 2.10). In the context of soils and water, an unusually high or low pH is a good indication of presence of some pollutants in the soil/water.

Table 2.9 Characterization of aqueous solutions on the basis of pH

pH class	pH
Ultra acid	<3.5
Extremely acid	3.5-4.4
Very strongly acid	4.5-5.0
Strongly acid	5.1-5.5
Moderately acid	5.6-6.0
Slightly acid	6.1-6.5
Neutral	6.6-7.3
Slightly alkaline	7.4-7.8
Moderately alkaline	7.9-8.4
Strongly alkaline	8.5-9.0
Very strongly alkaline	>9.0

The pH measurement involves the use of a pH sensitive electrode (usually glass), a reference electrode, and a temperature element to provide a temperature signal to the pH analyzer schematically shown in Fig. 2.3. The glass bulb is filled with a stable pH solution, usually pH 7. When the outside of the bulb is exposed to a solution or paste where hydrogen ions vary, it creates a potential that is read versus the stable potential of the reference electrode at any given temperature. The reference electrode also completes the pH measuring circuit within the solution. The signal is sent via the silver chloride coated wire to the instrument, where it is measured as a voltage and then converted to pH units. Most pH sensors are designed to produce a 0 mV signal at 7.0 pH, with a (theoretically ideal) slope of -59.16 mV/pH at 25°C. Recently, units that combine both the electrodes into one have been introduced but the basic principle of measurement remains the same.

Before operating a pH meter, it is warmed up for some time usually about 30 minutes or to minimize the drift. It is also calibrated. For this purpose buffer solutions, which contain salts that create pre-determined hydrogen ion activity in aqueous solutions, are used. The accuracy range of commercial buffer solutions is between ±0.01 and ±0.04 pH. Using this as reference, sensors are checked and readings of the measuring transmitter (meter) are adjusted. Usually, buffers of pH close to the pH values likely to be encountered in sample analysis are used. If it is not known or wide deviations are expected, a three-point calibration starting with buffer 7 may be required followed by a buffer in the acidic and basic range.

Table 2.10 pH of some common items

Item	pH	Acid/base
Hydrochloric Acid (HCl)	0.0	Acid
Lemon juice	2.2	Acid
Vinegar	2.8	Acid
Apples	3.0	Acid
Soft drinks	2.5-3.0	Acid
Tomatoes	4.5	Acid
Black coffee	5.0	Acid
Human saliva	6.4	Acid
Rain water	6.5	Acid
Acid rain	< 5.6	Acid
Milk	6.6-7.0	Acid
Pure water	7.0	Neutral
Human blood	7.34 – 7.45 (7.4)	Base
Sea water	7.7 – 8.3	Base
Baking Soda (Sodium bicarbonate)	8.3	Base
Milk of Magnesia	10.5	Base
Ammonia	11.0	Base
Lime (Calcium hydroxide)	12.4	Base
Sodium hydroxide (NaOH)	14.0	Base

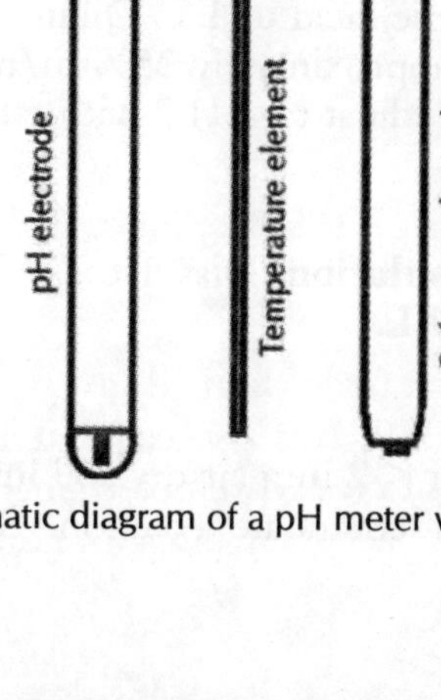

Fig. 2.3 A schematic diagram of a pH meter with two electrodes

Like EC meters, pH meters also require some form of temperature compensation to ensure standardized pH values. Meters with automatic temperature compensation receive a continuous signal from a temperature element and automatically correct the pH for temperature of the solution. Manual temperature compensation requires the user to enter the temperature. pH buffers are also labeled with temperature correction tables. For maximum accuracy and in conditions where the buffers are being brought to *in-situ* temperatures, these tables are used to assess the actual pH.

From categorization point of view, the soil is categorized as 'normal to most of the common crops' if pH_2 is 8.5, 'tending to become sodic' if pH_2 is between 8.5 and 9.0; and 'sodic soil', harmful for growing crop' if pH_2 is >9.0; (Muhr *et al.,* 1963). Gupta (1972) suggested that last two limits of pH_2 should be changed to 8.5-9.5 and >9.5 respectively. Also, in order to relatively distinguish the degree of sodicity in the soils, the soils having pH_2 values < 9.3, 9.3-10.0, 10.0-10.5 and >10.5, may be called low, medium, high and very high sodic soils respectively.

Precautions

Since probe is the key of the instrument that measures pH, it is always stored in a moist environment to prevent it from drying out. If it dries out, most likely the probe has to be replaced. The best way to store the probe is to keep it submerged in pH 4 buffer or pH storage solution. Never keep the probe submerged in distilled water. It will dilute the internal electrolyte and shorten the probe life. While drying and cleaning, avoid using any material that might cause scratches on the surface of the glass bulb.

EXCHANGEABLE CATIONS

Reagents for Preparation of Extracts

- **Alcoholic ammonium chloride, 1M:** Dissolve 53.5 g NH_4Cl in about 300 ml of water and add 600 ml of ethanol (95% v/v). Adjust the pH to 8.5-8.6 with concentrated ammonia solution and make up to 1 L. Store in a well stoppard bottle to prevent access of CO_2.
- **Ammonium ethanoate solution, 1M**: Dilute approximately 230 ml of glacial ethanoic (acetic) acid to 1 L. Dilute approximately 220 ml of ammonium solution (approximately 35% m/m NH_3) to 1 L. Mix the solutions together and adjust to pH 7 with ethanoic acid or ammonia. Dilute to 4 L.
- **Ethanol**, 95%
- **Potassium chloride solution**: Dissolve approximately 100 g KCl in water and make up to 1 L.

Procedure

- Weigh 5 g (±0.01 g) of < 2 mm air dry soil into a 100 ml beaker. Add 20 ml of ammonium ethanoate solution. Stir and allow to stand overnight. Transfer the suspension to a filter funnel fitted with a Whatman filter paper no. 44 standing in a 250 ml volumetric flask. Thoroughly wash the beaker with ammonium ethanoate solution into the funnel. Leach the soil with successive 25 ml volume of ammonium ethanoate solution. Continue until 250 ml of filtrate is collected. Make up to the mark with ammonium ethanoate solution.

- Use the extract so collected to determine potassium, calcium and magnesium.
- Follow the same procedure with alcoholic ammonium chloride and use the leachate to determine sodium.
- Sodium so determined minus the soluble sodium present in the soil saturation extract would give the exchangeable sodium.

Determination of Cation Exchange Capacity (CEC)

The dominant residual charge on most soil colloids (clay and humus) is negative. Any electrically charged colloidal surface area attracts mobile substances of opposite charge i.e. the cations the positively charged ions in the soil water. Cation exchange sites immobilize cations but keep them exchangeable so as to make them available to plant roots. Cation exchange capacity has thus been defined as the 'sum total of the exchangeable cations that a soil can adsorb'. It can be determined by several methods. The basic principle is that all the adsorbed cations in a soil are replaced by a common ion, such as barium, potassium, sodium, or ammonium and then the amount of adsorbed barium, potassium, sodium, or ammonium is determined (Richards, 1954). The cation exchange capacity is expressed in terms of moles of positive charge per unit mass. For the convenience, cation exchange capacity is expressed in centimoles of positive charge per kilogram [cmol (p^+) kg^{-1}] of soil or in meq $100g^{-1}$ of soil. A rough estimate of CEC can be made by the following equation (http://www.agriculture.gov.sk.ca/Default.aspx?DN=2d20bb89-4290-4eea-b265-dfd3a155cc51):

$$\text{CEC } (\text{cmol}(+)\ \text{kg}^{-1} \text{ or meq } 100\ \text{g}^{-1}) = 0.5 \times (\%\ \text{clay}) + 2.0 \times (\%\ \text{organic matter}) \quad (2.12)$$

Verma *et al.* (2006) showed that CEC in a black sodic soil is significantly correlated with clay content as per the following relation:

$$\text{CEC} = 2.83 + 0.772\ \text{Clay}\ (\%) \quad (2.13)$$

Laboratory Method

The soil is leached with sodium acetate solution (pH 8.2) to replace all exchangeable cations by Na^+ ions. The excess salt is washed down by alcohol and adsorbed Na^+ ions are replaced by NH_4^+, using neutral normal NH_4OAc solution. The Na^+ ions released from the exchanged sites are measured in NH_4OAc leachate with the help of flame photometer. The following reactions are involved.

$$CH_3COONa \rightleftarrows CH_3COO^- + Na^+ \quad (2.14)$$

$$X\text{-Soil} + CH_3COO^- + Na^+ \rightleftarrows Na\text{-Soil} + CH_3COO\text{-}X \quad (2.15)$$

$$CH_3COONH_4 \rightleftarrows CH_3COO^- + NH_4^+ \quad (2.16)$$

$$Na\text{-Soil} + NH_4^+ \rightleftarrows NH_4^+\text{-} Soil + Na^+ \quad (2.17)$$

Apparatus/Reagents

- Funnel, beaker, filter paper, flame photometer, volumetric flask, centrifuge tube.
- **Sodium acetate (1 N):** Dissolve 136 g of sodium acetate trihydrate in distilled water and dilute to a volume of 1 L (adjust the pH to 8.2 using dilute NaOH or CH_3COOH).
- **Ammonium acetate (1 N)**: Add 57 ml of 99.5% acetic acid and 68 ml concentrated ammonium hydroxide, specific gravity 0.90, to about 800 ml water and dilute to 1 L (adjust the pH 7.0 using dilute NH_4OH or CH_3COOH). Optionally prepare from NH_4OAc reagent salt (dissolve 77.08 g of ammonium acetate in distilled water and dilute to a volume of 1 L; (adjust the pH to 7.0).
- **Ethanol:** 60%
- **Standard sodium solution**: Dissolve 2.338 g of dry sodium chloride in 1 L distilled water. This will give a stock solution of 40 meq L^{-1} Na. Pipette 25 ml solutions from the 40 meq L^{-1} stock solution of Na into a 250 ml volumetric flask, and make volume to capacity. This will give 4 meq L^{-1} Na solution. Pipette out 0, 5, 10, 15, 20, 25, 50 and 75 ml portions of this solution to 100 ml volumetric flasks and make the volume to 100 ml. These working standards are prepared by diluting the aliquot with 1N NH_4OAc solution. These will give 0, 0.2, 0.4, 0.6, 0.8, 1.2 and 3.0 meq L^{-1} Na concentration. Calibrate the flame photometer for operation over the concentration range of 0 to 4 meq L^{-1} from above mentioned solutions. Use the first and the last solution to calibrate the instrument for operation over the concentration range of 0 to 4 meq L^{-1} of solution.

Procedure

Weigh 5 g soil in a beaker and add about 33 ml of NaOAc (preferably keep overnight to saturate the soil with Na), and transfer the soil on to the funnel fitted with moist Whatman filter paper No. 42. Add 33 ml NaOAc on soil to leach it. Add another 33 ml NaOAc. Repeat one more washing with NaOAc and discard the leachate. Add about 30 ml ethanol to soil and leach it to remove excess NaOAc. Continue washing with ethanol until the electrical conductivity of leachate from last washing is between 40 to 55 μmhos cm^{-1} and discard the leachate of ethanol. Replace the adsorbed sodium from the sample by extracting with three 33 ml portions of NH_4OAc solution and collect it in 100 ml volumetric flask. Make the volume of flask to 100 ml and determine the sodium concentration using flame photometer.

Alternative method (Hesse, 1971)

Weigh 5 g soil and place in a centrifuge tube. Add 33 ml NaOAc, stopper the tube, and shake it for 5 minutes. Remove stopper and centrifuge until the supernatant liquid is clear (usually 10 minutes at 2000 rpm). Decant the supernatant liquid as possible and discard. Repeat same procedure bringing the soil in centrifuge in suspension two more times, discarding the supernatant liquid each time. After the last saturation, wash the stopper and use adsorbent paper to remove any acetate crystals remaining on the lip of the centrifuge tube. Add about 30 ml ethanol to centrifuge tube, put stopper, shake for 5 minutes, remove stopper and centrifuge until the supernatant liquid is clear. Decant and discard the supernatant liquid. Continue washing with ethanol until the electrical conductivity of supernatant liquid from the last washing is between 40 to 55 μmhos cm^{-1}. Replace the adsorbed sodium from the sample by extracting with three 33 ml portions of NH_4OAc solution. Make the volume 100 ml. Determine Na concentration in the extract by using flame photometer.

Calculations

$$\text{CEC [meq 100 g}^{-1}\text{ soil or Cmol(P}^{+}\text{) kg}^{-1}] = \frac{\text{Meq L}^{-1}\text{ Na from standard curve}}{\text{Weight of soil sample (g)}} \times 100 \quad (2.18)$$

Or

concentration of Na in the leachate against reading from standard curve = X ppm (2.19)

$$\text{CEC [meq 100}^{-1}\text{ soil or Cmol(P}^{+}\text{) kg}^{-1}] = \frac{X}{23} \times 2 \quad (2.20)$$

Example

Weight of soil taken	= 5 g
Volume of NH_4OAc extractant used to replace Na^+	= 100 ml
Reading of flame photometer in sample	= 93
Na^+ concentration in meq L^{-1} from the standard curve	= 3.72
1000 ml contains	= 3.72 me Na^+
But it was extracted with 100 ml extract	
So 100 ml solution contains	= 3.72X 100/1000
	= 0.372 meq Na^+
100 ml was extracted from 5 g soil so 5 g soil contains	= 0.372 meq Na^+
100 g soil contains	= 0.372 X 100/5
CEC me100g^{-1} soil	= 7.44

Approximate method

Soils with pH above 5 contain very little exchangeable aluminium, and so the exchange capacity of the saline and sodic soils is dominated by Na^+, Ca^{2+} and Mg^{2+} with small amounts of K^+. Thus, the effective CEC is the sum of the four cations calcium, magnesium, potassium and sodium. If these are known, the CEC can be determined by adding the exchangeable cations. Other cations (manganese, iron, copper and zinc) are usually present in only trace amounts and so don't contribute significantly to the total. In the example shown in Table 2.11, the effective CEC is 34.9 Cmol (P^+) kg^{-1}.

Table 2.11 Example of calculating exchangeable cations as % of effective CEC

Cation	Amount		% of effective CEC
	mg kg^{-1}	cmol (+) kg^{-1}	
Calcium (Ca)	3000	3000/200=15.0	100 x 15.0/34.9=43.0
Magnesium (Mg)	2020	2020/120=16.8	100 x 16.8/34.9=48.1
Potassium (K)	351	351/390=0.9	100 x 0.9/34.9=2.6
Sodium (Na)	512	512/230=2.2	100 x 2.2/34.9=6.3
Aluminium (Al)	nil	0	0
Total (effective CEC)		34.9	34.90

Calcium Carbonate in Soil

Calcium carbonate occurs in soils in various forms such as soluble, crystalline or amorphous. With the translocation of sodium to the exchange complex, the soil pH increases resulting in decreased solubility of calcium (Table 2.12). Calcium under such situations precipitates as calcium carbonate nodules or concretions in the B or C horizon of a soil profile. While the soluble form is measured by titration with sulphuric acid and versenate solution, the crystalline and amorphous forms of calcium carbonate are measured in-situ by its shape, size and release of effervescence on application of dilute hydrochloric acid on the soil profile. A calcimeter is widely used for geological explorations of carbonate material at field scale. The application of dilute HCl to soils produces carbon-dioxide that replaces water. The volume of carbon-dioxide is measured by the equivalent volume of water displaced. It is calculated with reference to the volume of CO_2 released for 0.1 g of crystalline $CaCO_3$.

Table 2.12. Effect of pH on the solubility of $CaCO_3$ in water

pH	solubility of $CaCO_3$ (meq L^{-1})
6.21	19.3
6.50	14.4
7.12	7.1
7.85	2.7
8.60	1.1
9.20	0.8
10.12	0.4

According to Soil Taxonomy (Soil Survey Staff, 1998) a calcic horizon is an illuvial horizon in which calcium carbonates and other alkaline earth carbonates have accumulated to a significant extent. Normally, it contains 5% or more $CaCO_3$ equivalent which is higher than the underlying horizon. It is commonly associated with sandy, coarse loamy or loamy particle size class. In general, calcium carbonate equivalent is used as a parameter to measure calcium carbonate in soils. It is defined as the quantity of carbonate (CO_3) in the soil expressed as $CaCO_3$ and as a weight percentage of the less than 2 mm size fraction. In sodic soils, presence of $CaCO_3$ is used to decide whether an acid or an acid forming amendment or waste products such as organic compost, organic garbage, waste acidic products and rejected dairy waste etc. will be useful to reclaim the soil. In the absence of $CaCO_3$ such amendments will not be effective.

Procedure

Place one g of soil sample in a 100 ml conical flask. A 25 ml vial containing 20 ml HCl (1:3) is carefully inserted into the conical flask. The flask is closed with a corked inlet tube of the calcimeter installed in a chamber full of water. The acid is carefully mixed with the soil sample and shaken properly for a few minutes. The volume of water displaced due to release of CO_2 is noted. Similar technique is followed by taking 0.1g $CaCO_3$ instead of 1g soil sample and the volume of CO_2 released is noted. The $CaCO_3$ equivalent is calculated as follows:

Calculations

$$CaCO_3 \text{ equivalent (\%)} = \text{Volume of } CO_2 \text{ released for 1 g soil X 100) / (Volume of } CO_2 \text{ released for 0.1g } CaCO_3) \quad (2.21)$$

Calcium carbonate (Puri, 1930)

Principle

Calcium carbonate of the soil is determined by titrating the soil suspension with 0.5N H_2SO_4 in the presence of bromothymol blue and bromocresol green indicators. Calcium sulphate and aluminium chloride are added so that appearance of colour is distinct. Aluminium chloride ($AlCl_3$) is added to lower the pH of the soil suspension to bring the color change just at the point, where no carbonates are present. Without its use, even a soil free from $CaCO_3$ gives green color. Thus, $AlCl_3$ helps to make sharp distinction between the soils containing carbonates and no carbonates.

Reactions involved

$$AlCl_3 + 3H_2O \longrightarrow Al(OH)_3 + 3\,HCl \quad (2.22)$$

$$CaCO_3 + H_2SO_4 \longrightarrow CaSO_4 + H_2O + CO_2 \uparrow \quad (2.23)$$

Reagents

- **0.5N H_2SO_4**
- **0.1N $AlCl_3.6H_2O$**: Dissolve 8.05 g of aluminum chloride in a small quantity of water and dilute to 1 L with distilled water
- **Bromothymol blue (1% alcoholic solution)**: Dissolve 1 g of Bromothymol blue in 100 ml of alcohol
- **Bromocresol green (1% alcoholic solution)**: Dissolve 1 g of Bromocresol green in 100 ml of alcohol

Procedure

Weigh 10 g of soil and stir it with 100 ml of distilled water in a conical flask. Add 0.2-0.5 g of $CaSO_4$ and bring the suspension to boiling and add 10 ml of 0.1N $AlCl_3.6H_2O$ solution. Shake the suspension and add 10 drops of bromothymol blue and bromocresol green indicators and note down the color. Green color shows the amount of $CaCO_3$ present (more than 1%) and a golden yellow color is indicative of the absence of $CaCO_3$. If carbonate is present, bring the contents of the flask to boiling point and titrate against 0.5N H_2SO_4. Boil for a minute or two after each addition and allow to settle for a couple of minutes. Continue the titration till golden yellow color persists on boiling for a couple of minutes. Note down the volume of 0.5N H_2SO_4 used for titration.

Calculations

Weight of soil taken = 10 g

Amount of 0.5N H_2SO_4 used = Y ml

Amount of H_2SO_4 used is equivalent to total $CaCO_3$ in soil.

1 ml of 0.5 N H_2SO_4 = 1 ml of 0.5 N $CaCO_3$ = 0.5 meq of $CaCO_3$

= 50/2 x 1/1000 = 0.025 g $CaCO_3$

$$\% \ CaCO_3 \text{ in soil} = \frac{0.025 \times Y \times 100}{10}$$

$$= 0.25 \, Y$$

or

$$\% \ CaCO_3 \text{ in soil} = \frac{\text{Volume of } 0.5\,N\,H_2SO_4 \text{ used}}{4}$$

Interpretation of results

From crop production point of view, following guidelines are used to rate the soil.

$CaCO_3$ (%)	Soil rating
<5% $CaCO_3$	Fit for all crops
5-10% $CaCO_3$	Fit for all crops except citrus
>10% $CaCO_3$	Unfit for all crops

Reporting of the Results

To avoid any major change in the physico-chemical properties of the samples at high temperatures, air dried samples are commonly used for analysis. On the other hand, it is customary to report the results on oven dried basis. Therefore, the results need to be multiplied by a moisture factor (F), that could be calculated as per the following procedure.

Take exactly 10 g of air dry soil passed through 2 mm sieve. Keep the sample in an aluminum can and dry it at 105°C for 24 h. Cool in a dessicator and weigh the can immediately to the nearest 1/100 of a gram. The F then can be given as:

$$F = 10/w \tag{2.24}$$

such that w is the wt of the oven dry soil (w). Since w would be less than 10, F would always be more than 1.0. The analytical results are thus given as:

Analytical result on oven dry basis = Analytical result on air dry basis* F

Preparation of Sodic/Alkali Soil

To conduct experiments on sodic lands, many times it is necessary to prepare a sodic soil artificially. The following procedure is used to prepare a sodic soil of known pH or ESP.

- Estimate the amount of soil required for the experiment / study
- Properly sieved and dried soil from normal (non-saline, non-sodic) fields is spread on polythene sheets in uniform layer, not thicker than 0.5 cm
- Find out the initial ESP and pH_2 of the soil. (Approximate relationship between ESP and pH_2 for alluvial soils of the Indo-Gangetic plains is given in Table 2.13.
- Calculate the amount of salt required to increase the pH_2 of the soil for pre-decided sodicity level

Table 2.13 Approximate relationship between ESP and pH_2 of soils

ESP	Approx pH_2
20	9.1
25	9.2
35	9.4
40	9.5
45	9.6
50	9.7
65	9.8
75	9.9
85	10.0
100	10.6

Quantity of salt required

Let us assume that a soil of ESP 50 is to be prepared using a medium textured alluvial soil of the Indo-Gangetic plains having a CEC of 12.

ESP = Ex. Na (meq 100 g^{-1}) x 100/CEC

50*12 = Ex. Na x 100

If Ex. Na is denoted by X, then

X = 50*12 /100 = 6.0 meq Na

6 x 23 = 138 mg Na100 gm^{-1} soil or 1.38 g Na 1000 gm^{-1} soil

Using $NaHCO_3$ (Equivalent weight 84)

23 gm of Na is available in 84 gm $NaHCO_3$

1.38 gm of Na is available in = 84 x 1.38/23 = 5.04 g kg^{-1} soil

- Dissolve the total amount of salt for the given quantity of soil in water
- Spray the salt solution on the soil with spray pump, till it becomes saturated
- Cover the soil with polythene sheet
- The remaining salt water can be sprayed on following days
- After spraying the total salt solution, the soil is kept moist for 15-20 days to allow the complete reaction of salts
- After the soil comes in condition, mix it thoroughly 4-5 times. After 4-6 weeks, the soil pH_2 may be estimated before using it for experiment.

CHARACTERISTICS OF SODIC SOILS

A sodic soil has the following characteristics.

- Characterized by high pH_s (>8.2)
- Occurs in regions with flat topography with slope varying from 0.1 to 0.2% with poor drainage conditions
- Low infiltration rate, < 5% of the rate in normal lands in the vicinity
- Poor internal drainage resulting in dry subsoil and a moist or wet surface layer. Crops fail because of excess surface water or for lack of water even though there may have been adequate rainfall or irrigation.
- Black patches may appear at the soil surface because of the dissolution of organic matter at high pH, which is transported to the soil surface through capillary action. It is why these soils are sometimes referred as black alkali soils
- Presence of *Kankar* layer ($CaCO_3$ concretions) at depth ranging from 0.5 m to 1.8 m depth
- Commonly encountered in semiarid and sub-humid agro-climatic conditions with annual rainfall of more than 500 mm.
- Underlain by good or sodic ground water

A saline-sodic soil on the other hand has the following features:

- Contains salts in excess of 4 dS m^{-1} and ESP in excess of 15. On the basis of EC_e, the soils are characterized as in Table 2.14.

Table 2.14 EC_e and degree of salinity hazard

Soil salinity class	EC_e (dS m^{-1})	Effect on crop plants
Non-saline	0 - 2	Salinity effect negligible
Slightly saline	2 – 4	Yields of sensitive crops restricted
Moderately saline	4 – 8	Yields of many crops restricted
Strongly saline	8 – 16	Tolerant crops give satisfactory yields
Very strongly saline	> 16	Only a few very tolerant crops yield satisfactorily

- Have all the features of the saline soil, and if reclamation procedures are used that do not include gypsum, they become sodic upon leaching of salts.
- Reclamation takes several years (2 or more), depending upon the time required to get about two pore volumes of good quality water to pass through the soil.

Based on soil ESP, the impact on crop productivity is graded from slight or negligible to extremely high or severe (Table 2.15).

Table 2.15 ESP and degree of sodic hazard

Approximate ESP	Sodic hazard
Up to 15	None to slight
15 – 30	Slight to moderate
30 – 50	Moderate to high
50 – 70	High to very high
> 70	Extremely high

Diagnosis of Sodic Soil : Visual

The visual characters that serve as a diagnostic criterion to identify the sodic soils are reported in Table 2.16. Exact analysis should however, be based on the chemical characteristics as revealed by the sample analysis. The differentiating parameters have already been described in Table 1.1.

Table 2.16 The visual observations to assess sodic problem during field visits

Visual observation	Possible conclusions
Surface stagnation of rain/irrigation water	Sodic soil, compaction of the soil and/or a depressed land
Hard dense subsoil	Compaction due to machinery or sodic subsoil
Surface salt crust	Sodic: if dark/brown/ash coloured clay crust and black spots visible
$CaCO_3$ nodules present	May be sodic
Unusual leaf colour	Poor nutrition, water logging, imbalance in pH
Presence of sodicity tolerant weeds/vegetation	*Sporobolus marginatus, Desmostachya bipinata, Suaeda maritma, Diplachne fusca, Dichanthium annulatum, Cynodon dactylon, Chloris barbata, Brachiaria mutica, Kochia indica, Panicum antidotale*

FACTORS AFFECTING PLANT GROWTH IN SODIC SOILS

Plant growth, yield and/or quality of crops in sodic soils are affected by one or a combination of the following factors, the extent depending upon the amount of exchangeable sodium, pH, nature and kind of crops grown, macro and micro-environmental conditions and management interventions (Gupta *et al.*, 2007).

Chemical Factors

- High pH due to high concentrations of HCO_3 and CO_3; although high pH has no direct adverse effect on plant growth *per se*, it frequently results in altering the availability of some essential plant nutrients. For example, the concentration of the elements calcium and magnesium in the soil solution is reduced with increasing pH.
- Toxicity of Na, HCO_3 and CO_3 ions
- Excess of Na causing imbalance in the uptake of other ions due to antagonistic effect such as K and Ca nutrition. Deficiency of Ca is encountered since nearly all the soluble Ca is precipitated as insoluble $CaCO_3$
- Decreased solubility and availability of micronutrients like Zn, Fe and Mn due to high pH and presence of $CaCO_3$ and soluble CO_3 and HCO_3 ions
- Increased solubility and accumulation of toxic elements like F, Se and Mo in plants that may affect crop yield and/or health of the animals feeding on such crops
- Presence of hard $CaCO_3$ layer, which acts as a physical barrier for the movement of water, salts and roots of crops and trees.

Physical Factors

- Poor water and air permeability as a result of high dispersion of soil aggregates and clay particles
- Low water availability due to poor conductance from the lower to the upper soil layers and restricted effective root zone
- Hard crust on the surface layer which greatly hinders the seedling emergence and reduces the germination percentage and thus plant population
- Poor drainage resulting in water logging and aeration problems
- Difficulties in preparing seedbed as all parcels of the land do not attain proper moisture conditions at the same time; this also results in clod formation and poor tilth of the soils.
- Low activity of useful microbes due to high pH and high exchangeable Na
- Continuous loss of fertile soil due to air and water erosion
- Low organic matter content.

In saline sodic soils, several other effects come into play to affect plant growth because of excess salinity. First and the most important is the availability of water to the plant, which gets reduced resulting in a situation where plant is starved of water even though the soil is moist (physiological drought condition). It is because the osmotic pressure of the soil solution increases with the increase in salt concentration and the plants are unable to extract water as readily as they can from a relatively non-saline soil. In addition to the osmotic effect of salts in the soil solution, at high concentration, absorption of an individual ion may prove toxic to the plants. For example, some constituents of salts e.g. chloride, which at moderate concentration is harmless to most crops, may be toxic at higher concentrations to some crops e.g. citrus and grapes. In addition, some elements like lithium and boron etc. may prove toxic to some crops e.g. citrus and grapes even though they may be present in very less quantities. Besides, preferential absorption of one ion may also retard the absorption of other essential plant nutrients necessary for the normal growth of plants. The conditions like periodic submergence or water logging and the presence of shallow brackish ground water also present severe limitations, as these factors further accentuate soil salinization. Plant growth under field conditions may be adversely affected due to a combination of two or more of these factors, depending on the level of exchangeable sodium, soil salinity status, nature of the crops and the overall level of management.

PHYSICO-CHEMICAL CHARACTERISTICS

The saline and sodic soils can be distinguished from normal soils as well as from each other on the basis of several physical, chemical, biological and mineralogical characteristics and soil-water relationships that get modified upon chemical degradation. Some of such properties are discussed in the following sections. The discussions center around parameters that help differentiate sodic from the normal lands.

Physical Characteristics

Effective soil depth

The effective soil depth that can be exploited by plant roots is an important consideration in plant growth. Generally, excellent yields are produced on normal soils having an effective root zone depth varying from 45 to 90 cm although shallow-rooted crops do fairly well even with depths up to 30 cm. Root penetration is seriously inhibited in sodic soils due to high ESP and presence of an impermeable layer of *kankar* (impure $CaCO_3$), limiting the effective soil depth to 10-15 cm. Though some crops adapt to edaphic environment by modifying their root system, it is desirable to grow only shallow rooted crops at least in the early phase of reclamation. In saline soils, the zones of *kankar* and gypsum accumulation, if any, restrict the effective soil depth.

Bulk density

Bulk density is defined as the dry weight of a unit volume of soil and is usually expressed in g cm^{-3} or Mg m^{-3}. The bulk densities of highly productive normal soils usually range from 1.1 to 1.5 (medium to fine texture) and 1.2 to 1.65 (coarse texture). The sodic and saline soils (medium texture) have a high bulk density (1.5 to 1.65 g cm^{-3}) inhibiting root penetration and proliferation as well as impeded drainage. Because of dispersion effect of sodium coupled with impact of rain drops, bulk density in top surface layer of sodic soils can go up to 1.7 g cm^{-3}. The rates of infiltration rates and permeability are usually low in medium or fine textured soil with bulk densities exceeding 1.65 g/cm^3. The bulk density decreases with application of amendments, tillage and cropping.

Pore space distribution

Pore space and bulk density are closely and inversely related. Soils with a low porosity generally contain little air space at field capacity. In highly productive normal, medium to fine textured soils, the non-capillary porosity in the plough-zone is usually 10-30%. The sodic soils may have a non-capillary porosity as low as 2% at which normal aeration of the plants is adversely affected. Gupta and Narain (1971) observed that the total porosity and non-capillary porosity in normal soil profile (0-110 cm) varied from 45.8 to 38.5% and 8.6 to 5.7% as compared to 39.5 to 38.9% and 4.9 to 2.5% in saline-sodic soils.

Soil tilth

The soil tilth is a product of several parameters such as organic matter content, texture, consistency etc., which under good management combine to form a surface or near-surface physical condition of the soil favourable to plant growth. The highly productive normal soils have a very good tilth. In sodic soils, the soil tilth is quite poor as a result of high pH and ESP. The soils are very hard and compact when dry and very sticky when wet. These soils therefore, have to be tilled carefully at optimum moisture content. The saline soils generally have satisfactory tillage properties.

Cracking behavior

Sodic soils are highly dispersed and crack upon drying. In alluvial sodic soils cracks are not very wide. Even sodic Vertisols, which swell on wetting and shrink on drying, do not show wide cracks as per the characteristics of Vertisols. In fact depth and width of cracks increase with increasing ESP (Table 2.17). Upon continuous drying, thin flakes gets peeled off and upon further drying narrow, shallow and more number of micro-cracks are formed with increasing ESP (Table 2.17).

Table 2.17. Effect of soil ESP on cracking of sodic Vertisols

Soli ESP	Depth of cracks (cm)	Width of cracks (cm)	No. of flakes (m^{-2})
6	>90	5-6	-
10	50	2-3	2-5
15	30	1-2	5-10
22	10	0.5-1.0	10-30
38	<2	0.2-0.3	30-50
58	0.1-0.2	0.05-0.10	80-100
>60	Negligible	Absent	NA

Source: Verma *et al.* (2006), NA = Not available

Crust formation

Sodic soils have low structural stability. These soils disperse and slake when they are wet and may develop a hard crust upon drying. The crust presents a serious barrier to the emergence of seedlings and in some crops it may be the main cause of poor stand. The crust formation is not only limited to sodic soils alone but may also be formed in normal or saline soils. Typically, sodic soil crust having a total salt content of 22.8% was found to contain 48.3% Na_2CO_3, 33.4% $NaHCO_3$, 13.3% Na_2SO_4 and 3.9% NaCl. However, hard surface crusts are typically associated more with high ESP, a characteristic of sodic soils, low organic matter, puddling and wetting the soil to zero tension. Modulus of rupture, a property determined to infer the strength of the crust formed on the soil surface upon drying, is found to increase with increasing ESP of the soil as per the following relation (Sandhu *et al.*, 1980):

$$Y = 3.92 + 0.204 \text{ ESP} \qquad (r = 0.92) \qquad (2.29)$$

Here, Y is the modulus of rupture in bars.

Dispersion and swelling

Dispersion is the breakdown of soil clay minerals into single particles in solution when the bonds in a soil are not strong enough to keep the particles together. The opposite of dispersion is flocculation. Dispersion occurs because a soil has too great a proportion of exchangeable sodium or is too low in organic matter. However, working a soil under too wet or too dry conditions can also increase its tendency to disperse on wetting especially soils that are low in organic matter. Salinity can mask the tendency of a soil to disperse. As the sodium on the exchange complex increases, the forces that bind clay particles together are disrupted. It happens because of distortion of the diffuse double layer of the adjacent clay particles leading to their electrostatic dispersion and due to the size of the hydrated ions in the diffuse double layer (Fig. 2.4). The excess of singly charged particles increases and doubly charged particles decreases the thickness of diffuse double layer. Similarly, large cations in the diffuse double layer push clay particles apart. Sodium ions being larger in size than potassium and magnesium ions being larger than calcium ions, sodium ions are more damaging to clay structure than potassium and magnesium more damaging than calcium.

As water moves into soil, dispersed colloids are oriented to plug soil pores that results in reduced soil permeability.

Cations bring together negatively charged clay particles to flocculate soil clays (making clumps or "aggregates"). Sodium (Na) is a poor flocculator than calcium (Ca) and magnesium (Mg) because it has less charge and its ionic size in water is much larger. Relative to Na^+, Ca^{2+} and Mg^{2+} has 43 and 27 times greater floccultaing power, respectively. Once the separation occurs, the clay particles expand, causing swelling and soil dispersion.

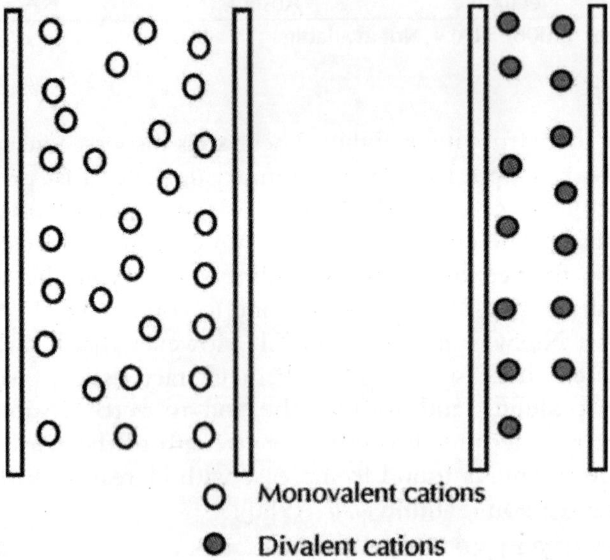

Fig. 2.4 Effect of cation type on the thickness of diffuse double layer

Dispersion coefficient defined as the ratio of water dispersible clay to the total clay expressed as percent is usually high in sodic soils. It may exceed 90% in soils having pH more than 10.0 compared to 20-30% in normal soils. The dispersion coefficient increases with increasing ESP (Fig. 2.5). On the contrary, in sodic Vertisols, dispersion ratio increased up to 15 ESP and thereafter remained almost constant up to 60 ESP. It is argued that a smectitic sodic soils present similar structural degradation at 15 ESP as at 60 ESP because of high buffering nature of the clay mineral.

The soil dispersion is one of the major factors that increases sediment content in the run-off waters from sodic soil catchment. The formation of a layer of high clay content in the profile at some depth is also attributed to high dispersivity. Flocculation, on the other hand is beneficial in terms of soil aeration, root penetration, and root growth. Although high electrolyte concentration of the soil solution in a sodic soil minimizes dispersion, yet the salinity cannot be increased to maintain soil structure without considering its potential impacts on plant health.

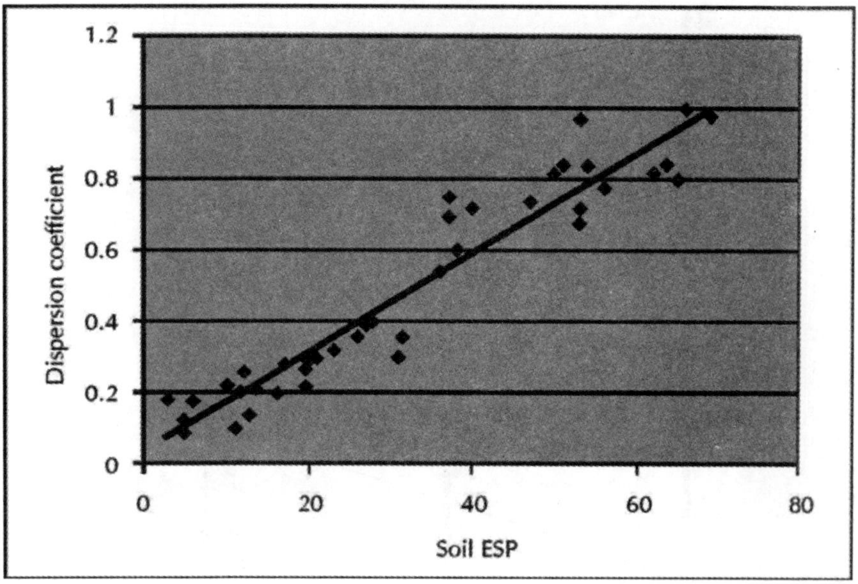

Fig. 2.5 Effect of soil ESP on dispersion coefficient

The three main problems caused by sodium-induced dispersion of clay platelet and aggregate swelling are surface crusting, reduced infiltration and reduced hydraulic conductivity.

Surface crusting, a characteristic of sodic soils is primarily caused by 1) physical dispersion caused by impact of raindrops or irrigation water, and 2) chemical dispersion, which depends on the ratio of salinity and sodicity of the applied water. Surface crusting due to rainfall is greatly enhanced by sodium induced clay dispersion. When clay particles disperse within soil water, they plug macropores in surface soil by two means. First, they block avenues for water and roots to move through the soil. Second, with repeated wetting and drying, clay dispersion occurs, it then reforms and solidifies into almost cement-like soil with little or no structure. The ratio of salinity (EC) to sodicity (SAR/ESP) determines the effects of salts and sodium on soils. While salinity promotes soil flocculation, sodicity promotes soil dispersion. The combined effect of salinity and sodicity of soils is measured by the swelling factor, which is the amount a soil is likely to swell with different combinations of salinity and sodicity (Pearson, 2009). Essentially, the swelling factor predicts whether sodium-induced dispersion or salinity-induced flocculation will influence soil physical properties. Swelling factor as a function of sodium content (adjusted ESP) of soil and salt concentration of soil water has been given by McNeal (1968). The figure proposed by him helps to show how the dispersive effects of soils with high ESP can be lessened with the flocculating effects of irrigation water with high EC (Fig. 2.6).

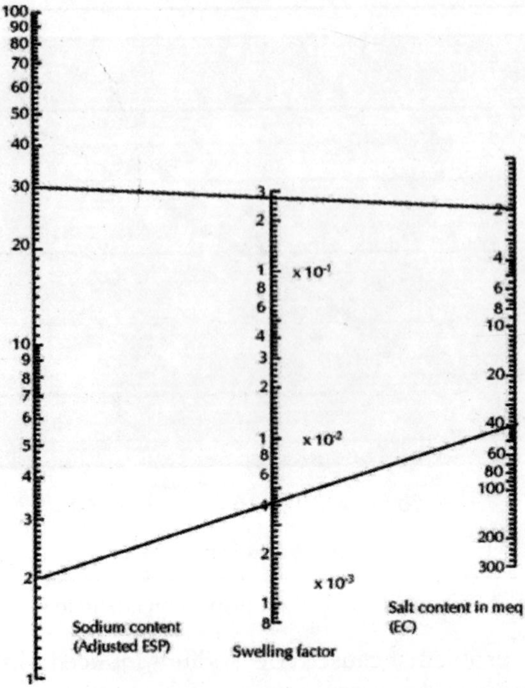

Fig. 2.6 Swelling factor as a function of sodium content (adjusted ESP) of the soil and salt concentration of soil water (After McNeal, 1968).

Test for Dispersion

To check whether soil is dispersive or not, drop small air-dry aggregates (3 to 5 mm diameter) into a saucer of rain (distilled) water carefully so as not to bump the dish to avoid additional dispersion. If the water turns cloudy within two hours, the soil is highly dispersive and will probably respond to gypsum or lime. This property has been used to develop a soil sodicity kit for diagnosis of sodic soils in Australia and India (See Box).

Soil Sodicity Kit: The Diagnosis Procedure

- Take about 100 g of soil to be tested. Pour it in a jar/beaker of about 600 ml capacity.
- Take 500 ml of rain/distilled water and pour it gently in the jar/beaker containing the soil.
- Ensure that soil is not unusually disturbed.
- Invert the jar/beaker once gently and bring to its original position.
- Keep it in the original position for about 4 hours.
- Check the suspension above the soil using a white spatula.

Clearly visible: Normal soil
Partially visible: Slightly sodic
Poorly visible: Sodic soil

Water stable aggregates

The cumulative percentage of water stable aggregates of 1.0 mm size decreased with increasing ESP in a Vertisol, from 61% at an ESP of 6.0 to 4.0 at an ESP of 58 (Table 2.18). The mean weight diameter also reduced from 1.52 mm for a normal soil of ESP 6.0 to 0.23 mm for a soil of ESP 58 (Table 2.18).

Table 2.18 Cumulative percentage of water stable aggregates and mean weight diameter at different ESP in a sodic Vertisol

ESP	Water stable aggregates (%)						Mean weight diameter (mm)
	>2 mm	>1 mm	>0.4 mm	>0.2 mm	>0.08 mm	>0.05 mm	
6	40.0	61.0	77.8	88.2	92.0	92.7	1.52
10	35.8	54.8	74.0	83.2	86.6	88.3	1.39
15	20.6	44.6	70.7	83.9	90.1	92.0	1.09
22	18.2	36.0	63.6	79.1	84.7	86.2	0.97
38	18.0	38.3	49.7	64.6	71.1	74.2	0.88
58	0.9	4.0	19.4	34.3	45.9	51.4	0.23

Source: Verma *et al.* (2006)

Shear Strength

Shear strength of a soil is quite important in studies pertaining to construction of buildings and dykes *(bunds)* as well as in studies related to soil erosion. The shear strength decreases with increase in pH and moisture content. Arubnandam and Heinzer (1977) reported that the critical shear stress which is defined as the stress at which erosion starts, reduces from 25 dyne cm^{-2} to 5 dyne cm^{-2} as the SAR increases from 5 to 15 for the same CEC of the soils. At still higher SAR, the critical shear stress further reduces to 1 dyne cm^{-2} at SAR of 60 and is independent of CEC. As the shear stress caused by rainfall and run-off in a region is similar, sodic lands are likely to be more eroded compared to a normal soil and hence more sediment is likely to occur in the run-off arising from watersheds that have large sodic lands. As a consequence, the dykes cross-sections and heights need to be considerably larger in sodic than that of normal lands.

Chemical Characteristics

Soil reaction (pH)

Sodic soils having carbonates and bicarbonates show high pH. The reaction between Na_2CO_3 and H_2O can be represented as follows:

$$2Na^+ + CO_3^{2-} + 2H^+ + 2OH^- => 2Na^+ + 2OH^- + H_2CO_3 \quad (2.30)$$

The acid H_2CO_3 is unstable and produces H_2O (water) and CO_2 (carbondioxide gas, escaping into the atmosphere). This explains the remaining alkalinity (or rather basicity) in the form of soluble sodium hydroxide and the high pH or low pOH.

The pH of the saturated soil paste (pH_s) has been widely recognized as an index, distinguishing sodic soils relative to the normal or saline soils. Bhargava and Abrol (1978) and Abrol *et al.* (1980) showed that the pH of the saturated soil paste and the exchangeable sodium percentage of the soils studied by them are correlated. Several other workers (Agarwal and Yadav, 1956; Chang and Dregne, 1955; Kovda, 1965, Agarwal *et al.*, 1979) also reported that pH and ESP are correlated. Gupta *et al.* (1981) pointed out that for soils containing sodium-carbonate type of salts the exchangeable sodium ratio (ESR) and pH were quantitatively related and that the relationship governing their dependence can be derived from Na^+ - $(Ca + Mg)^{2+}$ exchange equilibrium in soils.

For a rapid and convenient appraisal, pH of the 1:2 soil water suspension (pH_2) has also been reliably used for the same purpose in specific cases. The pH_2 values are usually higher than pH_s, the most striking difference being in sodic soils. In sodic soils, pH_2 values are higher than pH_s by about 1.0 unit while in normal or saline soils, it is higher by 0.2-0.5 units. This is ascribed to the nature of electrolytes present in sodic and saline soils. Sodic soils contain predominantly Na_2CO_3, a weak electrolyte, whereas the saline soils contain NaCl, a strong electrolyte. On hydrolysis, the pH of weak electrolytes increases sharply with dilution, whereas changes in the pH with dilution are less pronounced in strong electrolytes (Gupta, 1972). The increase in the soil pH_2 by about 1.0 unit from pH_s is useful to identify sodic soils.

Based on these considerations and on the basis of analysis of large number of samples from sodic soils of Uttar Pradesh (India), Bhargava and Abrol (1978) suggested an approximate relationship between pH_2 and ESP, which is given as follows (Fig. 2. 7):

$$ESP = 49.4\, pH_2 - 424.3 \qquad (2.31)$$

It can be used to assess the approximate ESP of soil from pH_2 measurements. It may be noted that this equation gives a computed value of zero at pH_2 of 8.6 and absolute sodium saturation at a pH_2 of 10.6. On the other hand, a sigmoid relationship between pH_2 and ESP may be more appropriate (Fig. 2.7). It showed that the soils having pH_2 of about 9.2 have ESP of about 20 and the increase in ESP beyond 10.2 slowly attains an absolute saturation at a pH_2 of about 10.6 as given by the linear relation. It may be noted that such a relationship is not universal and therefore, is applicable for specific and similar conditions. From management point of view, a semi or partially reclaimed sodic soil is one whose original pH_2 value > 10 has decreased to < 9.3 in the root zone (0-20 cm) but higher values persist in the lower depth (Fig. 2.8).

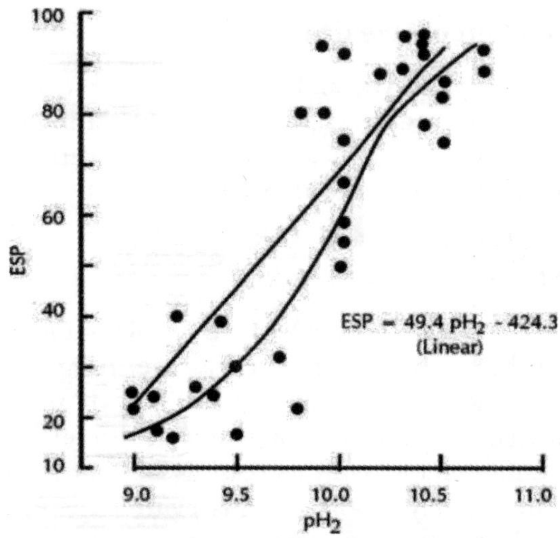

Fig. 2.7 Relationship between pH2 and ESP of soils in indo-Gangetic alluvial plain (India)

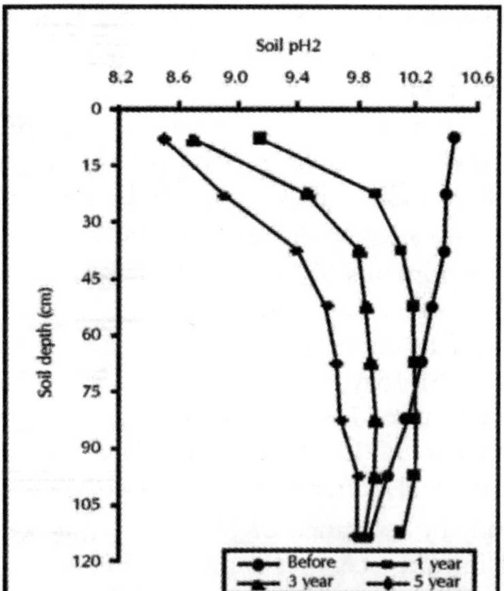

Fig. 2.8 Changes in soil pH2 at different soil depths before and after 1, 3 and 5 years of gypsum application on farmer's fields

In addition to dilution, other factors which affect the pH of the salt affected soils are: the total soluble salt concentration, nature of salts, the composition of exchangeable cations, the nature of cation-exchange materials and the presence or absence of gypsum and alkaline earth carbonates. Sodium carbonate causes strong sodication of the soils. Whereas pH of saturated solution of sodium bicarbonate is only 8.4, that of sodium carbonate is as high as 12.0. The excessive soluble salt concentration, particularly those of neutral salts, depresses the soil pH. The highly saline soils having an EC_e of about 100 dS m^{-1} usually have a pH close to 7.

Exchangeable Sodium Percentage (ESP)

The sodium adsorption ratio (SAR) and the exchangeable sodium ratio (ESR), which is determined as the cation exchange capacity minus the exchangeable sodium at the saturation percentage of the soil are significantly correlated. Based upon this relationship, ESP is related with SAR by the following equation:

$$\text{ESP} = \frac{100\ (-0.0126 + 0.01475\ \text{SAR})}{1+(-0.0126 + 0.01475\ \text{SAR})} \tag{2.32}$$

Thus, ESP can be calculated if the SAR of the soil solution is known. Either ESP or ES (exchangeable sodium) expressed as me/100g or SAR is usually a good indicator of the structural stability of a soil and of the physical response that might be anticipated when water is applied. Although physical properties become increasingly unfavourable with increasing ESP, at a given level of ESP, physical properties are usually poor in soils with expanding type 2:1 clay minerals, containing a low electrolyte concentration of salts capable of alkaline hydrolysis, *viz*, sodium carbonate, sodium bicarbonate, silica, etc. Compared to this, soils with clay minerals of the non-swelling 1:1 type (kaolinite) or mixed types, containing a high electrolyte concentration of salts, viz. sodium chloride or sodium sulphate have better physical properties. The sodic soils containing expanding type 2:1 clay minerals (illite, vermiculite, montmorillonite) exhibit unfavourable physical properties at levels of ESP>15, ES>3.0 meq $100g^{-1}$ or SAR>15. ESP or SAR of sodic soils is generally linearly and positively related with pH whereas there is no such relationship in saline soils.

The most productive normal soils have ESP or SAR less than 5 and ES less than 1.0 meq $100g^{-1}$. The saline soils have ESP or SAR usually less than 15 but SAR values up to 100 are not uncommon. The sodic soils have ESP >15 but in highly deteriorated soils, ESP may be as high as 100 and SAR may be as high as 500. Qualitatively on the basis of ESP, the sodic hazard can be anticipated as per information provided in Table 2.15. As the soil ESP increases from 5 to 70, the relative productivity of the soil decreases from almost 100 to 0.

The soil texture and the tolerance of plants to ESP are other two factors, which are of practical importance. At a given ESP, the coarse-textured soils exhibit less deterioration in physical properties than the fine-textured soils. Horticultural plants are extremely sensitive (ESP 2-10) followed by pulses (ESP 10-15). Wheat, barley, oat, raya (mustard), *senji*, *berseem*, *shaftal*, *dhaincha*, sugar cane, millets and cotton are semi-tolerant (ESP 15-30), and transplanted rice, sugar beet, para grass and Karnal grass are tolerant (ESP 30-60).

SAR

Although SAR has been recognized as a parameter to assess sodic conditions defining a natric horizon as the one having SAR >= 13. The major problem arises that even saline soils have high SAR but may not show high ESP because of the predominance of chlorides and sulphates or high pH as the excessive soluble salt concentration of neutral salts depresses the soil pH. Since the sodic soils have sizeable amounts of carbonates/bicarbonates, calcium and magnesium tend to precipitate resulting in high SAR values sometimes even exceeding 100. A relationship between SAR and ESP of the sodic soils of Uttar Pradesh revealed a positive correlation between the SAR and ESP of the soils for all samples having SAR less than 100 (Bhargava, 2003; Fig. 2.9). The following linear relation is fitted to the 24 data sets having SAR less than 100 out of 41 data sets. For a SAR of 10, the calculated ESP is about 21.

$$ESP = 0.809\ SAR + 12.78 \quad (2.33)$$

The exchangeable sodium status of soils can be predicted fairly well from the SAR of the saturated soil extract since the two are related by the expression:

$$ESP = 100\ [Na_{ex}/(Ca_{ex}+Mg_{ex})] = K_G\ SAR \quad (2.34)$$

where the exchangeable ion concentrations are in cmol (+) kg^{-1} (the sub-script ex indicates exchangeable), and K_G is the exchange constant or Gapon's constant. Several studies have shown that there is a linear relationship between SAR of the soil solution and ESR up to an ESP of about 50 so that SAR of the soil solution can be used as a fair measure of the exchangeable sodium status of soils. For a better estimate of exchangeable sodium, the value of constant K_G needs to be determined experimentally for each major group of soils.

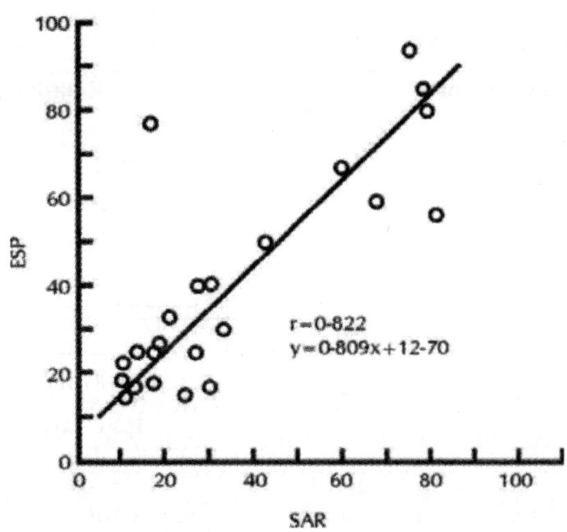

Fig. 2.9 Relationship between SAR and ESP in sodic soils of Uttar Pradesh

Salinity (Soluble salts)

Most productive normal soils have EC_e less than 1 dS m^{-1}. Conventionally, a soil having EC_e more than 4 dS m^{-1} has been designated as saline and less than 4 dS m^{-1} as non-saline. The Soil Science Society of America (1974) has recommended that soils having EC_e greater than 2 dS m^{-1} should be called as saline soils. This seems to be more logical as the growth of most sensitive plants, particularly of the horticultural and pulse crops, begins to experience the adverse effect of salinity at an EC_e of 2 dS m^{-1}. It would be still more appropriate if this limit were further reduced to 1.5 dS m^{-1}, which would roughly correspond with 1000 mg L^{-1} salt concentration.

The saline-sodic soils, thus may be associated with high ESP but their physical properties are not affected adversely because of the high electrolyte concentration of the soil solution due to presence of neutral salts. The soluble salts in saline soils mainly comprise the chlorides and sulphates of sodium, though calcium and magnesium are also present in appreciable quantities. These salts are highly deliquescent and therefore, absorb moisture from the atmosphere, causing the soil to appear moist. Higher concentrations of magnesium than that of calcium cause harmful effects on the physical properties of the soils and crop growth (Yadav and Girdhar, 1980, 1981). Bicarbonates occur in very low amounts and carbonates are usually absent or occur only in traces.

The kind of soluble salts, the texture of the soil and nature of plants are some of the important factors, which determine the plant growth on saline soils. Chloride salts are more harmful than sulphate salts (Table 2.20). The effective salinity is often described as [½ SO_4 + Cl]. As the EC_e of the soil increases from 4 to 20 dS m^{-1} in case of chloride salts and 6 to 40 dS m^{-1} in the case of sulphate salts, the relative productivity of the soil in most cases decreases from almost 100 to 0.

Table 2.20. Relationship between nature of salts in the soil and plant growth

EC_e (dS m^{-1})		Relative plant growth (%)
Chlorides	Sulphate	
4	6	100
4-8	6-12	80-90
8-12	12-20	50-70
12-20	20-40	20-50
20+	40+	0

Source : Kovda et al., (1973)

At a given salinity, the adverse effects on plant growth are more in fine than in coarse textured soils. The plants also differ widely in their tolerance to salts. The horticultural plants except date palm and pulses are sensitive to salinity (EC_e 1.5–3.0 dS m^{-1}). Sunflower, groundnut, maize, sugar cane and rice comprise the semi-tolerant crops (EC_e 3-5 dS m^{-1}) while cotton, sorghum, pearl millet, wheat, barley, mustard, safflower and sugar beet comprise the tolerant group (EC_e 5-10 dS m^{-1}) (Gupta and Yadav, 1986). Amongst vegetable crops, kale, asparagus, spinach are tolerant; tomato, broccoli, cabbage, cauliflower, lettuce, potato, carrot, onion, pea, squash and cucumber are semi-tolerant; and radish, celery, and green bean are sensitive.

The sodic soils, although relatively low in soluble salts, have a composition that differs considerably from that of normal and saline soils, containing mainly carbonates and bicarbonates of sodium. Chlorides and sulphates also occur but in relatively small amounts. The calcium is precipitated in the soil in the presence of bicarbonate ions and therefore, it is commonly deficient in sodic soils. In presence of sodium bicarbonate, calcium bicarbonate still retains some-degree of solubility but in the presence of sodium carbonate even when the total alkalinity level drops to 0.1-0.2 g L^{-1}, calcium ions are virtually eliminated from the solution. Similarly, if concentration of HCO_3 ions exceeds 0.5 g L^{-1}, there would be practically no calcium salts in the soil extract. Absolute concentration of bicarbonate ions in amounts greater than 10.0 mmol L^{-1} may affect mineral nutrition and tend to reduce the availability of iron in many plants, thereby causing chlorosis.

Calcium carbonate

Sodic soils of the Indo-Gangetic plains invariably contain free $CaCO_3$ concretions at some depth in the soil profile. These concretions play a significant role in the self reclamation of sodic soil due to liberation of CO_2 and other biological activities of the crop roots. As more and more $CaCO_3$ is dissolved, Ca replaces sodium on the exchange complex. As the solubility of $CaCO_3$ increases with decrease in pH, $CaCO_3$ (< 2 mm) decreased from 1.85 to 0.48% in 0-15 cm layer after 5 years of gypsum application but in deeper layers change was quite less (Table 2.21).

Table 2.21. Effect of gypsum application on soil properties under rice/wheat cropping

Year after gypsum application	Depth (cm)	EC_2 (dS m^{-1})	$CaCO_3$ (%)	Organic carbon (%)
1	0-15	0.96	1.60	0.15
	15-30	1.00	2.00	0.11
	30-60	1.23	2.60	0.09
3	0-15	0.38	1.00	0.19
	15-30	0.61	1.90	0.15
	30-60	1.00	2.52	0.12
5	0-15	0.22	0.48	0.23
	15-30	0.25	1.82	0.18
	30-60	0.40	2.50	0.13
Before gypsum application	0-15	3.25	1.85	0.10
	15-30	3.00	2.10	0.09
	30-60	2.46	2.62	0.08

Source: Sharma (1986)

Biological Characteristics

Organic matter

Soil organic matter or humus comprises of non-humic and humic substances. While the non-humic substances are the unaltered remains of plant and animal tissues including the organic compounds, the humic substances are chemically and biologically modified substances, being the most active fraction of the organic matter. Humic substances are resistant to microbial degradation. Based on their solubility, these are separated into four fractions namely (i) fulvic acid soluble in both alkali and acid (ii) humic acid, soluble in alkali but insoluble in acid and alcohol; (iii) hymatomelanic acid, soluble in alkali and alcohol and (iv) humin, insoluble in both alkali and acid. Humus found in small amounts in productive agricultural soils; generally representing less than 5% of the dry weight, imparts unique and remarkable properties to the inorganic matrix of sand, silt and clay.

The alkaline solutions in sodic soils is capable to dissolve soil humus becoming highly dispersed in the soil solution. On evaporation, a dark coloured residue is left in the form of a thin crust or rings around small depressions, thereby imparting a dark black colour to the surface of the soil. As a result, the organic matter content in sodic soils is low, generally in the range of 0.1-0.5%.

Micro-organisms

Micro-organisms in soils include bacteria, actinomycetes, fungi, algae and protozoa. There are hundreds of millions of micro-organisms in a spoonful of soil, most of them of benefit to plants and soil. They are important for soil structure, as they form organic bonds that hold soil particles together. They also release carbon dioxide and energy as well as mineral nutrients that plants can use. They do this because they derive their energy by breaking down plant and animal material. When they die and get decomposed, they release more nutrients for plant use.

The physical and chemical properties of a soil determine the nature of the environment in which micro-organisms are found. The soil reaction plays a vital role in the growth of microorganisms. Singh (1950) listed the species of blue-green algae found in sodic soils of northern India. Saxena (1960) collected the algal flora of sodic soils and observed that if pH is high (9.0-10.9) and the total nitrogen content of the soil is low (below 0.2%), the population of algal flora is poor. In ditches with stagnant water, where pH varied from 9.0 to 10.2 and total nitrogen content was about 0.03%, green and blue algae were in equal proportions. As the pH was lowered (7.9-8.6), the proportion of green algae increased. The development of the algal flora is the most prominent during monsoon. The total number of bacteria, actinomycetes and fungi are usually the highest in productive normal soil followed by saline and sodic soils (Table 2.22). These environmental characteristics in turn affect the composition of the microbial population both qualitatively and quantitatively.

Table 2.22 Microbial counts in different soils

Nature of the soil	Organisms x 10^5 g^{-1} soil		
	Bacteria	Actinomycetes	Fungi
Normal	96.00	11.6	0.21
Saline (EC_2 6 dS m^{-1})	11.30	0.4	0.045
Sodic (pH_2 10)	2.33	0.04	0.0003

Source : Neelkantan and Pahwa (1977)

Total organic carbon (%), dehydrogenase activity (μg TPF g^{-1} soil), microbial biomass carbon (mg kg^{-1} soil), % microbial biomass carbon and DHA, all were low in barren sodic soils (pH = 10.6, ESP = 95). The values were 0.5, 4.5, 56.7, 2.3 and 0.08 respectively (Batra *et al.,* 1997). All values improved to 0.44, 112.1, 203.3, 4.6, 30.55 upon reclamation with Karnal grass and gypsum applied at the rate of 50% GR (pH - 9.3, ESP = 40).

Mineralogical Characteristics

The kinds of clay minerals present in the soil determine several of the physical and chemical characteristics of a soil. The clay soils containing montmorrillonitic or smectitic minerals (swelling clays) are more subject to alkalinity problems than illitic or kaolinitic clay soils. The reason being that the former types of clay have larger specific surface areas (i.e. the surface area of the soil particles divided by their volume). Thus, mineralogical characteristics of sodic soils determine at what ESP the physical properties of the soil begins to deteriorate. For example; Balpande *et al.* (1996) suggested that ESP 5 should be used as the lower limit for sodic subgroups of Vertisols of central India rather than ESP 15 as given in Keys to Soil Taxonomy (Soil Survey Staff, 1998). It is because there are severe limitations to the use of such soils owing to the development of adverse physical conditions in terms of very poor drainage even at such low ESPs. Vertisols of Western India with an ESP around 15 cause extreme difficulty in tillage operations (Bhattacharyya *et al.,* 1994). Recent evidences show that dispersion of soil colloids is a determinant of a range of soil properties and among them clay mineral type appears to be one of most impo-rtant factors (Gupta and Abrol, 1990). The 1:1 layer clay minerals (kaolinite) have only a slight stickiness, a small amount of shrinkage on drying, a small surface area and a low cation exchange capacity. Most 2:1 type clay minerals (illite, vermiculite, montmorillonite) have opposite properties. In general, the most productive normal soils, which are dominated by 1:1 layer clay minerals together with iron and aluminum oxides, have excellent soil-water relationships and easy workability associated with their high degree of aggregation and non-swelling nature. At the same level of ESP, these soils are more permeable than those containing 2:1 layer clay minerals. However, 1:1 layer clays have low cation exchange capacity and a low capacity for moisture retention. The saline soils in India have been identified to contain high amount of muscovite, biotite, hornblende and chlorite. The sodic soils are dominated with illite, mixed layer minerals, vermiculite and chlorite, etc. The normal soils in association contain the same minerals as adjacent saline or sodic soils, but differ in the contents of their constituent minerals and degree of crystallinity (Chatterji and Sachdev, 1979).

Smectites are the highly reactive group of clay minerals because of their very high specific surface area. Although soils containing all other clays swell with changes in moisture content, changes are particularly extreme in smectite (Borchardt, 1989), especially when Na- saturated (Low, 1980). Smectites with low charge density swell more than smectites with high charge density

(Alperovitch *et al.* 1986; Slade and Quirk, 1991). In a comparative study of sodic Haplusterts and Typic Natrustalfs, the total clay (<2μ m) is composed of smectite, chlorite, vermiculite, mica, 10-14A° mineral and kaolinite with additional small amounts of quartz and feldspar. This fraction contains considerable amount of smectite (700 g kg^{-1}). The former soils contain 491 g kg^{-1} of smectite whereas the later only 46 g kg^{-1} in the soil control section (Table 2.22). This suggests that the observed impairment of hydraulic properties of Sodic Haplusterts even at a low level of ESP is due to large amount of clay, fine clay in particular which is dominated by low charge smectite whereas similar impairment occurs at a much higher level of ESP (40-50) in Typic Natrustalf because of loamy texture and low CEC caused by very small amount of smectite. It suggests that soils which do not contain smectite may not have the problem of drainage even at an ESP exceeding 50. Thus, the role of content and nature of smectite in fixing the lower limit of ESP in soils assumes importance (Balapande *et al.*, 1997).

Table 2.22 Semi-quantitative estimates of the clay minerals

Horizon	Clay minerals in < 2μm clay fraction (g kg^{-1})						Smectite on fine earth basis (g kg^{-1})	smectites
	Sm[1]	Vm	ML	M	K	Ch		
Sodic Haplustert' (Purna Valley, Maharashtra soil), Basaltic alluvium								
Ap	700	60	60	8	Tr2	100	504	491
A	710	60	60	70	Tr	100	520	
Bss1	720	50	50	50	50	80	490	
Bss2	750	50	50	50	Tr	100	446	
BC	730	60	60	60	Tr	100	390	
Typic Natrustalf (CSSRI, Zarifa Viran soil), Indo-Gangetic alluvium								
A1	120	70	220	460	90	60	15	46
A2	160	90	210	420	70	50	29	
A2 B	280	20	160	360	60	120	62	
Bt	330	50	150	340	50	80	78	
2Bk	170	Tr	190	370	160	116	32	

1 Sm = smectite, Vm = vermiculite, ML-mixed layer minerals (1O-14A 0 minerals), M = mica, K = kaolinite, Ch-chlorite; 2 Tr=Trace

Kankar (impure $CaCO_3$) pan

The *kankar* or impure $CaCO_3$ accumulates in saline or sodic soils that develop under arid and semiarid climatic regions, the accumulation is either diffused throughout the soil profile or soft concretions or nodules are concentrated in a continuous horizon (caliche, cemented hard clay pan) at different soil depths. In sodic soils, a distinct, regular and thick zone of calcium carbonate accumulation, mainly dolomitic exists about 1 to 1.5 m below the surface (Sharma and Bhargava, 1981). The smooth boundary of the calcic horizon, irregular shape of individual concretions and regularity of occurrence underneath vast areas highlight their pedogenic origin (Sehgal and Stoop, 1972). The zone of calcic horizon possibly indicates the zone of shallow water table as the recession of water table causes soil solution to concentrate resulting in precipitation of calcium and magnesium as carbonates. Investigations made at the Usar Reclamation Farm, Chakeri, Kanpur revealed the occurrence of a compact layer 30-40 cm thick and 30-40 cm below the surface, beyond which there is generally a zone of *kankar*, often as a thick pan of 1.2-1.5 m depth with typical characteristics (Table 2.23).

Table 2.23 Physical and chemical characteristics of clay pans

Depth (cm)	Clay (%)	Bulk density (g cc^{-1})	Aeration porosity (%)	Hydraulic conductivity (cm h^{-1})
0-12	18	1.5	5.4	0.039
12-35	22	1.6	3.6	0.014
35-60	25	1.7	2.5	0.004
60-110	24	1.6	2.9	0.009
110-180	19	1.6	-	0.021

Source : Singh (1980)

The presence of $CaCO_3$ restricts the water movement as well as root penetration and proliferation. The calcareous soils ($CaCO_3$ 2%) usually have pH_s in the range of 7.6-8.4. The presence of $CaCO_3$ in the soil profile in moderate amounts, particularly in the root zone depth, prevents excessive deterioration of saline or sodic soils by limiting the rise in ESP. The rain water owing to its acidic nature helps in converting $CaCO_3$ to $Ca(HCO_3)_2$ and replace sodium by the calcium released from $Ca(HCO_3)_2$. Highly calcareous soils suffer seriously due to nutrient and micro-nutrient deficiencies.

Gypsum

The small amount (2%) of gypsum ($CaSO_4.2H_2O$) in the soils is considered favourable for crop growth because it helps to maintain the soil structure as calcium replaces sodium on the exchange complex. Excessive gypsum (25%) on the other hand is harmful to plant growth. Unfortunately, sodic soils do not contain sufficient reserves of gypsum in India. The saline soils, particularly along the sea coast have been identified to possess gypsum at some depth in the soil profile. Gypsum has also been found to occur in the saline soil profiles of the Rann of Kutch in Gujarat and arid portions of Rajasthan. As much as 0.25-1.0% of $CaSO_4$ is reported to occur in these soils (Satyanarayana, 1961).

SOIL-WATER RELATIONS

Infiltration Rate

Infiltration rate is defined as the rate at which a soil, in a given condition at a given time, can absorb water. It is also the rate at which a soil will absorb water ponded on the surface as a thin sheet when the ponded area is infinitely large or when adequate precautions are taken to minimize the effect of horizontal flow at the borders. It is the volume of water passing from the soil surface into the soil per unit area per unit time. The initial intake rate rapidly decreases as the soil is wetted and the intake reaches a steady state after some hours, which is known as the basic or equilibrium infiltration rate. Both the initial infiltration rate and the basic infiltration rate decrease with the increase in ESP (Table 2.24). The most productive normal soils have basic infiltration rates between 1.0 and 3.5 cm h^{-1}. The sodic soils generally have basic infiltration rate below 0.1 cm h^{-1} characterizing the soils as non-arable except for rice crop. In a sodic Vertisol, cumulative infiltration decreased sharply with increase in ESP up to 15 and was almost negligible at an ESP of 38. If the cumulative infiltration rate curve is described by Kostiakov's power function relation (cumulative infiltration, $I = at^b$), then the constant, a and exponent, b of the power function change with increasing ESP (Table 2.25). While exponent b has been found to decrease, constant a increased in the alluvial soils (Acharya and Abrol, 1991) while it was about the same in black sodic soils (Ranade and Gupta, 1987). Similar results have been reported by Tiwari *et al.,* (1981). The data on cumulative intake reveals that only limited amount of water per irrigation need to be applied to sodic lands to avoid water stagnation (Table 2.26, Fig. 2.10).

Table 2.24 Infiltration rate as affected by ESP

Soil ESP	Initial infiltration rate (cm h^{-1})	Basic infiltration rate (cm h^{-1})
5	22.1	3.7
14	20.3	2.7
22	11.9	1.4
30	9.9	1.0
44	8.6	0.7
49	5.9	0.3
62	5.4	0.2
65	2.2	0.1
70	2.0	0.06
72	1.8	0.05
74	1.7	0.04

Source : Singh (1980)

Table 2.25 Impact of ESP on constants of the Kostiakov's infiltration rate equation

ESP	Cumulative infiltration equation for Vertisols*	Cumulative infiltration equation for alluvial soils@
5	-	$I = 3.95\,t^{0.58}$
6.2	$I = 6.1\,t^{0.65}$	-
10.0	$I = 5.7\,t^{0.50}$	-
14.0	-	$I = 4.26\,t^{0.55}$
15.0	$I = 5.7\,t^{0.36}$	-
22	$I = 4.9\,t^{0.31}$	$I = 5.09\,t^{0.33}$
30	-	$I = 5.86\,t^{0.27}$
70	-	$I = 6.01\,t^{0.17}$

-Not determined, *microplot study, @ laboratory disturbed column study

Table 2.26 Cumulative intake (cm) in a normal and sodic land amended with gypsum

Soil	Gypsum added ($t\ ha^{-1}$)	Time (hours)			
		1	10	15	20
Normal	-	2.4	7.4	9.1	10.4
Sodic	0	0.6	1.9	2.4	2.8
	7.5	0.8	2.6	3.2	3.7
	15.0	1.1	3.4	4.2	4.8
	22.5	1.2	4.1	5.1	6.0
	30.0	1.2	4.2	5.3	6.2

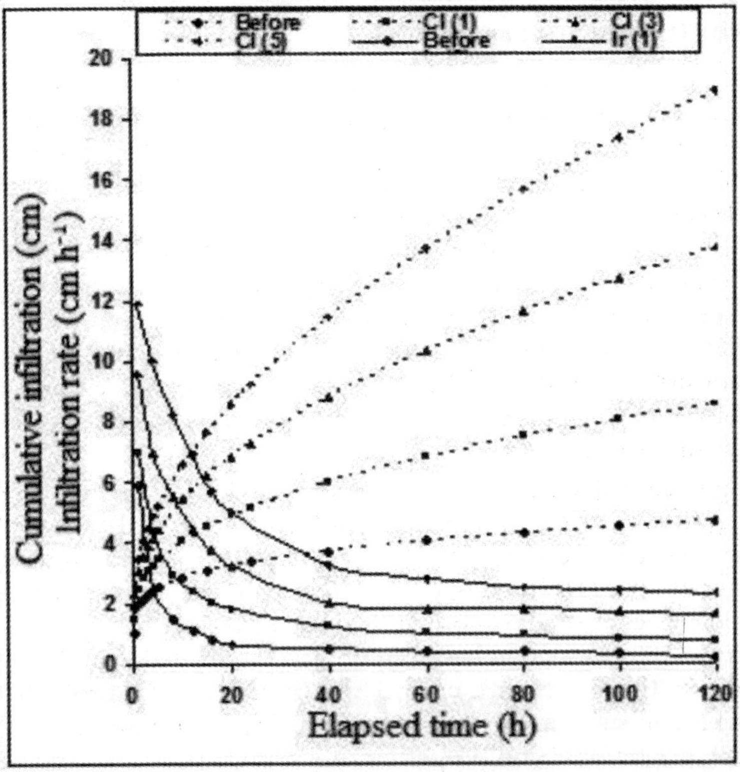

Fig. 2.10 Infiltration rate (Ir, cm h^{-1}) and cumulative infiltration (CI, cm) before and after 1, 3, 5 years of gypsum application on farmer's field

As a corollary to what has been said, it can be visualized that both the cumulative infiltration and basic infiltration rate increase with reclamation. The impact of number of years after reclamation on the infiltration rate curve and the cumulative infiltration curve is shown in Fig. 2.10, which clearly reveals the increasing trend. Since infiltration rate of the soil is affected by the interaction of irrigation water salinity and soil ESP, the infiltration rate has been found to decrease with ESP but increase with irrigation water salinity (Ranade and Gupta, 1987, Table 2. 27).

Table 2.27 Effect of irrigation water salinity and ESP on infiltration rates

Water concentration (meq L^{-1})	Cumulative infiltration at various soil ESP levels (mm)					
	6	10	15	22	37	58
5	31.1	27.2	11.0	-	-	-
50	38.0	34.3	30.2	24.1	10.2	-
10	46.8	44.0	34.4	27.8	16.9	9.8
200	71.9	62.8	49.3	35.3	26.9	15.9
300	80.7	67.0	54.4	41.6	32.1	22.0
500	77.3	67.2	55.7	45.2	34.5	24.8

Source: Verma *et al.* (2006)

Permeability (Hydraulic conductivity)

Permeability has been defined qualitatively as porous medium's readiness with which it conducts or transmits fluids. When defined quantitatively, the term hydraulic conductivity is used. It is defined as the constant of proportionality between the flux and the total driving force in Darcy's law or as the flux caused by a unit driving force. The most productive normal soils have reasonably good hydraulic conductivity such that saturated condition, whether from rainfall, irrigation or both, is unlikely to occur for more than 48 hour in the upper root zone. In soils, hydraulic conductivity primarily depends on the soil texture (Table 2.28), although other factors, viz. texture, structure and structural stability, consistency, $CaCO_3$, cleavage planes, visible pores and depth to impermeable strata such as bedrock or a hardpan also influence the hydraulic conductivity of the soils.

The saturated hydraulic conductivity decreases with increasing ESP. Abrol *et al.* (1978) reported that while an increase in ESP from 5 to 14 resulted in slight decrease in hydraulic conductivity, further increase in ESP brought about a sharp decline (Fig. 2.11). In consequence, the hydraulic conductivity values are lower in sodic as compared to normal or saline soils. The unsaturated hydraulic conductivity at 100 cm suction has been observed to be 0.05 cm d^{-1} in sodic soil as compared to 0.6 cm d^{-1} in a normal soil (Acharya and Abrol, 1975). With further increase in suction, the difference in hydraulic conductivity of two soils decreased. Water transmission properties of a normal (ESP 6.0) and sodic (ESP 58.0) clay (Vertisol) soil reveal that while lower layers are able to contribute to water availability to plants in a normal soil, it is not the case with sodic soil that show poor hydraulic conductivity, water diffusivity and differential water capacity as calculated with the help of moisture release characteristics curves for the two soil groups. Certain clay minerals with almost 100% ESP (fully sodium saturated) are called bentonite, which are used in civil engineering to place impermeable curtains in the soil, e.g. below dams, to prevent seepage of water or even to reduce seepage losses in tanks/reservoirs.

Table 2.28 Permeability of the soil as related to soil texture

Texture	Hydraulic Conductivity, K (cm h^{-1})
Clay	0.13
Clay loam	0.38
Sandy clay – loam	0.50
Sandy loam	4.50
Loamy sand	9.50
Sand	14.00

Source : Hermsmeier (1966, 1978)

Table 2.29 Water transmission properties of normal and sodic clay soils at different moisture contents

Moisture content ($cm^3 cm^{-3}$)	Hydraulic conductivity (cm h^{-1} * 10^3)		Water diffusivity ($cm^2 h^{-1}$)		Differential water capacity (cm*10^4)	
	Normal	Sodic	Normal	Sodic	Normal	Sodic
0.40	900.0	20.0	-	-	-	-
0.35	600.0	6.0	350.0	16.2	17.0	3.7
0.30	133.5	2.1	188.8	7.5	7.1	2.8
0.25	27.9	0.6	84.5	3.5	3.3	1.6
0.20	9.0	0.1	44.0	1.5	2.0	0.8
0.15	3.0	0.07	28.7	1.0	1.4	0.5
	0.8	0.03	10.0	0.04	0.8	0.1

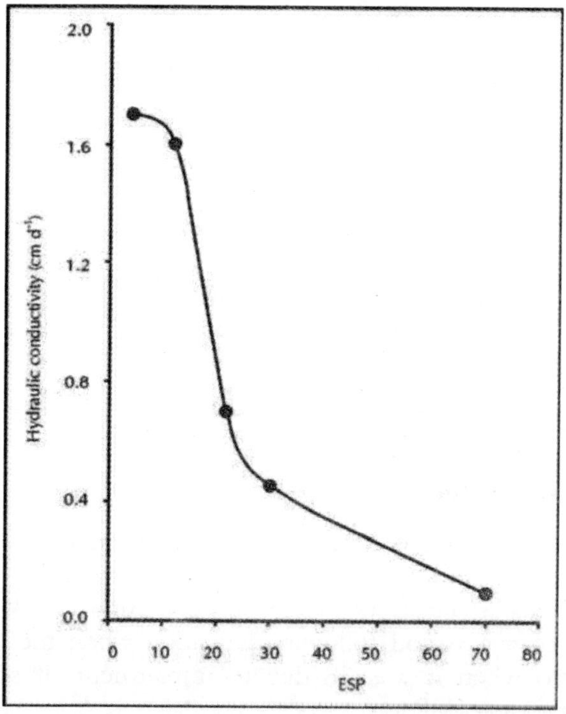

Fig. 2.11 Relationship between saturated hydraulic conductivity and soil ESP

Dispersion and hydraulic conductivity

The amount of dispersible clay regulates various physical and mechanical properties of the soil such as swelling, porosity, water retention capacity, hydraulic conductivity and modulus of rupture. In sodic soils due to high ESP nearly 90% silt + clay is found in water dispersible form in the entire profile. The dispersion decreased to 72, 66 and 58% in 0-15 cm layer after 1, 3 and 5 years of gypsum application (Table 2.30). This reduction continued in the lower depths also but at a slower rate. The higher reduction in 0-15 cm layer was due to the replacement of exchangeable sodium with calcium and also due to slight increase in organic carbon content which acts as the binding agent of soil particles (Sharma *et al.,* 1981). Similarly, saturated hydraulic conductivity of 0-15 cm layer increased from 2 mm d^{-1} to 12, 21 and 38 mm d^{-1} after 1, 3 and 5 years of gypsum application. Chawla and Chhabra (1991) reported that continuous cultivation of rice-wheat in rotation for 11 years and application of fertilizers significantly improved the soil physical properties.

Table 2.30. Changes in soil dispersion and hydraulic conductivity after 1-5 year of gypsum application

Year after gypsum application (12.5 t ha^{-1})	Depth (cm)	Dispersion (%)	Hydraulic conductivity (mm d^{-1})
1	0-15	72	12
	15-30	90	4
	30-60	95	2
3	0-15	66	21
	15-30	82	8
	30-60	84	5
5	0-15	58	38
	15-30	75	15
	30-60	78	9
Before gypsum application	0-15	88	2
	15-30	92	1
	30-60	94	0

Source: Sharma (1986)

Redistribution of soil water

Redistribution is the process of water movement within the soil profile under the influence of gravity and suction gradients and is governed by the properties of the surface and subsurface layers. The water content in the top 0-15 cm layer sharply decreased over a period following irrigation when the ESP of the soil was 4 compared to when it was 38 due to impairment of soil transmission properties with increasing ESP.

Soil water availability

The capacity of a soil to retain water available to plants has a direct bearing on the required depth and the frequency of irrigation. Although water retention increases with decreasing ESP, the differences in moisture retention decrease (Table 2.31). The readily available water, which is readily absorbed by the roots (about 50-75% of the total available moisture) may be 20% less in sodic than normal soils. In saline soils, the moisture availability to the plants decreases due to increase in the osmotic pressure of soil water solution. The osmotic pressure increases at the rate of 0.36 bar per unit increase in the electrical conductivity of the soil solution measured in dS m^{-1}.

Root Proliferation

The root distribution pattern is quite different in normal and sodic soils. Roots penetrate deeper and are able to exploit more soil mass in the former than in the latter (Table 2.32). Similar results for saline and sodic sub-soils (0.60-1.00 m) in the southern Mallee have been reported to have big impact on the root growth and water extraction by wheat (Salinity & Sodicity Subsoils Manual – 2006). Very little water extraction occurred when ESP >19% or EC_e > 8 dS m^{-1}, however, for sub-soils with lesser salinity and sodicity (ESP <19% or EC_e < dS m^{-1}) water use was significantly higher. Root density was 26 and 21 g m^{-3} in the former cases (ESP >19% or EC_e > 8 dS m^{-1}), while it was 45 and 49 g m^{-3} in the latter (ESP <19% or EC_e < 8 dS m^{-1}).

Table 2.31 Soil moisture retention characteristics of an sodic and gypsum treated soil

Soil moisture tension (cm)	Soil water content (cm^3 cm^{-3})	
	Gypsum treated soil	Sodic soil
50	0.35	0.27
100	0.26	0.24
150	0.24	0.22
200	0.22	0.21
300	0.21	0.20
400	0.20	0.19

Table 2.32 Soil mass (kg) exploited by roots of crops for a normal and sodic clay soil

Soil layer (cm)	Wheat		Cotton		Sorghum	
	Normal	Sodic	Normal	Sodic	Normal	Sodic
0-20	12.4	10.0	31.8	12.2	15.0	11.5
20-40	9.9	2.0	6.0	1.6	5.0	-
40-60	7.5	0.5	3.1	-	-	-
60-100	5.0		-	-	-	-
Total	34.8	12.5	40.9	13.8	20.0	11.5

HYDROLOGICAL PROPERTIES

Hydrological features of an area helps to appraise the water resources of an area. The important features relevant to the hydrological studies are: the topography, land use, crops, vegetation and engineering structures. The soils being the topmost medium on earth's surface through which the circulation and distribution of water takes place, the properties of soils influence the processes of run-off generation from rainfall, which in turn affects the formation, reclamation and management of salt affected soils. From hydrological point of view, a good catchment is one that has a good surface water yield potential. On the other hand, for ground water recharge, the same may prove to be bad. For salt affected soils where leaching is one of the most important step in the reclamation and management, it is desirable that the soil surface allows sufficient amount of rainfall to enter and move through the soil profile. Most of the salt affected soils in general and sodic soils in particular are located in flat terrain with restricted drainage. Sodic soils in general occupy lower portions in comparison to the normal soils of the region. They are not only capable of producing high run-off but run-off arising from adjoining areas also collects there resulting in flood like situations. Shallow ground water and poor hydraulic conductivity accentuate the run-off producing potential of these soils. Sodic soils from hydrological point of view are included in the Group D soils, having high run-off potential, shallow water table, clay pan near or at the surface with transmission rates less than 2.5 mm h^{-1}. Upon reclamation these soils need to be reclassified under a different group as the infiltration and hydraulic conductivity increase with time.

Hydrological Effects of Reclamation

The high dispersibility renders the soils prone to erosion and high runoff. It also causes prolonged water stagnation on soil surface. The soils are very hard when dry and produce big clods on ploughing. Thus, these soils pose management problems. On the other hand, package of reclamation along with land use changes affect the following hydrological components:

- Infiltration rate
- Catchment retention storage
- Runoff and peak rate of run-off
- Ground water recharge

Infiltration rate

Although soil infiltration in sodic soils and impact of reclamation on this parameter has been discussed in a previous section but herein discussion is limited to hydrological point of view. Singh (1980) used Horton (1940) equation to study the changes in infiltration characteristics of the soils under reclamation. Horton's equation given as follows was used to describe the changes in infiltration as a result of land reclamation.

$$ft = fc + (fo - fc)\, e^{-Kt} \tag{2.36}$$

Here ft is the infiltration rate at time *t*; fo is the initial infiltration rate or maximum infiltration rate, fc is the constant or equilibrium infiltration rate after the soil has been saturated or minimum infiltration rate and K is the decay constant specific to the soil. It was observed that three parameters of the Horton equations are functions of the soil ESP:

$$fc = \text{Exp.}\,(1.90 - 0.07\ \text{ESP}) \tag{2.37}$$

$$fo = \text{Exp}\,(3.46 - 0.04\ \text{ESP}) \tag{2.38}$$

$$K = 0.15 \ln(\text{ESP}) + 0.24 \tag{2.39}$$

Thus, it is apparent that as ESP decreases, both fc and fo increase while K decreases indicating increased infiltration rate with decrease in ESP.

Catchment retention storage

The watershed retention storage is found to increase owing to the increased infiltration rate and bunding of fields to store water. Based on an equation proposed by Linsley *et al.* (1949) for watershed storage and observed changes in basic infiltration rate, Singh *et al.* (1979) proposed that rain water retention in the catchment with rice crop can be given by the following relation. :

$$RW = a + b\,(1 - e^{-ct}) + Sd\,[1 - e^{-k(P - f(I) - E)}] \tag{2.40}$$

Here, RW is the rain water retention due to increased infiltration and bunding, a is the basic infiltration rate of the unreclaimed catchment, b and c are constants indicating increase in infiltration rate after complete reclamation and the rate of increase in basic infiltration rate during the time span of reclamation respectively, t is the time in years from the start of reclamation, Sd is the maximum storage capacity, k is a constant, P is the gross rainfall, E is the evaporation rate and f (I) is the basic infiltration rate in the ith year of reclamation. Thus, the term (P-f(I)-E) can be described as the excess rainfall. While evaluating catchment retention storage for a 1200 ha sodic watershed, which was being progressively reclaimed, for the low magnitude of storms there was not much difference in the water retention values corresponding to percentage of the area reclaimed (Singh *et al.,* 1979). However, for high magnitude storms, there was considerable difference in the amount of water stored within the watershed. In general, the water retention capacity of the watershed was found to increase with increase in the area reclaimed.

Runoff and peak rate of runoff

The reclamation of sodic watersheds considerably alters the time distribution characteristics and the total amount of runoff. In a study on identical plots of 1.26 ha each, the runoff volume was reduced by 75% while peak rate of runoff was reduced by 50% from a reclaimed plot compared to an unreclaimed one (Singh *et al.,* 1979). Based on Instantaneous Unit Hydrograph Technique, Singh (1980) showed that the flow duration characteristics change with reclamation resulting in reduced runoff and delayed peak discharge. Multiple regression analysis with variables such as depth of rainfall, intensity of rainfall, antecedent precipitation index, slope, area and shape of the watershed and reclamation factor, revealed that shape of the watershed and reclamation factor are the two most influencing parameters for runoff estimation. Reclam-ation factor defined as (1 + reclaimed area/ total area) varied from 1 for a completely unreclaimed watershed to 2 for a completely reclaimed watershed. It has been shown that with complete reclamation, runoff volume reduced to 45% of the total volume expected from the unreclaimed watershed (Singh, 1980).

MACRO-NUTRIENTS

The testing soil for major nutrients, viz. nitrogen, phosphorus and potassium is the only proven practical method of evaluating the fertility status of the soils. These tests provide a sound basis for making recommendations on fertilizer application. The most productive normal soils are high in organic carbon (above 0.75%), which is used as a measure of available nitrogen, high in available P_2O_5 (above 25 kg ha^{-1}) and available K_2O (above 280 kg ha^{-1}). A soil is rated as low in organic carbon, available P_2O_5 and available K_2O if the respective values are below 0.5%, 10 kg ha^{-1} and 110 kg ha^{-1}, respectively. Although one may anticipate that saline/sodic soils lying barren for several years might be fertile as the nutrients build- up is expected in the absence of their removal by crops. On the contrary, it has been observed that barren sodic and saline soils are usually low in nitrogen, medium to high in phosphorus and high in potassium. Barren sodic soils usually have high P and K (available P being 5-10 mg L^{-1}, available K 500-750 mg L^{-1}) and therefore, one may not expect any response to the application of P and K fertilizers for few years till the soils are depleted of these nutrients by few crops in succession.

MICRO-NUTRIENTS

Deficiency symptoms of zinc, iron and manganese are sometimes seen on plants because of their low content (Bhumbla and Dhingra, 1964) coupled with high pH, low organic matter content and presence of calcium carbonate that strongly modify the availability of micronutrients to plants grown in sodic soils. The availability of iron and manganese in these soils can be increased substantially by submergence and application of FYM (Swarup, 1981, Swarup, 2000). Zinc deficiency may still persist and has been widely reported. The deficiency gets accentuated when an amendment is applied to a Zn-deficient sodic soil (Singh *et al.,* 1982). Several field studies have shown significant increase in crop yields due to application of zinc. Singh *et al.* (1982) observed that application of 10 kg $ZnSO_4$ ha^{-1} was sufficient to mitigate the Zn deficiency in rice grown in an amended highly sodic soil. In saline soils, Cu and Zn content is usually less whereas that of Fe and Mn is more than those of sodic soils.

Next to zinc, iron is often a limiting micronutrient in sodic soils due to high pH and presence of calcium carbonate. Addition of iron salts however, was not found useful unless it was used to correct the deficiency as it was accompanied by changes in the oxidation status of the soil brought about by prolonged submergence and addition of organic matter (Katyal and Sharma, 1980). Swarup (1980) showed a marked increase in the extractable Fe and Mn status of a sodic soil upon submergence up to 60 days; the increase was more when organic materials like rice husk or farmyard manure were incorporated in the soil.

TOXIC ELEMENTS

The best productive normal soils are generally free of toxic elements or contain less than the permissible limits. The saline and sodic soils tend to contain one or more of these elements in toxic amounts. Almost all the essential elements, which are otherwise recognized as micro-nutrients, if present in the soil beyond specific limits, act as toxic elements and adversely affect the growth of certain plants. In addition, elements such as Al, As, Be, Cl, Cr, F, Pb, Ni, Se and V are also known to cause toxicity to plant, if present beyond permissible limits. The elements recognized to be of practical importance from the toxicity point of view are B, Li, F, Se and Mo. Out of these five, B, Li and F are phytotoxic. Se and Mo may not adversely affect the plant growth *per se* but crops, particularly the forages grown on them, may contain an excess of these elements, which cause adverse effects to the animals grazing on them.

Boron and molybdenum are not likely to be limiting elements for plant nutrition in sodic soils. On the contrary, these elements may be present in the toxic range. Boron content of 1.0 mg L^{-1} in soils has been considered adequate for the normal growth of most plants. Kanwar and Singh (1961) observed a positive correlation between water soluble boron and the pH and EC of soils. Boron in the saturation extract of normal, saline and sodic soils is less than 1.0, 1 to 2 and 2 to 10 mg L^{-1}, respectively (Bhargava *et al.,* 1974). In a laboratory study Gupta and Chandra (1972) observed a marked reduction in the water soluble boron content of a highly sodic soil upon addition of gypsum. At high pH and sodicity, boron is present as highly soluble sodium metaborate which upon addition of gypsum is converted to relatively insoluble calcium metaborate. Soluble boron was reduced simultaneously with decrease in ESP with gypsum treatment (Table 2.33). Reduced uptake of boron by grasses (Table 2.34) with decreasing ESP due to gypsum application was also reported by Ashok Kumar and Abrol (1982). The available molybdenum in normal and sodic soils has been reported to be 0.40 and 0.80 mg L^{-1} respectively (Singh and Singh, 1966). As with B, solubility of Mo increases with pH (Pasricha and Randhawa, 1971) and for this reason forage crops grown on sodic soils are likely to accumulate Mo in excessive quantities. Chhabra *et al.* (1980) and Gupta *et al.* (1982) studied the effect of sodicity on the solubility of fluoride, again an important element from the animal nutrition viewpoint. Water extractable fluoride increased with increasing sodicity and pH (Fig. 2.12), the latter having a more important role in determining the behaviour of fluoride in soils. It was further shown that the F content of plants increased with increasing ESP and decreased with application of P fertilizer (Singh *et al.*, 1979, 1980).

Table 2.33 Boron and ESP as affected by gypsum treatment

Gypsum treatment	ESP		B (mg L^{-1})	
	0-15	15-30	0-15	15-20 cm
0	84	95	3.0	4.4
25% GR	52	87	2.1	3.9
50% GR	37	82	1.7	2.5
75% GR	17	83	1.4	1.9
100% GR	17	43	0.8	1.3

Source: Gupta (1979)

Table 2.34 Average boron content (ppm) of four grasses as influenced by level of gypsum application in a highly sodic soil

Grass	Gypsum level (t ha^{-1})				
	0	3.12	6.25	9.37	12.50
Bermuda grass	42.8	39.4	35.7	33.2	29.7
Para grass	39.4	37.6	34.7	33.8	30.3
Setaria grass	41.3	35.7	33.8	31.9	29.2
Hybrid Napier	51.3	48.8	46.3	41.9	41.3

Source: Ashok Kumar and Abrol (1982)

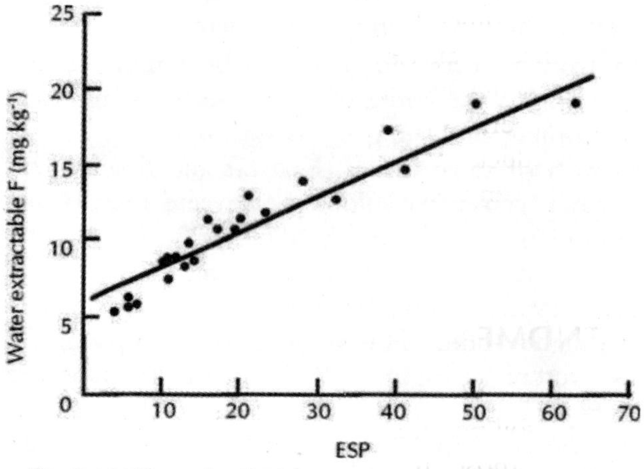

Fig. 2.12 Effect of soil ESP on water extractable fluorine

Similar to boron, significant reduction in Se accumulation by wheat was observed with gypsum application up to 0.8 t ha^{-1} and its residual effect was evident on the following crops for 2 years. Reduction in Se accumulation varied from 53 to 64% in wheat grain, 46 to 49% in wheat straw, 35 to 63% in rice grain and 36 to 51% in rice straw with an application of gypsum at the rate of 0.8 t ha^{-1}. A corresponding increase in S concentration was observed. In the gypsum-treated plots, the ratio of S: Se increased by 6 – 8 times in wheat and 3 – 6 times in rice (Dhillon and Dhillon, 2000). Reduction in Se accumulation by crop plants through gypsum application may help in lowering the risk of Se over-exposure of animals and humans that depend on diet materials grown on high selenium soils.

3

CHEMICAL AMENDMENTS

Sodic soils are characterized by $pH_s > 8.2$, ESP > 15 and variable which EC_e may or may not be greater than 4.0 dS m^{-1}, in the former case soil is designated as saline-sodic. The dominant cation on the exchange complex in both types of soils is Na^+ due to which these soils have poor physical conditions. Application of a chemical amendment that provides a source of Ca^{++} is therefore, essential to reclaim these soils, which replaces harmful Na^+ from the exchange complex. The kind and amount of chemical amendment to be used for reclaiming sodic/ saline-sodic soils depends on the soil characteristics, the desired rate and extent of exchangeable Na replacement and cost. Gypsum ($CaSO_4 . 2H_20$) has been the most commonly used amendment as it is cheap direct source of calcium and can be applied in the presence or absence of free lime. Over the years researchers have used other amendments such as elemental sulphur, sulfuric acid (H_2SO_4) and pyrites, which can be effective only in the presence of free lime in the soil. Several industrial by-products such as phospho-gypsum, distillery spent wash and press mud etc have been used both at the field as well as the experimental scales. In this chapter, principles involved in the use of several of these amendments, relative application rates and their relative merits are discussed.

CHEMICAL AMENDMENT: DEFINITION

Simply put, a chemical amendment is a material that is added to the soil to change and improve the soil physical or chemical properties to aid plant growth. Amendments, including those that improve soil physical properties work through chemical, biological, and physical modes of action. Even fertilizers are materials that contain useful amounts of one or more plant nutrients and help in better plant growth but in a practical sense, these are not categorized as amendments. For the purpose of this chapter, a chemical amendment is one that is used to reclaim sodic soils to overcome the adverse effects of high sodium and/or high pH. The chemical amendments neutralize the pH of the soil, react on the free sodium carbonate, and replace the exchangeable sodium by calcium. As an example, gypsum acts chemically by supplying calcium that counteracts the harmful effects of excessive sodium on soil physical structure. It may, however, be noted that even gypsum contains plant nutrients and may even be registered with the state as fertilizer. The degree of soil sodicity (ESP) and the presence or absence of *Kankar* (impure $CaCO_3$) in the soil help to decide the kind of amendment to be applied as well as decide the effectiveness of the chemical amendments.

Types of Chemical Amendments

The chemical amendments are broadly classified into three categories:

- Soluble calcium salts, viz. calcium chloride and gypsum which directly supply soluble Ca for the replacement of exchangeable Na.
- Acidifying materials, viz. sulphuric acid and elemental sulphur which due to their acidulating action and/or increasing the partial pressure of CO_2 supply Ca indirectly by dissolving calcite natively found in sodic soils. The Ca so mobilized is used to replace Na from the exchange complex
- Organic materials viz. rice husk, FYM and green manure etc. Compost and other organic materials improve soil physical structure by two distinct mechanisms – by stimulating microbial activity, which helps in soil aggregate formation, and at high application rates, by directly decreasing soil density. These also help to improve soil structure as well as dissolve the native $CaCO_3$ albeit weakly. The reclamation efficiency is usually low with these kinds of amendments.

Besides these many industrial wastes are used as amendments that work on similar principles depending upon their constituents. Amongst them pyrites, distillery spent wash, dairy waste, phospho-gypsum and press mud have been widely experimented and used.

Gypsum is by far, the most common amendment used in the reclamation of sodic soils, being in use for more than a century. Leather (1897) reported for the first time that the addition of small amounts of gypsum to reduce Na_2CO_3 to 0.1%, in the fine size, applied at the surface of sodic soils, was very beneficial. Pyrites, which came into use recently and became quite popular has gone out of use because of its use in other economic activities. Sulphuric acid seems to be the most effective chemical amendment but because of its prohibitive cost and operational difficulties, has not become very popular in India although it is commonly used in the USSR and the USA.

$CaCO_3$ as an Amendment

As a general rule, almost all sodic soils contain $CaCO_3$ in sufficient amount in the root zone depth commonly referred as *Kankar* ($CaCO_3$ concretions). Therefore, any additional application of $CaCO_3$ may not be useful. Moreover, ground limestone is an effective amendment only in soils having pH below about 7.0 because its solubility rapidly decreases with increasing soil pH (Table 3.1). Therefore, there is no point in adding this as an amendment for sodic land reclamation. On the other hand, efforts could be made to increase the solubility of slowly soluble $CaCO_3$ concretions (13.1 mg L^{-1}). It can be achieved through increase in the concentration of CO_2 so that $CaCO_3$ is converted into calcium bicarbonate whose solubility is much higher, being in the range of 60 to 140 mg

L^{-1}. Similarly, lowering the soil pH will also increase the solubility of $CaCO_3$ (Table 3.1). Thus, if by any means, CO_2 concentration in the soil can be increased or the pH of the soil can be decreased, the low and medium degree sodic soils can be reclaimed naturally. The most practical means of reducing pH and permanently increasing the CO_2 concentration in the soil is to intensify the microbiological processes of decomposition of organic substances ploughed into the soil in the form of green or farm yard manure. The application of small amounts of gypsum or other acidifying materials and the conservation of rain water during the monsoon season can also help to speed up the process of the natural reclamation. Gupta and Abichandani (1970) reported that in highly calcareous sodic soils, the high ESP was reduced from 59.5 to 18.0 and from 34.9 to 11.0 due to leaching alone caused by the rainfall (350 to 450 mm) in Western Rajasthan. In addition to the remarkable changes in ESP and pH, the total concentration and composition of the soluble salts also changed favourably.

Table 3.1 Solubility of $CaCO_3$ (mg L^{-1}) as affected by the CO_2 content and pH of the soil

CO_2 content in soil air (% by volume)	pH	Solubility of $CaCO_3$
0.00	10.23	13.1
0.03	8.48	63.4
0.30	7.81	133.4
1.00	7.47	202.9
10.00	6.80	470.0
100.00	6.13	1098.6

Source: Kovda *et al.* (1973)

GYPSUM

Three main commercial forms of calcium sulphate are: (i) anhydrite form, land plaster ($CaSO_4$); (ii) semi-hydrate form, commonly called plaster of Paris ($CaCO_4.1/2\ H_2O$); and (iii) di-hydrate form, generally known as gypsum ($CaSO_4.2H_2O$). It is the third kind of gypsum that is commonly used in agriculture. Gypsum reacts with sodium on the exchange complex as per the following reaction:

$$\underset{\text{Sodic soil}}{2\,Na - soil + 2CaSO_4} \rightarrow \underset{\text{Normal soil}}{Ca - soil + Na_2SO_4} \qquad (3.1)$$

Occurrence of Gypsum

Gypsum occurs extensively in natural deposits as a white coloured secondary mineral. In India, rich deposits of gypsum are found in Rajasthan. It also occurs in appreciable quantities in Tamil Nadu, Gujarat and Himachal Pradesh. The gypsum resources in India are estimated to be about 100.4 x 10^7 tonnes, 90% of which are available in the mines of Rajasthan only. Besides the gypsum from mines, it is also available as a by-product from some chemical industries manufacturing phosphoric acid, tartaric acid, formic acid, oxalic acid, citric acid and common salt. Major quantities of phospho-gypsum are produced during the manufacture of phosphoric acid by wet process in which rock phosphate is treated with sulphuric acid. For every tonne of P_2O_5 produced, 5.5 tonnes of phospho-gypsum containing 25% water is produced. More than 1.0 million tonnes of by-product gypsum is available from industries manufacturing chemical fertilizers in India.

Quality

Fertilizer Corporation of India Ltd. (FCI) mines three grades of gypsum having different purity levels: (i) Grade I: 85-90% $CaSO_4.2H_2O$, (ii) Grade II: 80-85% $CaSO_4.2H_2O$, and (iii) Grade III: 70-80% $CaSO_4.2H_2O$. The chemical analysis of various grades of gypsum is reported in Table 3.2.

Table 3.2 Chemical analysis of gypsum mined by the FCI

Constituents (%)	Gypsum grade		
	I	II	III
$CaSO_4.2H_2O$	88.67	83.32	77.56
SiO_2	4.84	9.75	12.37
Mixed oxides (R_2O_3)	0.68	0.85	1.59
MgO	1.12	0.88	0.81
CaO	30.59	28.08	28.56
Loss on ignition	21.17	20.40	20.44
Chlorides as NaCl	0.008	0.014	0.016

Bureau of Indian Standards (BIS) Specifications

According to the revised specifications as per IS:1290-1965, the mined gypsum is divided in five grades, viz. (i) Grade I for use in surgical plaster industry; (ii) Grade II for ammonium sulphate industry; (iii) Grade III for pottery industry; (iv) Grade IV for cement industry, and (V) Grade V for land reclamation programs with specifications reported in Table 3.3. The gypsum of grade III mined by the FCI conforms to the ISI specifications for grade V gypsum. The gypsum used in the reclamation should not contain more than 1.0% of Na_2O.

Table 3.3 BIS specifications for gypsum of different grades

Constituents	Gypsum grade				
	I	II	III	IV	V
Free water, maximum	1.0	–	1.0	–	–
CO_2, maximum	1.0	–	3.0	–	–
Silica and insolubles	1.0	0.5	6.5	–	–
Fe_2O_3 and Al_2O_3, maximum	1.0	1.5	1.0	–	–
MgO, maximum	0.5	1.0	1.5	3.0	–
$CaSO_4$,, minimum	96.0	86.0	83.8	80-85	70-80
Chlorides (as NaCl) maximum	0.1	0.01	0.01	0.5	–

Solubility

Gypsum is soluble in water to the extent of about 0.25% and is therefore, a direct source of soluble calcium. Obviously, the rate of reaction of gypsum is limited only by its moderate solubility. It has been conventionally recognized that an application of 90 to 120 cm of irrigation water is required to dissolve 10 to 12 t ha^{-1} of high-grade gypsum (Richards, 1954). Recent studies have shown that it is true only if applied to normal soils as the solubility of gypsum in sodic soil is higher depending upon pH, ESP and type of salts present in the soil. It has been observed that in sodic soils, the solubility of gypsum is ten times more (Oster and Halvorson, 1978). It is also reported that the amount of water required to dissolve the added gypsum to reclaim sodic soils does not increase with the amount of gypsum added provided it equals the gypsum requirement (Hira and Singh, 1980).

When mixtures of salts with dissimilar ions are present as in the case of saline soils, the solubility of gypsum gets altered when combined with chlorides (Table 3.4). The actual solubility increase is due to the effect of Na^+ and Cl^- in solution so that the concentration of the ions must increase to maintain the constancy of the solubility product. Redly *et al.* (1980) studied the solubility of $CaSO_4$ containing chemical amendments, semi-hydrates and mechanically activated anhydrites, in the saturation extracts of sodic soils differing in the concentration and composition of water-soluble salts and the degree of sodium saturation. The equilibrium concentration of Ca ions in the soil extracts saturated with these materials exceeded 10-50 times the Ca ion concentration in the original solution (Table 3.5). It increased with the increase in the ionic strength in the case of solutions low in sulphates. On the other hand, the relatively stable Ca ion concentration was found in the studied concentration range with the increasing ionic strength in solutions, where sulphates dominated. The thermodynamic solubility product of $CaSO_4$ calculated from the reduced activities of Ca and SO_4 ions is practically independent of the ionic strength of the solution but it depends on the kind of material, the average being as follows:

Table 3.4 Solubility of gypsum in water in the presence of NaCl and $CaCl_2$

	Chlorides (mmol L^{-1})	Solubility of $CaSO_4$ (g 100 g^{-1} water)
NaCl		
	0.0	0.204
	17.2	0.787
	20.2	0.823
	24.4	0.820
	29.3	0.614
	35.8	0.709
$CaCl_2$		
	0.0	0.204
	5.2	0.103
	9.9	0.086
	18.9	0.077
	40.8	0.035

Table 3.5 The average values of Ca^{++} and SO_4^- concentration (mmol L^{-1}) in the saturation extract of soil profiles

Soil characteristics			Treatments*	Ca^{++}	$SO_4^{=}$
pH	EC_e (dS m^{-1})	ESP			
5.9 - 9.6	3.0- 10.7	30 - 75	0	3.4	84.8
			1	29.8	107.1
			2	32.6	115.7
			3	38.7	119.8
7.4 – 9.5	0.4- 1.8	8 - 29	0	1.8	5.8
			1	31.4	42.5
			2	36.6	43.5
			3	43.5	49.7
6.6 – 8.7	1.8 – 9.9	2 – 32	0	3.0	44.0
			2	46.7	87.5
			3	48.8	92.4
7.5 – 8.9	0.9 – 1.3	3 – 12	0	1.2	5.2
			2	66.5	74.5
			3	67.3	72.7

*Treatments: 0 = Original soil, 1 = Analytical $CaSO_4$ semi-hydrate, 2 = Fine grained $CaSO_4$, 3 = Coarse grained $CaSO_4$

Analytical $CaSO_4$. ½ H_2O	3.03×10^{-5}
Fine grained $CaSO_4$	4.23×10^{-5}
Coarse grained $CaSO_4$	4.78×10^{-5}

The pH and the SAR/ESP of the soils are usually lower after saturation with $CaSO_4$. The amount of cations from the dissolution of $CaSO_4$ (given in $CaSO_4$ g kg^{-1} soil), resulting in the change of a unit value of ESP (in the range of 5-40) is 0.15 and 0.26 for samples with low and high SO_4 content, respectively.

The solubility of gypsum has been found to increase if it is applied in conjunction with farmyard manure or green manure (Maharudrappa *et al.*, 1972). Dargan *et al.* (1976 a,b) also observed that the use of FYM in conjunction with gypsum resulted in higher yield of the first crop of *berseem*, sugarcane and rice grown in separate fields (Table 3.6). A high order of interaction was observed when application of FYM was combined with gypsum.

Fineness

Gypsum is quite cheap; its cost at the mining site being negligible. The degree of fineness to which gypsum needs to be ground on the other hand would determine the cost of gypsum to the farmer. Conventionally, gypsum of about 2 mm particle size (10 mesh) has been used to reclaim sodic soils (Kovda *et al.*, 1973). According to the ISI specification, the whole of the material must pass through 10 mesh and at least 50% of it must pass through 60 mesh (0.25 mm) sieve. These standards pertain to 30 mesh fineness. About 15% of the total cost of gypsum is on account of grinding to a fineness of 30 mesh. The cost can be reduced if we adopt a standard where gypsum of 10 mesh fineness is used in reclamation programs.

Table 3.6 Effect of FYM and gypsum on the yield of crops (t ha^{-1})

Treatment	*Berseem* fodder	Sugarcane (Stripped cane)	Rice
Control	0.15	-	2.45
FYM 25 t ha^{-1}	0.83	0.49	3.59
FYM 50 t ha^{-1}	1.74	0.47	3.74
Gypsum 25% GR	–	14.32	4.23
FYM 25 t ha^{-1} + Gypsum 25% GR	–	21.37	5.29
FYM 50 t ha^{-1} + Gypsum 25% GR	–	35.77	5.46
Gypsum 50% GR	9.49	28.65	5.70
FYM 25 t ha^{-1} + Gypsum 25% GR	29.48	37.46	5.76
FYM 50 t ha^{-1} + Gypsum 50% GR	31.89	39.11	6.89
CD at p = 0.05	4.77	9.29	0.89
Initial soil pH_2	10.5	10.6	10.6

Source: Dargan (1979)

Quantity of Gypsum Required for Reclamation

The quantity of gypsum obviously depends upon the initial exchangeable sodium content of the soil and the level to which it has to be reduced, presuming that the irrigation water is of good quality. The levels to which the exchangeable sodium content of the soil is to be reduced depends upon the texture of the soil and the first crop to be grown following land reclamation. For example, if the soil is initially fully saturated with the exchangeable sodium (ES = 10 to 15 meq $100g^{-1}$, ESP = 95 to 100), it may suffice to reduce the ESP to about 50 if tolerant crop like rice (transplanted) is to be grown. Likewise, ESP has to be reduced to 30 if semi-tolerant crops like wheat, or to 10 if sensitive crops like gram, are to be cultivated. Similarly, for the same crop, coarse-textured soils can support optimum growth at relatively higher ESP than fine-textured soils. Bhumbla and Abrol (1975) reported that satisfactory yield of rice, wheat and barley can be obtained by adding gypsum @ 50% of gypsum requirement as determined in the laboratory by Schoonover's method (Fig. 3.1).

The quantity of gypsum also depends upon the depth of the soil to be reclaimed. It is not practical or even necessary, to remove all of the exchangeable sodium from the entire soil profile by the gypsum. A common practice is to apply sufficient gypsum to remove sodium from the top 10-20 cm of the soil. This improves the physical condition of the surface soil quickly and permits the growth of shallow rooted crops. The shallower mixing of gypsum @ 50% GR (calculated for 0-15 cm soil depth), in the top 10 cm depth has been reported to be superior to the mixing in the 0-30 cm depth (Table 3.7).

Fig. 3.1 Rice crop in a gypsum (50% of GR) treated plot (left) and without application of gypsum (right)

Table 3.7 Effect of depth of mixing gypsum on the yield of crops (t ha^{-1})*

Gysum (t ha^{-1})	Depth of mixing (cm)	Barley 1970-71	Rice 1971	Wheat 1971-72
0	10	0.00	5.18	2.04
	20	0.00	5.69	1.77
	30	0.00	4.62	0.93
13.5	10	2.64	7.02	3.28
	20	2.46	6.05	3.16
	30	0.53	5.64	2.06
27.0	10	2.83	6.71	3.67
	20	3.22	6.28	3.67
	30	2.30	5.98	3.53
C.D. at p = 0.5 for Gypsum and depth		0.35	0.73	0.41

* Initial soil pH_2 = 10.5; Source: Khosla *et al.* (1979)

Frequency of Application

Single application of gypsum at the rate of 25% GR (6.5 t ha^{-1}), before the first crop of rice, has proved as effective as the higher and repeated doses in subsequent years for a highly sodic soil (Table 3.8). The pH_2 came down from 10.7 in the first year before the rice crop to 9.5, 8.6 and 8.4 after the wheat crop in the first, second and third year, respectively. It shows therefore, that if reclamation has to proceed with the rice as the first crop, application of gypsum @ 25% GR would be enough, if average yields for 3 or more years are taken as the basis of recommendation. However, the yields may be low in the first year when gypsum is applied at the rate of 25% GR as compared to 50% GR.

Table 3.8 Effect of gypsum doses applied over three years on crop yields

Gypsum (t ha^{-1})			Grain yield (t ha^{-1})	
1st year	2nd year	3rd year	Rice	wheat
6.5	–	–	7.01	4.33
6.5	6.5	–	6.40	3.98
6.5	6.5	6.5	7.46	4.16
13.0	–	–	6.64	4.51
13.0	6.5	–	7.04	4.35
13.0	6.5	6.5	7.24	4.50
19.5	–	–	6.92	4.62
19.5	6.5	–	7.31	4.72
19.5	6.5	6.5	6.96	4.57
26.0	–	–	7.20	4.69
26.0	6.5	–	7.24	4.89
26.0	6.5	6.5	7.14	4.17

Self Reclamation

With continuous use of high quality irrigation water following good irrigation and cropping practices, reclamation in the sub-soil usually proceeds with time because the following processes come into operation:

- Almost all sodic soils contain sufficient reserves of *Kankar,* whose solubility increases over the years through various processes.
- The irrigation water of high quality, usually available from the canals carrying fresh water, contains Ca eight to ten times higher than Na and this leads to constant replacement of Na with Ca on the exchange complex particularly after the application of gypsum
- The use of nitrogenous fertilizers, particularly $(NH_4)_2SO_4$ is very effective in lowering down the soil pH. This also exerts a marked influence in releasing more calcium from $CaCO_3$
- Consequent upon the start of the process of reclamation through cultivation of crops, the plant roots evolve CO_2 which helps in the conversion of $CaCO_3$ into $Ca(HCO_3)_2$. The solubility of the latter being more than the former
- Because rain water is acidic in nature due to its CO_2 content, the natural reclamation of the soil also takes place to some extent.

In view of the foregoing analysis, it can be generalized that in a particular situation small amounts of gypsum (25% GR) can help to initiate the process of reclamation of sodic soils provided

- Good quality water is available for irrigation
- The rains are appreciable and effective
- The soils contain sufficient reserves of $CaCO_3$
- Adequate amount of nitrogenous fertilizers are used
- Rice (transplanted) is grown as the first crop.

The process of reclamation proceeds with time provided the land is cropped continuously without leaving it fallow.

Assessing Gypsum Requirement

Gypsum requirement (GR) for reclamation of sodic soils is determined using several procedures. Theoretically, for every one milliequivalent of Na to be replaced one needs one me of soluble Ca. This will amount to 0.02 g Ca 100 g^{-1} soil or 0.86 g of pure $CaSO_4.2H_2O$ kg^{-1}. Taking 2 x 10^6 kg as weight of 0-15 cm soil layer (or the actual value can be calculated by multiplying d_b x A x D such that d_b is the bulk density in kg m^{-2}, A is area in m^2 and d is the depth the soil profile to be reclaimed), the amount will be 1.72 tonnes of gypsum per ha per meq of Na to be replaced. Thus, the GR can be calculated by the following relationship:

$$GR\ (meq\ 100\ g^{-1}) = [CEC\ (initial\ ESP - final\ ESP)] / 100 \quad (3.2)$$

$$GR\ ha^{-1}\ (15\ cm\ soil\ depth) = GR\ (meq\ 100\ g^{-1}) * 860 \times 10^{-6} \times 2.10^{6}$$
$$= GR\ (meq\ 100\ g^{-1}) \times 1720\ (kg) \quad (3.3)$$

For reclaiming upper 30 cm soil one may need about double the amount of above calculated gypsum if the initial ESP of this layer is of similar magnitude as that of 0-15 cm layer. If the gypsum is less than 100% pure, the quantity of gypsum needs to be increased proportionately.

Laboratory procedure

Schoonover method is commonly used in the laboratories to work out the gypsum requirement (Schoonover, 1952). In this method, saturated gypsum solution of known calcium concentration is prepared by adding 5 g of chemically pure $CaSO_4 \cdot 2H_2O$ to one liter of distilled water. It is vigorously shaken for about 10 min on a mechanical shaker or 1 h by hand and then filtered through Whatman filter paper no. 1.

Take 5 g of air dry soil in 250 ml conical flask. Add 100 ml of the saturated gypsum solution. Firmly put a stopper and shake for 5 min on a mechanical shaker or 50 min by hand. Filter the contents through Whatman filter paper no.1. Transfer 5 ml of aliquot into 100 ml porcelain dish. Determine the calcium and magnesium concentration. Run a blank using 5 ml of the saturated gypsum solution prepared before.

$$GR\ (meq\ 100\ g^{-1}\ soil) = 2\ [(Ca\ conc.\ in\ blank,\ meq\ L^{-1}) - (Ca\ conc.\ in\ the\ filtrate,\ meq\ L^{-1})]$$

$$GR\ for\ 15\ cm\ soil\ layer\ (t\ ha^{-1}) = GR\ (meq\ 100\ g^{-1}\ soil) * 1.72 \quad (3.4)$$

Abrol *et al.* (1975) proposed an amendment to this method. They argued that the gypsum requirement as calculated by Schoonover method includes the gypsum required to neutralize the soluble sodium carbonate, if present in sodic soils. It is therefore, prudent to leach the sample of the soil (5 g) with 20 ml of 60% ethanol to leach down the soluble sodium carbonate so that it does not add to the gypsum requirement of the soil. Rest of the procedure is same as discussed in the foregoing paragraph.

Extensive work done at Central Soil Salinity Research Institute, Karnal, India, has shown that 10 to 15 t ha^{-1} of gypsum containing 70% $CaSO_4.2H_2O$, which approximates about half of the actual GR determined by Schoonover's method (0-15 cm depth), is sufficient to reclaim the surface 15 cm soil if one is to start with rice as the first crop (Table 3.9). This is keeping in view the high tolerance of transplanted rice to soil ESP and its shallow root system. Moreover, its roots capacity to facilitate the soil water to dissolve soil $CaCO_3$ also helps to reduce the soil exchangeable sodium. Over time, ESP of the surface as well as deeper layers gets reduced so as to allow cultivation of other crops. To grow sensitive crops like pulses in the initial years, one needs to add higher amounts of amendments for reclamation.

Table 3.9 Yield (t ha^{-1}) of crops as affected by gypsum treatment of a sodic soil

Gypsum (t ha^{-1})	*Sesbania** 1970	Wheat 1970-71	Rice 1971	Wheat 1971-72	Rice 1972	Wheat 1972-73	Mash 1973	Lentil 1973-74	Pearl millet 1974	Gram 1974-75
0.0	0.44	0.0	4.39	1.52	7.18	1.35	0.05	0.28	1.06	0.12
7.5	3.99	1.89	6.21	2.79	7.23	1.96	0.25	0.82	1.33	0.47
15.0	4.35	3.49	6.39	3.45	7.13	2.43	0.32	1.16	1.96	0.69
22.5	4.70	4.16	7.08	3.72	7.17	2.26	0.38	1.36	1.81	0.90
30.0	4.50	3.79	6.66	3.98	7.03	2.74	0.42	1.40	1.94	0.93
LSD at P = 0.05	0.51	0.64	0.81	0.70	NS	0.50	0.12	0.35	0.52	0.44
Normal yield	-	4.5	7.0	4.5	7.0	4.5	1.0	2.0	3.0	2.0

* Green matter yield; Initial soil pH = 10.3; ESP = 94 in surface 15 cm soil
Source: Abrol and Bhumbla (1979)

pH *versus GR graph*

The absolute quantity of exchangeable sodium percentage of a sodic soil will continue to be the basis to accurately determine the quantity of gypsum required for reclamation. In practice, pH of the saturated soil paste (pH_s) or pH of 1:2 soil water suspension (pH_2) may also be used since both are fairly correlated with each other and with ESP of the alluvial soils of the Indo-Gangetic plains. The method is especially useful where skilled staff and laboratory facilities to determine the exchangeable sodium and gypsum requirement are not available. Abrol *et al.* (1973) prepared a graph between pH_2 and GR of the soils for different soil textures (Fig. 3.2). This graph is a quick and fairly good method to estimate GR of sodic soils of the Indo-Gangetic plains. The GR calculated with this graph need not be reduced to half as for laboratory determined GR. It may however, be noted that this graph is not universal as pH_s and ESP of black cotton soils (Vertisols) are not well related. Therefore, use of such a graph is not advisable for such soils.

Once the GR is known, equivalent quantities of other amendments could be easily calculated using Table 3.10.

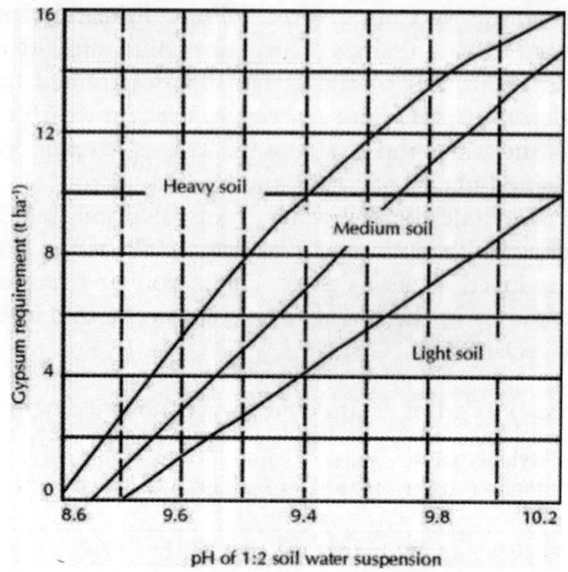

Fig. 3.2 pH_2 versus GR graph for alluvial soils of Indo-Gangetic plains

Table 3.10 Commonly used amendments and their equivalent amendment values

Amendments (Chemical formula)	Tonnes of amendment material equivalent to	
	1 tonne of pure gypsum	1 tonne of pure sulphur
Gypsum ($CaSO_4 . 2 H_2O$)	1.00	5.38
Calcium chloride ($CaCl_2 . 2 H_2O$)	0.86	–
Calcium nitrate (Ca $(NO_3) . 2 H_2O$)	1.06	–
Press mud (lime–sulphur, 9% Ca+24% S)	0.78	4.17
Sulphuric acid (H_2SO_4)	0.61	3.20
Iron sulphate ($FeSO_4 . 7 H_2O$)	1.62	
Ferric sulphate ($Fe_2SO_4)_3$.	1.09	5.85
Aluminium sulphate ($Al_2 (SO_4)_3 . 18 H_2O$)	1.29	6.34
Sulphur (S) **	0.19	1.00
Pyrites ((FeS_2, 30% S) **	0.63	

*The above quantities are based on 100% pure materials. If the purity is less than 100%, then necessary corrections must be made. Thus, if gypsum is only 80% pure then the equivalent quantity to be applied would be 1.25 times of quantities mentioned in the Table.
** For sulphur and pyrites, 100% oxidation is assumed.

PYRITES

A major supplier of the pyrites (FeS_2) in India, Ms. Pyrites Phosphates and Chemicals Limited, mines the pyrite mineral along with pyritepherous shells. Pyritepherous shells besides containing iron have sulphur, which acts as an amendment to reclaim sodic soils. Two main commercial grades of pyrites are:

(i) low grade pyrite shales, which contains 10% sulphur and (ii) high grade pyrites, which contain 40% sulphur. Intensive exploration in Amjhore block by Pyrites Phosphates and Chemicals Ltd. have shown a reserve of 24 million tonnes of 40% 'S' grade and 150 million tonnes of 10% 'S' grade. In the present method of mining, 0.6 MT of pyrites (40% S) and 1 MT of pyrite ferrous shale (10% S) are being mined simultaneously. The run off mine thus obtained contains 22% sulphur. This low-grade pyrites, containing about 22% sulphur, is used as a soil amendment (Jaggi, 1978).

Occurrence

Pyrites occur extensively in natural deposits as black coloured sediment. In India, rich deposits of pyrites are found in the Vindhya hills at Amjhore in Rohtas district in Jharkhand state. The pyrite bed having a thickness of about 0.6 m is overlaid by 5-6 m thick pyrites ferrous shales. Pyrites is also mined in Sikar district of Rajasthan. The pyrites resources in India are estimated to be 1500 million tonnes (10% S) excluding 350 million tonnes (40% S), available exclusively in Jharkhand.

Quality

The mineral pyrites occur in various physical forms, viz. crystalline, amorphous and powdery. The powdery is the most reactive form containing about 22-24% sulphur. The pyrite ferrous shale lying on the top of the pyrites bed contains, on an average of about 10% sulphur. The chemical composition of the pyrites of the two grades is given in Annexure III. The shale lying below the pyrites bed does not contain any sulphur. The mine water becomes acidic with a pH of around 3 as a result of pyrophoric nature of the pyrites.

Chemical Reaction

The sulphur present in the pyrites is a highly reduced compound in a higher oxidation state. The reaction rate depends upon several factors such as moisture content, available oxygen, temperature and presence of toxic constituents around the granules. The biological oxidation of sulphur in the amendment is carried out mainly by *Thiobacillus ferrooxidans* but only at a specific pH range normally obtainable in rice crop fields (Singh *et al.*, 1978). In buffer suspensions, it has been found that the oxidation of pyrites is maximum at pH 5.0. At a pH 2.5, oxidation is low and at pH 8.0 it is negligible (Venkat Krishnan, 1980). Pyrites first gets oxidized to H_2SO_4, which in turn reacts with soil lime to indirectly furnish soluble calcium for exchange.

$$2FeS_2 + 2HOH + 7O_2 \longrightarrow 2FeSO_4 + 2H_2SO_4 \tag{3.5}$$

$$CaCO_3 + 2H_2SO_4 \longrightarrow CaSO_4 + CO_2 + H_2O \tag{3.6}$$

$$2\,Na\text{–soil} + CaSO_4 \longrightarrow Ca\text{–soil} + Na_2SO_4 \downarrow \tag{3.7}$$

The oxidation of pyrites begins as soon as the material is mined gets exposed to the air. The reaction is slow but wasteful as some of the S escapes to the atmosphere as follows:

$$FeS_2 + 3O_2 \longrightarrow FeSO_4 + SO_2 \uparrow \quad (3.8)$$

For efficient oxidation, a warm, well aerated moist soil is required. When pyrites is added to a sodic soil, the acid already contained in it reacts with $CaCO_3$ of the soil to release Ca for exchange with Na on the clay complex. As such, pH in the vicinity of pyrites granules decreases to adversely affect the biological oxidation of pyrites (Sharma and Swarup, 1990). Thus, oxidation of S in pyrites remains incomplete. As such its efficiency remains below that of gypsum when applied on equivalent S basis. Many studies have shown that its efficiency is about 70 to 80% as compared to gypsum at the recommended dose of application. Singh *et al.* (1978) observed that more than 36% of the sulphur is extracted in water in the first two days from the pyrites containing 30% sulphur. Later on, the oxidation is slow. This aspect has to be kept in mind while judging the effectiveness of sulphuric acid produced due to the oxidation of pyrites in the reclamation of sodic soils, containing sufficient reserves of $CaCO_3$.

Later studies revealed that the soluble S content of the material at the time of field application rather than the total S content determines its reclamation efficiency. Freshly mined pyrites contains many times less soluble S than the pyrites stored under a rain shelter for few years. Depending upon storage conditions, the soluble S content of pyrites varies from 1 to 11%. As a thumb rule, for pyrites to be effective, it must contain at least 5 to 6% soluble S at the time of application. Reclamation efficiency of the pyrites can be increased as follows.

Method of Application

Not much information is available on the appropriate method of pyrites application to ensure its optimum efficacy. Singh *et al.* (1978) observed that the surface application of pyrites lowered pH and ESP of the soil to a greater extent as compared with its mixing in the surface 10 cm of the soil. Higher amounts of leachate were collected from column when pyrites was surface placed. Venkat Krishnan (1980) reported that mixing of the pyrites in the whole soil column was inferior to surface placement. Mixing the surface pyrites with the rest of the soil after 12 and 18 days of surface application performed comparably and resulted in better and more uniform improvement of the soil than retaining the pyrites on the surface unmixed. Mixing of the pyrites after 6 days of application led to very poor improvement. Pyrites showed signs of appreciable oxidation forming a rusty coloration after about 15 days. The efficiency of the amendment also reduced as the soil water content at the time of application reduced from full saturation to one-fourth saturation. Some guidelines on pyrites application are:

- The pyrites must be properly crushed before use. Its soluble S content should be more than 5%

- Apply pyrites only when the field moisture content is around field capacity or above
- For proper oxidation, it should be applied as thinly as possible. Besides, sufficient time should be allowed to oxidize the material before the leaching process is initiated
- The soils should be leached with good quality water for 7-15 days preferably in intermittent leaching mode before transplanting the first rice crop.

Pyrites containing 6-8% soluble sulphur has a pH of 2-2.5 when dissolved in water (1:100). As such, it is used as a field test to assess the quality of the pyrites for land reclamation. The test procedure is as follows:

Dissolve 1-g pyrite in 100 ml distilled water and check the pH. If the pH is more than 2.5, the pyrite must be stored and periodically moistened by sprinkling water over the bags. The material should be stored until the test reveals a pH of 2.5 or less. One may use a litmus paper or a pH meter for this test.

Precaution

Since pyrites contains free H_2SO_4, proper care must be exercised during field application. Gloves and gumboots should be used to avoid acid injury.

SULPHUR AND SULPHURIC ACID

Alike pyrites, mineral sulphur is also converted to H_2SO_4 through the process of biological oxidation by *Thiobacillus thiooxidens.* Being a very slow process, the relative efficiency of mineral S as an amendment is also low.

$$2S + 5O_2 + 2\,H_2O \longrightarrow 2H_2SO_4 \tag{3.9}$$

Further reaction will be as per eq. (3.6) and eq. (3.7) in the case of pyrites.

Sulphuric acid is applied directly to the soils. The farmers in the Western States of USA (California, Arizona etc.), USSR and Central Asian Countries have been using concentrated sulphuric acid as a means of reclaiming calcareous sodic soil under irrigated farming. It reacts rapidly with soil lime since it does not have to go through the oxidation process. However, it is highly corrosive and quite difficult to handle at the field level. The reclamation is based upon the following chemical reactions:

$$H_2SO_4 + CaCO_3 \rightarrow CaSO_4 + CO_2 + H_2O \tag{3.10}$$

$$2\text{Na-clay} + CaSO_4 \rightarrow \text{Ca-clay} + Na_2SO_4 \tag{3.11}$$

but if the soil is non-calcareous,

$$2\text{Na-clay} + H_2SO_4 \rightarrow 2\text{H-clay} + Na_2SO_4 \tag{3.12}$$

Although the initial cost of the sulphuric acid is high, its transportation cost is lower because only about one half of the quantity of gypsum is required on equivalent basis. By any standard, sulphuric acid is more efficacious and economical than gypsum in the highly calcareous sodic soils, provided calcium and organic matter are not depleted completely (Kovda *et al.*, 1973). Even the horticulture plants can be established well because the soil profile is reclaimed through a deeper depth, whereas, with gypsum, reclamation is a slow and proceeds only gradually with time through small depths. Overstreet *et al.* (1951) compared the effectiveness of gypsum, sulphur and sulphuric acid on a highly sodic soil (ESP 60-100). After seven months of experimentation, including leaching of the soils, sulphuric acid reduced soil ESP to 10, more effectively than gypsum.

Large scale reclamation programs of sodic soils are being executed in Armenia by the application of 1% sulphuric acid solution, heavy leaching (1500-30,000 m^3 ha^{-1}) sometimes combined with rice cultivation, and extensive deep drainage (horizontal drains combined with tube wells). Since the sodic horizons of these soils are very thick (as much as 60-100 cm), the sulphuric acid is applied layer by layer, with multi-tier ploughs or using a special ripper. The standard application of sulphuric acid is 30 t ha^{-1} (expressed as 80% acid), and the crop (cotton, fruits and vines) yields for the first two years are sufficient to pay for the reclamation cost.

In the Indian context, Shah (1975) used waste sulphuric acid (70% strength) in the field and achieved remarkable improvement in the sodic soils. However, farmers being more dependent on subsidies, and where adequate machinery for transport and application of H_2SO_4 are yet not available, probably use of sulphuric acid is not gaining acceptance.

Gypsum and pyrites would continue to dominate as amendments in land reclamation programs. Yadav (1973) presented data on the effect of gypsum and sulphuric acid applied at 80% of the laboratory estimated gypsum requirements and in chemically equivalent quantities on crop yields in a highly sodic soil. pH and ESP of the surface soil of the experimental field varied between 9.1 to 9.8 and 41 to 60, respectively. During the three years of study, crop yields (Table 3.11) were always higher in sulphuric acid than in the gypsum treated plots. There was also improvement in crop yields over the years which was apparently due to biological action of plant roots, etc. This aspect is discussed in detail in another section.

Table 3.11 Effect of gypsum and equivalent quantities of sulphuric acid on crop yields (t ha^{-1}) (Yadav, 1973)

Treatment	1969-70		1970-71		1971-72	
	Rice	Wheat	Rice	Wheat	Rice	Wheat
Control	1.51	0.72	2.12	1.71	3.56	1.46
Gypsum	3.67	1.92	4.36	2.61	5.13	2.74
Sulphuric acid	4.56	2.56	4.68	3.10	5.53	3.05

Some workers have postulated that sulphuric acid is more effective because calcium sulphate formed *in- situ* as a result of its reaction with soil calcium carbonate is more effective in neutralizing free sodium carbonate in soils and in replacing the adsorbed sodium. Miyamoto *et al.* (1975) presented equations to predict the changes in exchangeable sodium and dissolved salts from known applications of sulphuric acid. Prather *et al.* (1978) suggested that combining either $CaCl_2$ or H_2SO_4 with $CaSO_4$ as an amendment can appreciably shorten the time of reclamation and improve reclamation efficiency as compared to $CaSO_4$ alone under certain soil conditions.

PHOSPHO-GYPSUM

Phospho-gypsum (PG) is a byproduct of the phosphate fertilizer industry. The PG emanates when phosphoric acid is produced using rock phosphate through a process known as the wet process. The apatite (or rock phosphate) reacts with concentrated sulphuric acid to form phosphoric acid and phospho-gypsum as follows:

$$\underset{\text{Apatite (rock phosphate)}}{Ca_{10}(PO_4)_6F_2} + 10H_2SO_4 + 20H_2O \longrightarrow \underset{\text{Phospho-gypsum}}{10\ CaSO_4 \cdot 2H_2O} + 6H_3PO_4 + 2HF \quad (3.13)$$

For each tonne of P_2O_5, 5.5 tonnes of byproduct PG, a fine powder like material containing about 25% moisture is produced, about 1.0 million tonnes being available from industries manufacturing chemical fertilizers in India. It contains 80-90% calcium sulphate and 1-2% fluorine with pH in the range of 4.5 to 5.0. The detailed composition of phospho-gypsum from different industries is given in Table 3.12 and Table 3.13. The PG sourced from M/s Paradeep Phosphate limited, Odisha has S 16.4%, Ca 18.8%, Na 0.11%, Al 0.02%, Fe 0.30%, Si 3.35% and P 0.34% (Nayak *et al.*, 2011). In comparison, mineral gypsum has S 16.1%, Ca 18.3%, Na 0.18%, Al 0.06%, Fe 0.01%, Si 0.82% and P 0.01%.

Table 3.12 Chemical composition (%) of phospho-gypsum

Characteristics	Albright Morarji and Pandit Ltd., Ambernath	FCI, Trombay	Coromandel Fertilizer Ltd., Visakhapatnanm	FCI, Haldia	SPIC, Madras
Combined water	19.5	20.2	–	–	–
Total P_2O_5	1.0	0.15-1.0	0.8-1.0	–	–
CaO	31.1	30-52	30.8	–	32.0
Mixed oxides	0.9	–	0.45	–	0.5
SIO_2 and other insolubles	1.8	–	4.6	0.7	0.9
F	1.52	–	0.6	0.35	1.1
Na_2O	0.56	–	1.3	–	–
K_2O	0.06	–	–	–	–
Equivalent $CaSO_4.\ 2H_2O$	–	95-98	90.2	70.2	–

Table 3.13 Chemical composition of unprocessed phospho-gypsum

Constituents	Coromandal Fertilizer Ltd., Vishakhapatnam	Southern Petro-chemical Industries Corp., Tuticorn	BIS 12679-1989 limits	Fly Ash*
P_2O_5	1.15	1.08	0.40 max	
F	0.72	0.66	0.40 max	
Organic matter	0.31	0.36	0.15 max	
Cl	0.14	0.08	0.10 max	
SiO_2 + insoluble in HCl	5.14	2.86	-	70.60
$Fe_2O_3 + Al_2O_3$	0.85	1.40	-	24.40
CaO	31.57	31.80	-	2.60
MgO	0.43	0.08	-	0.73
SO_3	41.98	43.10	-	
$Na_2O + K_2O$	0.32	0.30	0.30 max	0.20 (Na_2O)
Free moisture	19.78	18.8		
Purity ($CaSO_4.2H_2O$)	88.51	92.66	85.0 min	
pH	3.0	3.5	5.0 min	

Values are in % except pH. Source: Singh and Garg (2002), *Singh and Garg (2005); max = maximum; min = minimum

The exact quality of PG depends upon the presence of impurities in the rock phosphate. Some of the impurities could be due to the presence of radioactive nuclide (Uranium and Radium) with minor amounts of toxic metals such as arsenic, barium, cadmium, chromium, lead, mercury, selenium, silver, fluoride and aluminum. The average composition of PG in USA is reported in Table 3.14. U.S. Environmental Protection Agency (Federal Register 6/3/92) has permitted the controlled use of phospho-gypsum in agriculture if radium-226 levels are <10 pCi g^{-1}. This restriction on the maximum radium radioactivity essentially eliminates the use of southern Florida phospho-gypsum because its radium-226 levels are commonly in the range of 15 to 25 pCi g^{-1}.

Table 3.14 Composition of phospho-gypsum

Major constituents	g kg^{-1}*
Ca	200–240
P	1-5
S	150-190
F	5-38
Minor constituents	mg kg^{-1}*
K	100-800
Mg	8-400
Mo	65
Cd	0.23
Radioactive element	
226Ra	10–25 pCi g^{-1} #

Values are averages of those presented in Mays and Mortvedt (1986), Pavan et *al.* (1987), Lin *et al.* (1988), Alva and Sumner (1989), Alva *et al.* (1990), and Sumner (1990). † Value is from Mays and Mortvedt (1986) and is for southern Florida phospho-gypsum only.

Current researches in India have proved that PG is a good amendment to reclaim sodic lands and sodic water. The studies also revealed that one time application of phospho-gypsum for land reclamation and/or small doses for water reclamation are unlikely to result in phytotoxic level of fluoride. Even then, the rice and wheat in fluoride affected areas should be consumed with care for first two years of reclamation with this amendment. Fears of radioactive elements have also been discounted although the exact quantum and nature of these elements may vary due to the source of rock phosphate used in the manufacturing process (Nayak *et al.*, 2011). The fate of radium-226 in Florida phospho-gypsum was investigated by Mays and Mortvedt (1986). They applied phospho-gypsum containing 25 pCi g^{-1} 226Ra at rates up to 112 Mg ha^{-1} to the surface of a silt loam soil and grew successive crops of corn (*Zea mays* L.), wheat (*Triticum aestivum* L.) and soybean (*Glycine max* L.). Application of phospho-gypsum even at the 112 Mg ha^{-1} rate had no effect on the radioactivity levels in grain of corn, wheat, or soybeans. The 112 Mg ha^{-1} rate was more than 200 times the normal rate of gypsum used for peanut fertilization. Additionally, they noted no increases in grain Cd levels although at the highest rate they found that corn growth was slow.

Mehta and Yadav (1977) reported that phospho-gypsum is a promising chemical amendment and can be used effectively for the reclamation of sodic soils. Reclaimed lands supported good crop stand. Singh *et al.* (1980) conducted detailed studies on the phospho-gypsum and its effect on the soil properties and crop growth. This study discounted the adverse effects of fluorine in the phospho-gypsum and explained that the adverse effects of fluorine to a great extent are nullified because of the phosphorus contained in this amendment.

DISTILLERY SPENT WASH

Distillery spent wash (DSW), a byproduct of the molasses based distilleries, is a nutrient rich wastewater. It has high biological oxygen and chemical oxygen demand (BOD and COD). It has 0.10-0.15% N, 0.01-0.03% P_2O_5, 0.80-1.20% K_2O and appreciable quantities of Ca, Mg, S and micronutrients (Table 3.15). It is estimated that for each liter of alcohol, about 15 liters of DSW is produced. In Tamil Nadu, under the AICRP on Management of Salt Affected Soils and Use of Saline Water in Agriculture, distillery spent wash (DSW) has been used to reclaim sodic lands. Around 0.5 million liters ha^{-1} of DSW needs to be applied for reclamation (Fig. 3.3; Haroon and Bose, 2004). Tests on this amendment in Madhya Pradesh, under the AICRP on Management of Salt Affected Soils and Use of Saline Water in Agriculture, have also proved the efficacy of this amendment in soybean-wheat crop rotation. Singh *et al.* (1980) reported that the addition of spent wash was very effective in increasing the water intake rate of a highly sodic soil. ESP was reduced from 100 to 2 in the top 15 cm of the soil used in the column. The application of spent wash followed by irrigation rather than the dilution of spent wash at the time of its application was very effective. DSW is also used as a source of nutrients, being rich in N, P and K.

Fig. 3.3 Application of DSW for reclamation of sodic lands

Table 3.15 Chemical Characteristics of pre-treated distillery spent wash (undiluted)

Characteristics	Value
pH	7.7
Electrical conductivity (dS m^{-1})	26.4
Total solids	47200
Total dissolved solids	37100
Total suspended solids	10240
Settleable solids	9880
BOD	16100
Carbonate	Nil
Bicarbonate	12200
Total phosphorus	40.5
Total potassium	7500
Calcium	900
Magnesium	1244
Sulphur	70
Sodium	520
Chlorides	6204
Iron	7.5
Manganese	980
Zinc	1.5
Copper	0.25
Cadmium	0.005
Lead	0.16
Chromium	0.05
Nickel	0.09
Ammonical nitrogen	750.8
Carbohydrates (%)	22.8

Units: All values are in mg L^{-1},except pH or otherwise specified; Source: Chandraju et *al.* (2013)

PRESS MUD

In India, 400 sugar factories rank as the second major agro-industry that produce several co-products of immense potential value. The co-products include press mud, molasses and spent wash. Out of which press mud produced during clarification of sugarcane juice is about 3.6% to 4% of sugarcane crushed i.e. 36 to 40 kg of press mud is produced from each tonne of cane crushed. Press mud is a soft, spongy, amorphous and dark brown material containing sugar, fiber and coagulated colloids including cane wax, albuminoids, inorganic salt, sand and soil particles (Table 3.16). It consists of 80% water and 0.9%-1.5% sugar organic matter, nitrogen, phosphorus, potassium, calcium, coagulated colloids and other materials in varying amounts. As such, it is a good substitute as a fertilizer, both in the unprocessed or processed form (Yaduvanshi and Yadav, 1990). Processes used to improve its fertilizer value include composting, treatment with microorganisms and mixing with distillery effluents (Nasir, 2006).

Sugar mills in India produce about 12 million tonnes of press mud (filter cake) as a waste from double sulphitation process. Its disposal is a major issue. In many cases, it is burnt in brick kilns. Such a use of the material is resulting in the loss and wastage of millions of tonnes of nutrients besides causing environmental degradation. Dry press mud can be exploited for generation of energy as it contains higher percentage of combustibles. It is also a good source of biogas as it contains about 5–15% of sugar.

Sugar mills either use the carbonation (calcium carbonate-sulphur) or the sulphitation process (Sulphur) to purify the sugarcane juice. Press mud coming out of the sugar mills with the carbonation process contains about 70% calcium carbonate and 11% organic matter. Since the solubility of calcium carbonate is quite low, the press mud available through this process cannot be used as an effective amendment. However, it is applied along with farmyard manure as an amendment on mild sodic soils or to improve soil physical properties. The press mud, produced in sugar mills using sulphitation process, contains 9.3% calcium sulphate and 35.8% organic matter. Since the percent calcium sulphate is too less, the quantities required to reclaim barren lands is quite high. Nonetheless, it has a good potential to reclaim mild sodic lands as well as to ameliorate soils where sodicity build-up is due to irrigation with sodic water. Evidences have been generated to prove its usefulness as an amendment. Addition of 10-20 t ha^{-1} press mud can reduce the gypsum requirement of soils by about half because of the addition of calcium sulphate and organic matter. For example; Isaac *et al.* (2010) reported that the highest yield was obtained when gypsum was applied @ 50% of GR, and it was at par with gypsum applied at 25% GR when press mud @ 10 t ha^{-1} was applied along with.

Table 3.16 Values of some selected physico-chemical parameters of press mud (PM) (The values are mean ± SE of 10 observations each)

Parameters	Press mud (PM)
Water holding capacity (%)	78.2± 1.323
pH	7.66± 0.047
Electrical conductivity (S cm^{-1})	1.77± 0.092
Sodicity (meq $100g^{-1}$)	2.89± 0.198
Carbon content (%)	29.67± 1.058
Organic matter (%)	51.2± 1.827
Available phosphorus (%)	1.80± 0.157
Sulphate (%)	2.297± 0.123
Nitrate nitrogen (%)	0.271± 0.044
Total nitrogen (%)	1.13± 0.095
Total calcium (%)	0.204± 0.015
Total magnesium(%)	0.088± 0.011
Total sodium (%)	0.005± 0.001
Total potassium (%)	0.037± 0.003
C:N	26.25
C:P	16.48

Source: Joshi and Sharma (2010)

FLY ASH

Fly ash, a by-product of coal based thermal power plants, besides many elements contains soluble phosphorus, magnesium, sodium and calcium. Around 10 million tonnes of fly ash is generated by various thermal power plants in India. Depending upon the origin of coal used in a particular power plant, fly ash could be alkali or acidic in nature. Few experiments have established its usefulness in sodic soils (Table 3.17). Guidelines for its application as an amendment are yet to be developed.

Table 3.17 Effect of fly ash and pyrites on rice and wheat yield in sodic soils

Amendment	Quantity (t ha^{-1})	Yield (t ha^{-1})			
		1979-80		1980-81	
		Rice	Wheat	Rice	Wheat
Control	0	1.02	0.57	1.98	0/91
Fly ash	10	2.85	2.40	4.03	0.75
Pyrite	10	3.03	2.87	4.33	2.79

Source: Tiwari *et al.* (1989)

ORGANIC AMENDMENTS

Most organic amendments like straw, rice husk (rice hulls), groundnut and safflower hulls, farm yard manure (FYM), compost, poultry dropping, green manure, trees leaves, saw dust etc. on their decomposition produce high partial pressure of CO_2 and organic acids, which help in increasing the electrolyte concentration and lowering the soil pH. Organic materials also feed soil microbes that produce polysaccharides and long chain aliphatic compounds capable of binding and stabilizing soil aggregates (Lado *et al.*, 2004), which tremendously helps to maintain soil structure. Such processes also increase the solubility of *kankar* to lower the ESP. Sodic soils are generally deficient in organic matter, and as such addition of crop residues and other organic materials in the soil benefits through processes such as improve and maintain soil structure, supply needed plant nutrients and hastening soil reclamation. Singh (1974) observed that when straw of barley and *Sesbania* was decomposed under aerobic conditions, their efficiency in terms of ESP reduction and improvement in soil physical properties was less as compared to when those were decomposed under saturated (anaerobic) conditions (Table 3.18). Thus, it is advisable to maintain continuous ponding during the course of decomposition of the organic matter so as to get maximum benefits from organic amendments.

Organic materials together with inorganic amendments are more cost effective, hasten the reclamation process, increase the yield and therefore, their use should be encouraged (Dargan *et al.*, 1976a). To be very effective alone, relatively large quantities (20 to 400 t ha^{-1} representing 1/5 to 1/3 of the soil volume) of organic amendments have to be used. If the C:N ratio of these materials is very high, as is the case with saw dust, rice husk and barley straw, then they decompose slowly and may be less effective as compared to *Sesbania* which has a low C:N ratio. Under these conditions, organic amendments may cause N deficiency which should be addressed.

Table 3.18 Effect of organic materials on soil physical properties

Treatment	pH_s	ESP	Modulus of rupture (bars)	Hydraulic conductivity (cm h^{-1})	Na in leachate (meq L^{-1})
Barley straw aerated	9.1	81.5	12.9	0.03	16.6
Sesbania straw aerated	8.8	68.5	11.8	0.04	26.3
Barley straw saturated	8.4	56.1	11.3	1.72	30.2
Sesbania straw saturated	8.3	47.2	7.5	1.29	43.0
Original soil	9.5	95.0	-	-	-

Source: Singh (1974)

RELATIVE PERFORMANCE OF VARIOUS AMENDMENTS

Application of amendment has profound positive effect on the physico-chemical properties of the soils and crops yield (Table 3.19). Since the replacement of cations in soils is governed by the type of cations in the soil solution (its valency, size and degree of hydration), ion concentration, nature of ions associated with the cation and nature of colloidal particles, wide variations in soil properties might be realized upon addition of amendments. If we compare the relative change in some of the soil properties, it is noted that the final soil ESP with gypsum and H_2SO_4 as amendments is similar (Table 3.19). The H ion having the same valency as Na but being smaller in size and non-hydrated, has greater replacing power and is preferred over Na on the exchange complex. Thus, H ions of the H_2SO_4 replace Na from the exchange complex to lower the soil ESP. In spite of ESP decrease being similar in H_2SO_4 and gypsum treatments, the saturated hydraulic conductivity of the soil was less and the water dispersible clay more in H_2SO_4 treated soil than gypsum treated soil (Table 3.19). Similarly, the ESP drop and improvement in soil physical properties was higher, when $Al_2(SO_4)_3$ was applied as an amendment. It could be due to Al of the $Al_2(SO_4)_3$, which is a more effective flocculent as compared to H of H_2SO_4. Similar comments are valid for Ca of $CaSO_4$. Besides solubility of $Al_2(SO_4)_3$ and $CaSO_4$ being low, major fraction of these amendments remain immobilized due to soluble $NaHCO_3$ and/or Na_2CO_3. As such, these amendments are quite effectively used to replace exchangeable sodium.

Table 3.19 Effect of different amendments (applied @100% GR) on physico-chemical properties of a sodic Vertisol

Amendment	pH_2	EC_e ($dS\ m^{-1}$)	ESP	Saturated hydraulic conductivity ($mm\ h^{-1}$)	Water dispersible clay (%)	Dispersion ratio
Control	8.81	0.80	65	0.06	37.2	0.41
Gypsum	7.9	0.72	14	4.77	8.0	0.09
Pyrites	8.0	0.31	20	1.64	32.4	0.36
H_2SO_4	7.5	0.18	14	2.98	30.4	0.54
$Al_2(SO_4)_3$	7.6	0.27	8	4.49	8.6	0.09
$FeSO_4$	7.9	0.85	21	1.59	33.7	0.37

Source: Sharma and Gupta (1986)

Pyrites versus Gypsum

Since the relative performance of pyrites is related to initial water soluble S, the relative decrease in soil ESP and yield of rice and wheat crops with pyrites containing more than 5.5% water soluble sulphur gave almost equivalent results to that of gypsum applied at equivalent levels (Table 3.20 and Table 3.21).

Table 3.20 Effect of gypsum and pyrite on yield and soil properties

Treatment	pH_2 (0-15 cm)		ESP (0-15 cm)	Yield ($t\ ha^{-1}$)	
	After paddy	After wheat		Paddy	Wheat
Control	10.0	10.1	69.1	3.3	0.0
Gypsum (50 % GR)	9.5	9.3	50.2	5.2	3.3
Gypsum (75 % GR)	9.3	9.3	32.9	5.3	4.2
Pyrite I* (50 % GR)	9.9	9.9	60.2	4.6	0.6
Pyrite I (75 % GR)	9.7	9.8	55.8	4.4	0.4
Pyrite II* (50 % GR)	9.7	9.5	48.5	5.1	2.9
Pyrite II (75 % GR)	9.2	9.2	30.9	5.3	4.1
Pyrite III* (50 % GR)	9.7	9.7	50.0	5.6	3.4
Pyrite III (75 % GR)	9.2	9.3	29.7	5.7	3.9

* Pyrite I, II and III contained 2.8, 5.5 and 8.0 soluble sulphur at the time of use

Source: Swarup and Yaduvanshi (2004)

Table 3.21 Effect of gypsum and pyrites on the yield of crops in sodic soils

Source	Soil type and pH	Crops	Grain yield ($t\ ha^{-1}$) at different levels (% GR) of amendments									
			0		25		50		75		100	
			G	P	G	P	G	P	G	P	G	P
Abrol and Verma (1978)	Alfisol, initial pH 10.5	Rice	3.86	3.86	6.71	5.71	6.81	6.09	7.44	6.71	7.24	6.91
		Wheat	0.02	0.02	0.15	0.15	3.15	0.54	3.60	1.35	4.22	1.35
Chauhan (1978)	Alfisol, initial pH 9.2	Rice	1.73	1.73	-	-	2.77	0.54	-	-	-	-
		Barley	2.62	2.62	-	-	3.67	3.64	-	-	-	-
Bhatt *et al.* (1978)	Alfisol, farmers field, pH varying from 9.0 to 10.5	Rice	2.93	2.93	3.23	3.49	3.57	3.87	3.13	4.06	4.09	4.87
		Barley	1.50	1.50	1.18	1.18	1.74	1.27	1.58	1.60	2.33	1.86
Singh *et al.* (1981)	Alfisol, initial pH 10.2	Rice	2.61	2.61	-	6.65[1]	6.87	-	-	7.59[2]	-	7.66
		Wheat	2.05	2.05	-	3.72	4.80	-	-	4.84	-	5.00
Sharma and Gupta (1986)	Vertisol, initial pH 8.5	Rice	0.14	0.14	0.97	0.84	1.77	1.60	2.50	2.15	-	-
		Wheat	0.50	0.50	1.00	0.88	1.67	1.51	2.40	3.22	-	-
Chhabra *et al.* (1989)	Alfisol, initial pH 10.2	Rice	1.75	1.75	2.93	2.09	3.06	2.37	-	-	3.43	2.60

G= Gypsum: P= Pyrites: 1- 33% GR, 2- 66% GR.

Gypsum versus Phospho-gypsum

The studies at the Lucknow center of CSSRI revealed that phospho-gypsum is a better reclaiming agent as compared to mined gypsum. The PG treatment resulted in greater reduction in pH and ESP of both the surface and the lower layers. It is attributed to higher solubility of PG over mined gypsum, which allows soluble Ca to permeate to lower depths (Table 3.22). Application of phospho-gypsum also gave comparatively higher yield over equivalent dose of mined gypsum i.e. phospho-gypsum @ 10 tha^{-1} (equivalent to 50% GR) resulted in slightly higher yield of 4.5 $t\ ha^{-1}$ over the treatment 50% GRm (4.22 $t\ ha^{-1}$) and 25% GRm+25% GRpg (4.32 $t\ ha^{-1}$) in the first year of reclamation. Similarly, application of phospho-gypsum @ 5 $t\ ha^{-1}$ resulted in comparatively higher yield over 12.5% GRm +12.5% GRpg. The yield of follow up wheat was also higher in the PG treated fields (Fig. 3.4).

Table 3.22 Effect of phospho-gypsum on ESP of soil after first and third year of rice-wheat cropping

Treatments	2006-07		2008-09	
	0-15 cm	15-30 cm	0-15 cm	15-30 cm
50 GRm	26.5	31.6	24.6	28.5
50 GRpg	20.4	27.3	18.3	25.7
50 GR_{PG}-KFC	18.3	25.2	17.9	23.4
25 GRm+ 25 GRpg	17.8	23.5	16.7	21.5
25 GRpg	19.4	25.6	17.9	24.3
12.5 GRm+ 12.5 GRpg	25.2	30.8	23.5	27.5
LSD (p=0.05)	1.5	2	1.5	2

50GRm = 50% gypsum requirement through mineral gypsum; 50 GRpg = 50% gypsum requirement through phospho-gypsum; 25 and 12.5 means 25% and 12.5% of GR

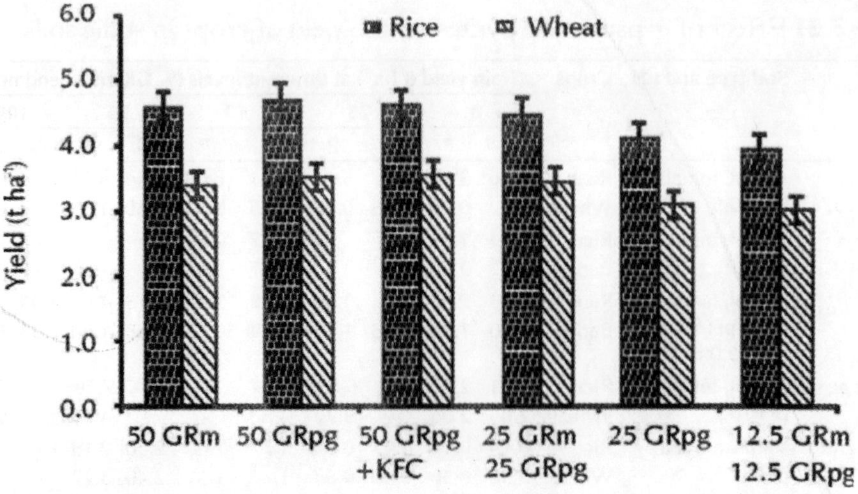

Fig. 3.4 Effect of gypsum and phospho-gypsum on rice and wheat yield at Shivri, Lucknow

PHYTOREMEDIATION AND BIO-RECLAMATION

Amelioration of sodic soils has been mainly achieved through the application of chemical amendments. Owing to gradual increase in amendment cost in some parts of the world during the last two decades, this amelioration strategy has become cost-intensive, particularly for the subsistence farmers in developing countries. There can be many other settings where it may not be possible to reclaim the soils using the amendments or resource constraints of the farming community may inhibit the adoption of technology. Thus, phytoremediation with low initial investment has emerged as a potential substitute of chemical amelioration. Cultivation of salt tolerant species for the purpose to reclaim sodic soils is known as bio-reclamation or phytoremediation hereinafter referred as phytoremediation. Phytoremediation has also been defined as the use of plants to remove pollutants from the environment and to render them harmless (Hasanuzzaman *et al.*, 2014). Phytoremediation works through plant root action that helps dissolve native soil calcite ($CaCO_3$) of low solubility to supply adequate levels of Ca^{2+} for an effective Na^+-Ca^{2+} exchange without the application of an amendment. The plants not only remediate the salt contaminated soils but also provide food, fodder, fuel wood, and industrial raw material and increase the income of the farmers owning salt affected lands. The process is a clean technology and is environmentally safe. The technology uses:

- Salt tolerant crops
- Grasses
- Forest plantations
- Halophytic plants
- Microbial consortium

Phytoremediation technology has many advantages (Qadir *et al.*, 2007), among them following are the major ones:

- Less investments as it either obviates or minimizes the need of chemical amendments
- Financial or other benefits from crops grown during amelioration
- Improves physical condition of the soil
- Higher organic matter and availability of plant-nutrient
- Amelioration is more uniform and proceeds to deeper depths than chemical amelioration
- No environmental issues rather carbon sequestration is achieved if plants are used as ameliorants

While the application of crops is discussed in Chapter 4 wherein the role of field crops such as rice, wheat and sugar beet etc has been described in bio-remediation of sodic soils, role of grasses/plants such as Karnal and Bermuda grasses, *Prosopis sp.* and others is discussed in chapter 5. A brief introduction to the application of microbial consortium is discussed in the following paragraphs.

Microbes, the most abundant entity in the universe, are increasingly being used in various bio-remediation processes. Halotolerant or sodic tolerant microbes can be used to reclaim sodic soils as these are able to dissolve the sodium and other cation complexes by producing organic acids (Paikray and Singh, 2012). A microbial technology package for the reclamation of sodic soils is disclosed in patent number WO 2012046908A1. The microbial culture used is the mixed enriched culture isolated from sodic soil (pH>10) of Kakori block, Lucknow, Uttar Pradesh selected by its ability to grow and survive on high stress of sodium ion concentration and pH. The microbial composition comprises of bacteria strains such as *Pseudomonas fluorescens* - MTCC 5525, *Bacillus* sp. -MTCC 5536, *Micrococcus* sp. - MTCC 5534 bacterial strain, *Kocuria* sp. - MTCC 5535, *Kocuria carniphila* - MTCC 5537 and *Trichoderma herzianum* - MTCC 5530 fungal strain. The Microbial consortia of 10^9 colony forming unit (cfu) along with jaggery/molasses/sugar and Farm Yard Manure (FYM)/organic manure/ agricultural waste are applied to the affected soil. The method, however, needs testing under field conditions for its effectiveness to reclaim sodic lands.

A growth promoting technology for sodic lands comprising of a microbial consortium of *Bacillus pumilus*, *Baciillus thurigensis* and *Trichoderma harzianum* in a dynamic media has been prepared by CSSRI, RRS, Lucknow. It has been patented and commercialized (Damodaran, 2014).

It is now emerging that phytoremediation or bio-remediation is an effective amelioration strategy for calcareous sodic and saline sodic soils having comparable performance to chemical amendments albeit the time scale as it may take 2-3 years before the land can be put to normal cropping.

ELECTRICAL METHOD

Investigations have shown that the treatment with electrical current may stimulate the reclamation of sodic soils, although it cannot replace conventional procedures involving controlled leaching. A remarkable result was obtained by the U.S. Bureau of Reclamation on a field trail in Montana where saline, boggy land that produced no crop for 19 years was improved to such an extent that a heavy growth of grasses and sweet clover covered the area 50 days later. In this work, after the current had been applied for a short time, the sodium content of the drainage water from the area increased from 1500 mg l^{-1} to 37,000 mg L^{-1}. The expenses for electrical current amounted to 6 dollars per hectare (Kovda *et al.*, 1973). The penetration of the chemical amendment at high densities of the electrical current (0.1 and 0.5 MA cm^{-2}) achieved the complete reclamation of the sodic soil and the treatment proved to be most efficacious. The results of these investigations have been used in field experiments also.

Vadjunina and El-Sawaby El-Koumi (1972) conducted a laboratory experiment on sodic soil monoliths (15 x 15 x 50 cm) to examine the effect of electrical reclamation as compared to gypsum and acid treatment. The following treatments were compared:

(i) Control
(ii) Gypsum @ 30 t ha^{-1}
(iii) H_2SO_4 @ 10 t ha^{-1}
(iv) H_2SO_4 @ 20 t ha^{-1}
(v) Electrical current for 3000 hr 0.01 mA cm^{-2}
0.10 mA cm^{-2}
(vi) Electrical current for 1000 hr 0.50 mA cm^{-2}

The soils were leached after the above treatments. The changes in the soil pH, ESP and infiltration rate shown in Table 3.23 reveal that electric current application at a density of 0.01 mA cm^{-2} was similar to the application of gypsum or H_2SO_4. The remarkable improvements in pH, ESP and the infiltration rate took place in the 0-30 cm soil depth.

Table 3.23 Effect of electrical and other conventional treatments on the soil pH, ESP and infiltration rate

Treatment	0-10 cm		10-30 cm		Average infiltration rate (mm d^{-1})
	pH	ESP	pH	ESP	
Original soil	10	72.8	10.09	89.9	–
Control	9.74	49.2	9.95	82.6	4.4
Gypsum (30 t ha^{-1})	8.90	19.2	9.72	62.2	13.0
H_2SO_4 (10 t ha^{-1})	9.46	31.1	9.88	79.3	10.0
H_2SO_4 (20 t ha^{-1})	7.56	2.9	9.11	39.4	13.0
C (0.01 mA cm^{-2})	8.68	10.3	9.85	78.5	14.4
C (0.10 mA cm^{-2})	8.20	1.7	8.26	1.2	57.0
C (0.5 mA cm^{-2})	7.98	1.0	8.09	1.2	86.4

THERMAL METHOD

Investigation have shown that the heating a sodic soil to a high temperature results in significant reclamation. Gupta and Bandyopadhya (1975) achieved remarkable improvement in the sodic soil, which was heated to high temperature in the muffle furnace. On heating, the sodic soil containing illite, possibly vermiculite, at 500°C for 2 hr, the EC_e dropped from 13.4 to 4.8 dS m^{-1}, ESP reduced from 92 to 37, boron (saturation extract) decreased from 9 to 1 mg L^{-1} and hydraulic conductivity (K) increased from 0.2 to 1.0 cm h^{-1} (Table 3.24).

Table 3.24 Effect of heating on properties of a sodic soil

Treatment	EC_e (dS m^{-1})	Soluble ions (m mol l^{-1})					CEC (meq100g^{-1})	ESP	B (mg l^{-1})	K (cm h^{-1})
		Na	Ca+Mg	CO_3	HCO_3	Cl				
Control	13.4	108	1.5	54	25	49	8.3	92	9	0.2
75°C for 24 h	12.0	102	1.5	44	25	45	8.3	91	–	–
100°C for 24 h	11.5	97	1.5	35	19	45	8.3	92	–	–
200°C for 6 h	10.4	88	1.5	16	9	45	8.3	91	–	–
400°C for 3 h	5.8	52	1.5	2	4	35	7.6	68	–	–
500°C for 2 h	4.8	30	1.5	0	3	35	6.8	37	1	1.0

The changes that take place due to heating are related with the clay mineral of the soil. Gupta and Bandyopadhya (1975) opined that on heating to 400-500°C, part of the sodium ions in the exchange complex of the soil penetrate into the vacant octahedral positions in the dioctahedral lattice of the illite, possibly vermiculite clay mineral. After entering the vacant octahedral positions, the sodium is rendered difficult to exchange or is fixed in a non-exchangeable form. However, the nature of the clay mineral will not undergo any change even after heating up to 500°C and illite possibly vermiculite dominated. The collapse of the interlayer space of the crystal lattice, follows the loss of swelling potential of the soil and increase in hydraulic conductivity from 0.2 to 1.0 cm h^{-1} is therefore, quite obvious. Further it seems that part of sodium carbonate is lost due to volatilization. Boron occurs in the organic form as well as inorganic form. Since the organic matter is destroyed beyond 200°C, some decrease in boron is quite probable. Volatilization of boron may also reduce its content to a certain extent.

Preliminary studies on germination and early seedling growth of wheat in the saturation extract of original soil and that of one heated at 500°C for 2 hours revealed 100% germination over nil in an original soil. The length of the seedlings was 14.3 cm in the heated soil extract after 12 days of growth. The data presented in Table 3.25 show excellent growth of seedlings in the extract of the heated soil.

Table 3.25 Germination percentage and seedling growth of wheat

Soil	Germination percentage	Length of seedlings (cm)
Original soil	0.0	–
Heated soil	100.0	14.3

It was further examined whether the type of clay minerals influences the reduction of ESP. It was observed that the illite mineral is more amenable to changes in ESP on heating to a high temperature followed by bentonite. The kaolinite mineral is not affected at all (Table 3.26).

Table 3.26 Effect of heating on ESP of different clay minerals

Clay mineral	ESP	
	Original	After heating at 500°C for 2 h
Illite	63.0	24.5
Bentonite	80.1	63.8
Kaolinite	79.6	79.6

Although, the experiments have been conducted on a laboratory scale but the results are very encouraging and promising. It is surmised that if a slow moving down draft furnace with its flame directed towards the ground on sodic lands is moved in the field, it should be possible to reclaim the soil instantaneously. Such type of down draft furnace is manufactured in Australia. One such machine was gifted by Australia to the Central Road Research, Institute, CSIR, New Delhi, for heating the black cotton soils around Bhopal and use the resulting hard substance, a kind of clinker, as a sub-base materials to construct roads. The machine heats the soil to about 15 cm depth to a temperature of about 700°C. The current results can be translated into field practice by the use of this type of machine. Alternatively, the soils could be heated in the field *in situ* by the use of electricity which may however, be prohibitive due to the high cost. Harnessing of solar energy in the near future seems to be more rational to implement this practice.

4

RECLAMATION PACKAGE

Reclamation and management of sodic soils are two distinct activities although practically both are applied simultaneously to achieve high efficiency or to hasten the process of land reclamation. Reclamation of sodic soils beginning with the application of a chemical amendment is a slow process. The efficiency of the reclamation process depends upon the management interventions comprising of physical/mechanical processes, hydraulic processes, chemical processes, biological processes, and reuse and/or safe disposal of drainage water, wherever necessary. The physical/mechanical processes are: on-farm land development including *bunding* (dyking), levelling, tillage to conserve rainfall or to increase infiltration rate and deep tillage/soil inversion, if beneficial (Table 4.1).

Table 4.1 Package for reclamation and management of sodic soils

Group/sub-group	Technology
Reclamation	On-farm land development including surface drainage
	Application of chemical amendment
	Flushing#/Leaching
Management	
Crop management	Selection of crops and cropping sequences
	Exploitation of varietal differences
	Improved agronomic practices
Soil management	Land forming/seeding
Irrigation water management	Shallow depth-high frequency irrigation regime
	Switchover to improved irrigation techniques
Chemical management	Application of additional nutrients/green manure/FYM
Rainwater management	*In-situ* rainwater conservation; Rainwater harvesting and reuse (helps in drainage)

#Flushing is adopted only in heavy texture soils with accumulation of excessive salts on the soil surface (http://www.biologydiscussion.com/soil/alkali-soils-factors-effects-and-reclamation/7188 and http://agricoop.nic.in/dacdivision/mmsoil280311.pdf) Opened on 17.07.2015

Land forming or modifying the planting technique allows plants to live with the salts. The hydraulic processes include flushing especially in heavy textured soils or where excessive salt accumulates at the soil surface, leaching, drainage improvements and environmentally safe disposal of drainage water. Amongst the chemical processes, application of a good mix of macro and micro nutrients assumes significance. Application of farm yard manure and green manuring helps to improve the soil physical and biological characteristics. Besides these processes, several improved cultural practices such as selection of salt/sodicity tolerant crops/varieties, increasing the crop stand through higher seed rate and improved irrigation practices provide a practical approach to manage these soils. Since rain water is the best quality water available for leaching, rain water management is also a key component in the whole scheme of land reclamation. Central Soil salinity Research Institute, Karnal has framed a package of practices to reclaim and manage sodic lands, which takes care of what has been stated in this paragraph (Table 4.1). This chapter discusses all these components of reclamation and management except the application of amendment that formed the part of Chapter 3.

PHYSICAL AMELIORATION

A study of the profile distribution of ESP determines whether physical manipulation of the soil will be helpful or not. The physical amelioration is possible only if the ESP decreases with depth. In such cases, it is useful to employ soil inversion practices to bring the low ESP layer on top while taking the high ESP top layer below. Clearly, there is no removal of sodium from the exchange complex but the GR of the soil (Usually determined for 0-15 cm layer) will be much less than without inversion. Profile studies in Sharda *Sahayak* canal command especially situated on both sides of the main canal revealed that such a situation prevails over large areas (Fig. 4.1). Therefore, it should be possible to partially reclaim the soil through profile inversion using a heavy duty moldboard or disc plough. It may also help to improve the soil physical condition. Sometimes, a hard pan is also noticed at shallow depth in these soils. In such cases, deep ploughing or deep ripping breaks the hard pan developed as a result of migrating dispersed clay. These practices benefit plant growth through greater water penetration and improved aeration. However, the benefits of ripping are short lived, especially on sodic soils.

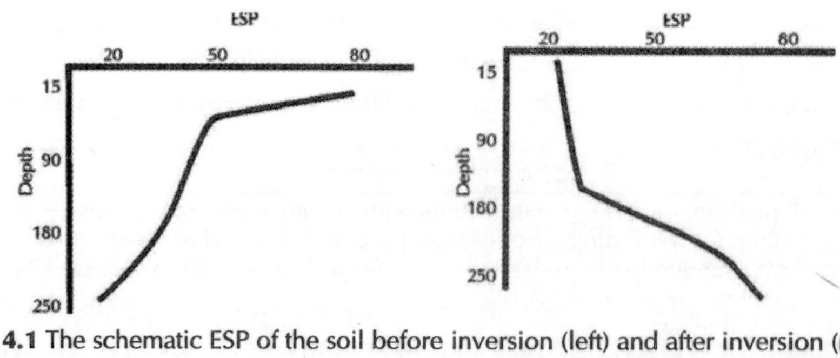

Fig. 4.1 The schematic ESP of the soil before inversion (left) and after inversion (right)

In Uttar Pradesh, this model was tested while implementing a multi-enterprise model. While digging an aquaculture pond, the upper high ESP soil layer went down while the low ESP soil came at the top. Thus, crops on the dykes and raised land could be cultivated without any chemical amendment. Soil breeding through application of sand to make the soil lighter in texture is another way of physical reclamation. Besides being costly and difficult to adopt, the anticipated results may not be forthcoming with this technique in sodic soils (Acharya and Abrol, 1991).

ON-FARM LAND DEVELOPMENT INCLUDING SURFACE DRAINAGE

Land Levelling and Construction of *Bunds*

Any reclamation program begins with appropriate land development. The first activity in this context is to dyke the fields (construction of bunds). It is essentially required for the two reasons. Firstly, the rainwater/irrigation water stored in the field after gypsum application should not go out of the fields. Secondly, runoff from the surrounding un-reclaimed areas should not enter the fields. As such, construction of 45 to 60 cm high dykes is crucial for an efficient reclamation program. During the first few years, these soils are prone to erosion by the flowing water and therefore, slightly wide and relatively high dykes need to be constructed (Fig. 4.2).

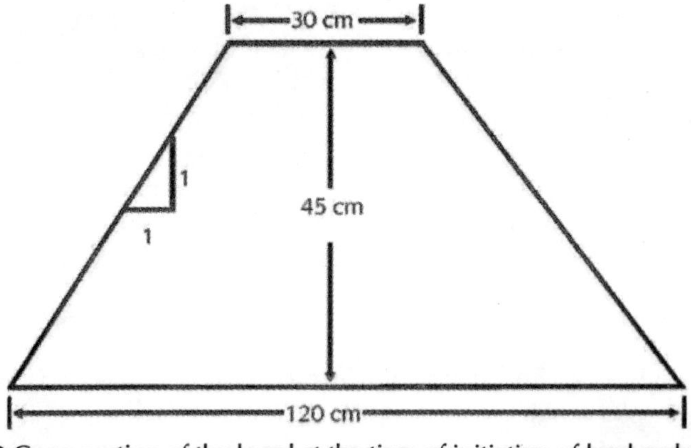

Fig. 4.2 Cross-section of the bund at the time of initiation of land reclamation

After the construction of bunds, fields are properly levelled. Even small differences in the level can result in non-uniform leaching causing patchy crop growth. However, the term levelling should not be misunderstood as tabletop levelling. Instead, it should be treated as grading in the direction of irrigation. Precise grading rather than tabletop levelling is essential to carry out efficient irrigation and surface drainage. As far as possible, field should be provided with a slope of 0.1% to drain excess water; the range being 0.05 to 0.20%. It is estimated that about 10 tractor hr ha^{-1} are enough to accomplish the land smoothening work. Studies have shown that a super leveller can effectively complete the task with sufficient accuracy (Table 4.2). Presently, laser land levelling is increasingly used which results in a more precise surface and grade (Table 4.3 and Fig. 4.3). The data reveal that the range within which the ground levels deviate from the mean improves with land levelling. The improvement is much superior in the laser than conventional land levelling. The levelling index (LI) with laser levelling can be brought to less than 1 while it remains more than 2.0 in conventional levelling (Table 4.3, Ambast *et al.*, 2015). The best time to undertake *bunding* and levelling operations is April-May. It helps to finish all the activities before the first crop of rice usually transplanted in the month of July.

1:Tripod; 2:Laser beam generator; 3:Laser beam receiver; 4:Control box; 5:Levelling equipment; 6:Tractor; 7: Laser beam

Fig. 4.3 A view of laser land levelling components and operation in the field

Table 4.2 Time, cost and land uniformity coefficient (LUC) for different levelling implements

Implement	Time of operation (hr)		Time for land shaping operation (h)	LUC
	Plowing	Discing		
Levelling blade	10	10	26.4	0.72
Levelling float	10	10	15.0	0.63
Levelling blade + float	10	10	18.5	0.90
Super leveller	10	12	19.5	0.96

Source: Tyagi and Singh (1979)

A well- planned field layout is then prepared to ensure that each field receives water and is connected to collector and/or main field drain and is fully trafficable. The irrigation system delivers water uniformly to all parts of the field, while the drainage channels are capable to remove excess irrigation or rainfall from all corners of the fields. In border irrigation, a drainage channel may run perpendicular to the borders at the tail end. The estimated length of irrigation and drainage channels for one ha of land is about 100 m each.

Table 4.3 Performance parameters in laser and conventional land levelling

Field	Levelling index (LI)	
	Before	After
Laser (Field 1)	1.56	0.80
Laser (Field 2)	2.29	0.90
Conventional	3.00	2.10

Assured Source of Irrigation Water

Timely application of water in adequate quantities is essential for the successful implementation of a reclamation program. In general, sodic areas are underlain by good quality ground water and the aquifer characteristics are good enough to install a tube well. Since, canal water is limited and undependable; it is advisable to install a tube well before embarking upon a reclamation program. Depending upon the aquifer characteristics, a cavity or a strainer type tube well can be installed. A tube well of 10 L s^{-1} capacity is sufficient to cover an area of 4 ha. A cavity well serves as a dual purpose structure. It provides water for irrigation and helps to lower the water table as land reclamation work progresses (Fig. 4.4). Currently, when water table is declining at a very fast rate; a submersible pump might have to be installed. It would naturally impact the cost of land reclamation.

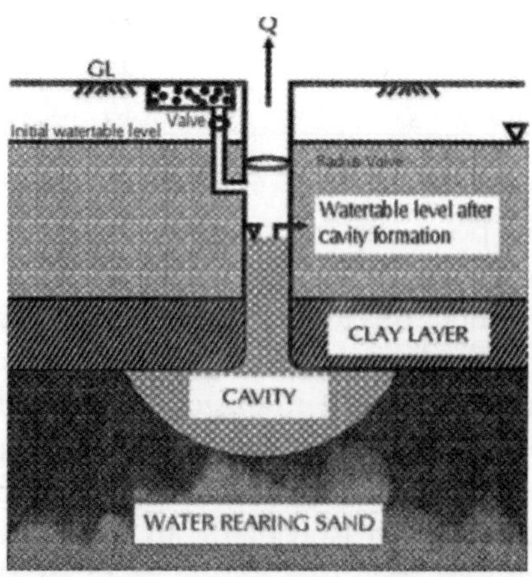

Fig. 4.4 A multi-purpose cavity well for irrigation, drainage and artificial ground water recharge

Surface and Subsurface Drainage

Since sodic soils are characterized by low infiltration rate, drainage is essential to ensure high productivity. The adverse effects of water logging in these soils appear due to a combination of poor aeration, reduced uptake of plant nutrients, imbalance between the uptake of various ions and toxicity effects of various elements/cations. Temporary water logging situation is encountered during monsoon season or following a low intensity shower if it follows a normal irrigation during the winter season. The adverse conditions continue to prevail even few days after the water has been removed from the soil surface. The reclaimed sodic lands often referred as semi-reclaimed sodic soils also require efficient and effective surface drainage system as revealed by yield losses from short-term water stagnation (Table 4.4). It may be noted that except for the barley, yield reduction exceeds 10% with water stagnation of 2 days in all the test crops (Gupta *et al.*, 2004). Therefore, in semiarid and sub-humid regions where sodic soils occur in abundance, improved surface drainage is an essential element of a reclamation plan. The three tier system of rainwater management discussed under the section rainwater management in this chapter is a good intervention to ensure adequate and timely surface drainage especially in land locked areas.

Table 4.4 Crops yield reduction (%) due to shallow submergence of various durations in semi-reclaimed sodic soils

Crop	Yield (t ha^{-1})	Yield reduction over drained with water stagnation (days)			
	Drained	1	2	4	6
Sorghum	4.13	3	11	16	20
Pearl millet	2.22	6	15	22	27
Pigeon pea	1.52	4	14	18	21
Wheat	4.20	8	17	27	39
Barley	3.65	4	7	13	25
Mustard	1.43	8	16	22	29
Berseem (seed)	0.48	2	21	35	48
Sunflower	1.86	13	19	26	30

In areas where water table exists at shallow depth and especially where ground water quality is poor, option of installing some kind of subsurface drainage needs to be explored. It has been established that areas underlain with fresh ground water are amenable to vertical drainage (tube well drainage). The sodic areas in Haryana, Punjab and Western Uttar Pradesh are typical examples where shallow water table has been managed with vertical drainage. The tube wells in these areas acted both as a source of irrigation and drainage. Although, vertical drainage has proved to be the most effective at these places yet at places where rainfall is more, canal water is available in relatively more quantity, ground water is sodic in nature and resource endowment of the farmers is relatively poor, it is not being adapted to the desired extent. Conventional pipe drainage to lower the water table may pose problems because of low hydraulic conductivity of these soils. As such, an integrated solution consisting of surface, pipe, vertical and bio-drainage needs to be adopted to hasten the land reclamation process (Gupta, 2016).

FLUSHING/LEACHING

Leaching process is greatly inhibited in heavy soils, sodic soils with low infiltration rates and lands with shallow water table. In such events, flushing is adopted as a means of leaching to remove the surface concentrated salts. Flushing proved effective in reclamation of saline-sodic soils of the central India. For effective flushing, water is kept ponded on the land surface for 2-3 days followed by puddling for effective dissolution of soluble salts followed by drainage of the ponded water (Fig. 4.5). Two flushing with an interval of about 7 days in between the two flushing is recommended. Since the soluble carbonates and bicarbonates of sodium are effectively flushed, gypsum application after the flushing proved relatively more effective than without flushing. Compared with conventional treatment, the surface flushing significantly reduced the EC_e and SAR of the soil and improved physical properties such as f-potential, dispersible clay content, water stable aggregates and saturated hydraulic conductivity (Nayak *et al.*, 2008). It has now been established that in relatively heavy textured highly saline-sodic soils of Uttar Pradesh, flushing along with leaching is a better option than leaching alone (Nayak *et al.*, 2008). With flushing, average yields of rice and wheat increased by 25% and 62%, respectively as compared to the conventional practice of leaching alone. Split application of gypsum between flushing was not beneficial and therefore, full dose of gypsum should be applied after second flush.

Fig. 4.5 Fields being prepared and filled with water for the flushing operation

As per the postulation made by Abrol *et al.* (1975), the gypsum requirement as calculated by Schoonover method includes the gypsum required to neutralize the soluble sodium carbonate, if present in sodic soils. It is therefore, prudent to leach the soil by applying a heavy irrigation before applying gypsum. Since flushing effectively accomplishes the same, Arora *et al.* (2015), observed that gypsum requirement is reduced significantly as determined after the flushing operation (Table 4.5). The GR after flushing reduced at all the places, being less than 50% at Dhaura.

Table 4.5 Effect of flushing on gypsum requirement

Location	pH_2	EC_2 (dS m^{-1})	Gypsum requirement (meq 100 gm^{-1})	
			Schoonover method	After washing
Kausambhi	9.2	0.74	3.70	2.90
Rasoolpur	8.9	2.43	1.60	1.10
Baratikhera	10.1	9.81	5.20	3.10
Lalganj	10.0	1.28	5.80	3.40
Sandila	10.3	2.50	5.30	3.70
Dhaura	10.1	4.04	4.50	2.20
Vishnupur	9.8	2.25	5.80	4.80

Besides what is stated in the previous paragraph, it is imperative to carryout leaching to transport the chemical constituents of an amendment to the exchange sites and to leach down the exchanged chemical constituents from the root zone. It is accomplished by ponding water on the soil surface so that percolating water through the root zone displaces them below the root zone.

Since the solubility of gypsum in pure water is quite less, it was thought that water required to leach gypsum amended soils is quite high. On the contrary, it is now known that depth of water required to dissolve gypsum in soils of high ESP is much less than calculated on the basis of its solubility in pure water. It is ascribed to the fact that soil in this case acts as a sink for the calcium ions. Laboratory studies revealed that gypsum dissolution increased with an increase in the exchangeable Na content of the soil. About 4 cm of water was sufficient to dissolve all the agricultural grade gypsum of <0.26-mm size (Hira and Singh, 1978). The amount of water required to dissolve the added gypsum and to reclaim a sodic soil did not increase with the amount of gypsum added, provided it equalled the gypsum requirement. The water required for complete dissolution of gypsum computed from a dissolution equation, increased from 2.8 cm to 15.9 cm as gypsum particle size increased from < 0.1 mm to 0.5–2.0 mm. Normally, 50 to 80% of the applied gypsum is dissolved within 7 to 10 days of continuous ponding and leaching. There is no need to leach down the soils beyond this period as the remaining leaching requirement is also met by the ponded water in the rice crop. Since rice is the first crop in land reclamation programs, this recommendation remains valid where rice is grown as the first crop. If crops other than rice are cultivated, leaching period might be increased.

In pyrites or S treated soils wetting and drying cycles help to maintain proper moisture and aerobic conditions necessary to ensure maximum oxidation and production of H_2SO_4. For pyrites amended soils, fields, after applying the amendment, are left as such for about 7-10 days. Afterwards, leaching is carried out for about 7-15 days either in continuous ponding mode or in intermittent ponding mode. Water standing in rice fields takes care of the remaining water requirement for leaching in this case as well.

CROP MANAGEMENT

Selection of Crops and Cropping Sequences

Crops differ widely in their tolerance to sodic environment of the root zone. Even crop varieties respond differently to sodic conditions. Therefore, selection of crops and cropping sequences is of practical and great economic importance to manage sodic lands. Generally, four criteria are used to assess relative tolerance of crops:

- Germination of crops
- Plant survival
- Absolute plant growth or yield
- Relative growth or yield

Large numbers of investigations on differential response of crops or varieties have been reported at germination stage and attempts made to extrapolate the tolerance limit for final performance. More than often-such attempts proved frustrating because tolerance characteristics differ from one stage of growth to another. Plant survival has been used as a criterion by ecologists but without yield, plant survival alone has little value to the farmer. On the other hand, it can prove to be useful criterion for plant breeders. Absolute plant growth or yield is of greatest interest to the farmers. However, this criterion does not permit comparison between crops because yields of different crops are not expressed in comparable terms.

The relative growth or yield, defined as the yield on a sodic soil as a fraction of the yield on a non-sodic soil under similar environment and nutritional conditions, seems to be the most desirable and popular criteria used in crop tolerance studies. This criterion is reliable provided the level of other essential factors such as nutritional factor or water does not affect yield reductions. The main difficulty with this method, however, is to decide at what level of yield reduction, different plants should be compared. A researcher comparing the salt tolerance at 10% yield reduction may essentially end up differently than the other who is comparing the salt tolerance at 50% yield reduction. Nevertheless, this criterion is favoured world over and is widely used for relative comparison.

In addition to these four criteria, attempts have also been made to evolve a criterion based on metabolic parameters. The sodium and potassium contents in plant tissues and particularly their ratio have been found to be suitable parameters to adjudge the relative sodicity tolerance. Studies on few varieties known for their differential affinity to sodic conditions revealed that at tillering stage, high K/Na ratio characterizes a tolerant variety compared to a variety which has low K/Na ratio (Joshi *et al.*, 1980). Free proline status of leaves at seedling stage has also been found a good indicator of sodicity tolerance (Bal, 1976). But in later tests, it turned out that these are not very much reliable. The change in pH of the cell sap as a result of sodic stress also has been found to be useful to compare relative/tolerance of crops and varieties (Dwivedi, 1979). The sodic halophytes showed only a nominal rise in pH of their cell sap at ESP 40 as compared to ESP 10. The tolerant *Kharchia* variety of wheat similarly exhibited a rise of only 1% in its sap pH at ESP 40 at 35 days growth stage. In spite of the concerted efforts that have been made in this direction, more information needs to be generated on such tests so as to arrive at a sound criterion.

Pearson as early as in 1960 listed the crop as per their tolerance to soil ESP (Table 4.6). Using the data from numerous studies conducted at CSSRI, Karnal, Gupta and Sharma (1990) for the first time used to describe the crop ESP tolerance through the piecewise linear model of Maas and Hoffman (1977) to describe the relative sodicity tolerance of various crops (Table 4.7). The threshold ESP, ESP at 50% yield reduction and slope (% yield reduction per unit increase in ESP) are reported in Table 4.7.

Table 4.6 Tolerance of various crops to exchangeable sodium (ESP) under non-saline conditions (Pearson, 1960)

Tolerance to ESP and range at which affected	Crops	Growth response under field conditions
Extremely sensitive (ESP = 2-10)	Deciduous fruits, Nuts, Citrus (*Citrus* spp.), Avocado (*Persea americana* Mill.)	Sodium toxicity symptoms even at low ESP values
Sensitive (ESP - 10-20)	Beans (*Phaseolus vulgaris* L.)	Stunted growth at these ESP values even though the physical condition of the soilis good
Moderately tolerant (ESP - 20-40)	Clover (*Trifolium* spp.) Oats (*Avena saliva* L.) Tall fescue (*Festuca arundinacea* Schreb.) Rice (*Oryza saliva* L.) Dallis grass (*Paspalum dilatum* Poir.)	Stunted growth due to both nutritional factors and adverse soil conditions
Tolerant (ESP - 40-60)	Wheat (*Triticum aestivum* L.) Cotton (*Gossypium hirsutum* L.) Alfalfa (*Medicago sativa* L.) Barley (*Hordeum vulgare* L.) Tomatoes (*Lycopersicon esculentum* Mill.) Beet, garden (*Beta vulgaris* L.)	Stunted growth usually due to adverse physical conditions of soil
Most tolerant (ESP more than 60)	Crested and Fairway wheatgrass (*Agropyron* spp.) Tall wheatgrass (*Agropyron elongatum Host* Beau.) Rhodes grass (*Chloris gayana* Kunth.)	Stunted growth usually due to adverse physical conditions of soil

Table 4.7 Threshold, slope and ESP_{50} of crops for medium textured soils (sandy loam)

Crop	Botanical name	Piecewise linear model		
		ESP_t	(%) Slope	ESP_{50}
Barley	*Hordeum vulgare* (P)	8.5	3.5	22.3
Cowpea	*Vigna unguiculata*	13.5	9.1	19.0
Sesbania	*Sesbania aculeata*	46.9	2.4	67.7
Gram	*Cicer arietinum*	7.7	5.0	17.7
Groundnut (Peanut)	*Arachis hypogaea*	8.0	2.2	29.7
Guar (Cluster bean)	*Cyamopsis psoraloides* D.C.	11.9	3.2	27.5
Lentil	*Lens esculentum*	4.9	5.5	14.0
Linseed	*Linum unitatissinum* L.	13.3	2.3	25.0
Pear millet	*Pennisetum typhoideum*	13.6	2.6	32.8
Raya (Indian mustard)	*Brassica juncea* L.	7.6	0.8	70.1
Rice	*Oryza sativa*	24.4	0.9	80.0+
		32.7	1.3	71.2*
Rye	*Secale cereale*	11.0	2.6	30.2
Sesamum	*Sesamum indicum*	9.0	3.6	22.8
Soybean	*Glycine max*	8.0	3.5	22.3
Sunflower	*Helianthis annus* L.	11.3	1.1	56.8
Saffflower	*Carthamus tinctorius*	7.6	5.2	17.2
Wheat	*Triticum aestivum* (P)	16.4	2.1	40.2+
		20.4	2.6	39.6*
Pea	*Pisum sativum*	7.7	4.1	19.9
Onion	*Allium cepa*	9.8	2.2	32.5
Garlic	*Allium sativum*	9.5	1.8	37.3

ESP values indicate average soil ESP of 0-15cm layer, data from pot (*) and field (+) experiments

Although the crop tolerance to soil sodicity might vary with the stage of growth, variety, environmental factors and level of management, the relative tolerance is summarized in Table 4.8 and 4.9.

Table 4.8 Relative tolerance of crops and grasses to exchangeable sodium

Tolerant	Moderately tolerant	Sensitive
Karnal grass (*Leptochloa fusca*)	Wheat (*Triticum astivum*)	Gram (*Cicar arielimun*)
Rhodes grass (*Chloris gayana*)	Barley (*Hordecum valgaro*)	*Mash* (*Phaseolus mungo*)
Para grass (*Brachiaria mutica*)	Oat (*Avena sativa*)	Chickpea (*Cajanus cajan*)
Bermuda grass (*Cynodon dactylon*)	*Shaftal* (*Torifolium resupinatum*)	Lentil (*Lens esculenta*)
Rice (Oryza sativa)	Lucern (*Medicago sativa*)	Soybean (*Giycino max*)
Dhaincha (*Sesbania aculeate*)	Turnip (*Brassica rapa*)	Groundnut (*Archis hypogaea*)
Sugar beet (*Beta vulgaris*)	Sunflower (*Helianthus annus*)	Sesamum (*Sesamum oriental*)
Teosinte (*Euchlaona maxicana*)	Safflower (*Carthamus tinctorius*)	*Moong* (*Phaseolus aureus*)
	Linseed (*Linum usitaissum*)	Pea (*Pisum saccharatum*)
	Onion (*allium cepa*)	Cowpea (*Vigna unguiculata*)
	Garlic (*Allium sativum*)	Maize (*Zea mays*)
	Pearl millet (*Benisetum typhoites*)	Cotton (*Gosstypium fndicum*)
	Cotton (*Gosssypium indicum*)	

Table 4.9 Tolerance of crops to various levels of ESP

Range of ESP	Crops
10 – 15	Safflower, mash, peas, lentil, pigeon pea, *urd* bean
15 –20	Bengal gram, soybean, maize
20 – 25	Groundnut, cow pea, onion, pearl millet
25 – 30	Linseed, garlic, guar, lemon grass, pamarosa, sugarcane, cotton
30 – 50	Wheat, *raya*, sorghum, sunflower, *berseem*, *shaftal*, *senji*, blue panic, gutton panic
50 – 60	Barley, *sesbania*, para grass, Rhodes grass, matricaria
60 - 70	Rice, sugar beet, Karnal grass

Note: Relative crop yields are only 50% of the maximum in the indicated ESP range

Intra-crop variability, besides its role in crop selection, reflects on the future success of any breeding program for stressed environments. Based on the work at CSSRI, Karnal, the range of intra-crop variability is reported in Table 4.10. It is apparent that rice has the maximum range of genetic variability for sodicity tolerance while barley has the maximum for salinity tolerance.

Table 4.10 Intra-crop variability of salt tolerance in some important crops

Crop	Alkalinity (pH)	Salinity (EC_e, $dS\ m^{-1}$)
Rice	9.2 - 10.2	4.0 - 10.0
Wheat	8.8 - 9.3	6.5 - 11.0
Barley	9.0 - 9.3	4.0 - 6.5
Indian mustard	8.9 - 9.2	4.0 - 5.5
Toria	8.8 - 9.0	4.0 - 6.0
Taramira	8.7 - 8.9	4.0 - 6.0
Linseed	8.7 - 8.9	4.0 - 6.0
Safflower	8.7 - 8.8	4.0 - 6.0
Sunflower	8.6 - 8.8	4.0 - 6.0
Sugarcane	8.8 - 9.0	-
Cowpea	8.6 - 8.8	4.0 - 5.0
Green gram	8.6 - 8.8	-
Pigeon pea	8.3 - 8.6	-
Pea	8.6 - 8.8	4.0 - 5.0
Gram	8.6 - 8.8	4.0 - 5.0
Bakla	8.6 - 8.8	4.0 - 7.0
Sugar beet	9.2 - 9.8	4.2 - 10.0
Pearl millet	8.7 - 8.9	4.0 - 8.0

Genotypic Differences

Large genotypic differences in sodicity tolerance have been reported and as such crop varieties widely differ in tolerance. Therefore, besides selection of crops, selection of crop varieties is equally important. Choice of the tolerant varieties within a crop can result in more production and better economic returns from sodic soils (Misra, 1996). Comparison of salt tolerance indices for 5 genotypes of rice indicated that CSR 3 (a natural selection from indica types) is the most tolerant to sodic with an ESP_t of 26.16 while M 1-48 is the least tolerant with an ESP_t threshold of only 8.02 (Table 4.11). Similar test with 6 genotypes of wheat indicated that Kharchia 65 was the most tolerant both in respect of ESP_t and ESP_{50}. The order of sodicity tolerance is Kharchia 65 >KRL 4-13 > KRL 3-23 > KRL 1-4 = HD 2009 > HD 4502. There are not much significant differences among the last four genotypes. A view of the salt tolerant genotypes CSR 10 (rice) and KRL 19 (wheat) is shown in Fig. 4.6. Some of the promising cultivars and tolerance of few of them to soil pH are given in Table 4.12.

Fig. 4.6 A view of the rice tolerant variety CSR 30 (Left) and wheat variety KRL 210 (right)

Table 4.11 Sodicity tolerance of rice genotypes

	CSR-1 Mutant	M1-48	*Basmati* 370	CSR-1	CSR-3
Threshold ESP_t	20.36	8.02	15.92	20.63	26.16
Slope	2.03	2.22	1.84	0.82	0.73
ESP_{50}	45.00	30.60	43.10	81.60	94.65
Mean yield (t ha^{-1})	1.49	1.62	1.32	1.49	1.64

Table 4.12 Promising varieties of important crops for cultivation in sodic soils

Crop	Tolerant varieties	Soil pH_2
Rice	CSR 10*, CSR 11, CSR 12, CSR 13*	9.8 -10.2
	CSR19, CSR23*, CSR27*, Basmati CSR30*, CSR36*	9.4-9.8
Wheat	KRL 1-4*, WH157, KRL 210, KRL 213	<9.3
	Raj3077, KRL19*	<9.3
Barley	DL200, Ratna, BH97, DL348	8.8-9.3
Indian mustard	Pusa Bold, Varuna	8.8-9.2
(Raya)	Kranti, CS52*, CSTR330-1,	8.8-9.3
	CST609-B 10, CS54*, CS56*	8.8-9.3
Gram	Karnal Chana 1	<9.0
Sugar beet*	Ramonskaaya 06, Maribo Resistapoly, Polyrava-E	9.5-10
Sugarcane	Co453, Co1341	<9.0
Pearl millet	MH 269, 280, 427, HHB 392	NA
Cotton	HY 6, Sarvottam, LRA 5166	NA
Safflower	Manjira, APRR 3, A 300	NA
Sorghum	SPV 475, 1010, CSH 1, 11, 14	NA
Tomato	Angurlata, Azad T2	NA
Chillies	Jwala, Chaman	NA
Garlic	Gattar Gola, Hansa	NA

*CSSRI varieties released by Central Varietal Release Committee, NA = Not available,
* Varieties are tolerant to both sodic and saline environment

In a recent study it has been observed that the rice genotypes CSR 10-M27, Bulk 212, CSR 2K 232, Bulk 18, CSR 11-117, CSR10 M2-27, NDRK 11-2, CSR 11-121, CSR 11-143 and CSR 27 -192 are most stable in salt affected soils. The genotype Bulk 212 was found to be the highest yielding variety in most situations across the salt stress environments. These rice genotypes can be used for commercial cultivation in salt affected soils and also used in the further breeding program to transfer the salt tolerance to high yield varieties (Krishnamurthy *et al.*, 2015).

It may be noted that some variations might occur in different tables reported by different workers. It is mainly because of the variations in the experimental conditions, varieties used and other agro-climatic conditions. For example, transplanted lowland rice is more tolerant to sodic stress than direct seeded rice. Many times it has caused confusion on tolerance of rice to sodic conditions. Crops that can withstand excess moisture conditions are generally more tolerant to sodic conditions.

A major difficulties with salt tolerant varieties is that the window in which these varieties yield more than high yielding varieties is quite small (Fig. 4.7). Thus, if the soil gets reclaimed, a salt tolerant variety yields less than the conventional high yielding variety. Moreover, any further deterioration in the soil condition of un-reclaimed soils may not allow the variety to give economic yield.

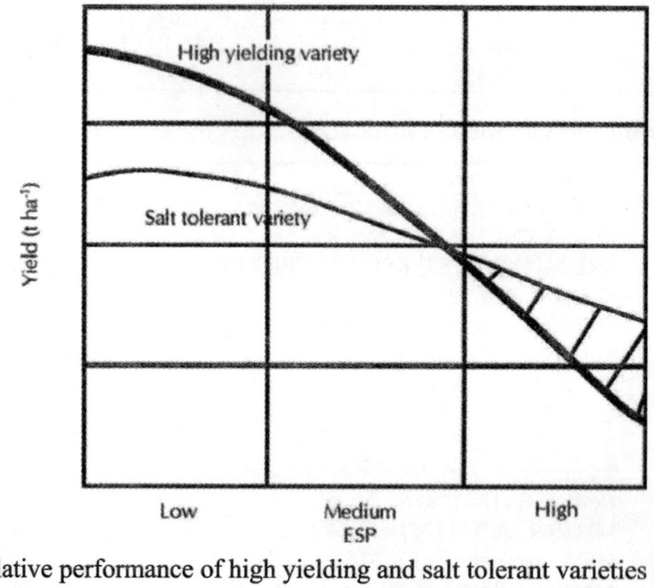

Fig. 4.7 Relative performance of high yielding and salt tolerant varieties in sodic soils

Cropping Pattern

In order to get more returns from the investments and to initiate a process that helps to reclaim the soils faster, an appropriate cropping sequence must be selected. Results of various experiments on comparison of different cropping systems such as rice-wheat, rice-*berseem*, rice-mustard, sorghum-wheat, sorghum-mustard and sorghum-*berseem* at Karnal indicated that the rice based cropping patterns are more promising and profitable than sorghum based sequences (Table 4.13).

Table 4.13 Cropping patterns recommended for sodic soils (reclamation phase)

Season	Year of reclamation		
	First	Second	Third
Summer (April–May)	Leaching/ *Sesbania* green manuring	*Sesbania* green manuring	*Sesbania* green manuring/ Fallow/ Cow pea (fodder)/ Maize (fodder)
Kharif	Rice	Rice	Rice
Rabi	Wheat/ *berseem*	Wheat/ *Raya* /*Shaftal*/ *berseem* Senji/ Sugar beet	Wheat/Barley/Raya/*Berseem*/*Senji*/Sugar beet

Why rice?

Rice is an ideal crop during the first two-three years of reclamation because of the following positive features of this crop:

- Being a relatively shallow rooted crop, gypsum requirement is less as only shallow soil layer (0-15 cm) needs to be reclaimed
- The transplanted rice exhibits high tolerance to exchangeable sodium chiefly because of the fact that rice requires a layer of water in the field. Under flooding conditions the high pH of sodic soils is reduced. Ponnamperuma and his colleagues observed a decline in pH from 8.8 to 7 twelve weeks after flooding. It was ascribed to evolution of large quantities of carbon dioxide from bacterial action and its accumulation in the root zone because of restricted diffusion of gases in flooded soils (Ponnamperuma, 1965; Ponnamperuma *et al.*, 1966).
- Although, rice can withstand water stagnation throughout the growing season, some drainage especially at tillering may prove beneficial.
- The low permeability of sodic soils is another advantage to rice because losses of water due to deep percolation are minimal. Deep percolation helps to leach soluble salts and exchanged products present in the root zone.
- Apart from being tolerant to high sodicity, rice has proved to enhance the reclamation process considerably resulting in continuous soil improvement.
- Up to 12 to 15 cm of rain water can be stored in the rice fields reducing the drainage need of the sodic soil areas on the one hand and encouraging ground water recharge on the other (Narayana, 1980).

The effect of growing rice on changes in soil sodicity in fields and other controlled laboratory studies (Abrol and Bhumbla, 1979; Chhabra and Abrol, 1977) showed that rice growth resulted in high cumulative removal of exchangeable sodium caused by mobilization of native insoluble calcium carbonate as a result of increased hydrolysis and CO_2 liberated by plant roots (Table 4.14). Long-term field studies (unpublished) showed that when rice was included in the crop sequence in a highly sodic soil with entire profile highly sodic in nature, there is a gradual reduction in sodicity making the upper 100 to 125 cm of soil nearly free of sodicity problems in about 10 years (Fig. 4.8). Moreover, the spectacular improvement even in the non gypsum treated plots in about the same time is largely attributed to rice. McNeal *et al.* (1966) experimenting with sodic soils in Pakistan concluded that rice cultivation indirectly facilitates the removal of exchangeable sodium by increasing the cross-sectional area of conducting pores, resulting in increased permeability. On the contrary, they questioned the rice crop's appreciable direct effect resulting from the dissolving action of carbon dioxide on the release of calcium from calcium carbonate.

Table 4.14 Effect of growing rice on changes in soil ESP (Abrol and Bhumbla, 1979)

Treatment gypsum (t ha^{-1})	Exchangeable sodium percentage (ESP)			
	Before rice (April 1971)		After rice (November 1971)	
	0-15 cm	15-30 cm	0-15 cm	15-30 cm
0	87.0	94.0	50.0	63.0
7.5	67.0	87.0	29.5	63.0
15.0	33.0	83.0	25.5	56.0
22.5	15.5	64.5	16.0	32.5
30.0	14.0	57.0	12.5	36.0

*Gypsum was applied in June 1970

Complete reclamation of sodic soils takes several years. After rice-wheat or rice-*berseem* sequence for about 3 years, cropping can be diversified to other semi-tolerant to sensitive crops depending upon the extent of land reclamation (Table 4.15). With passage of time more sensitive crops can be cultivated such that any crop can be grown after 8-10 years.

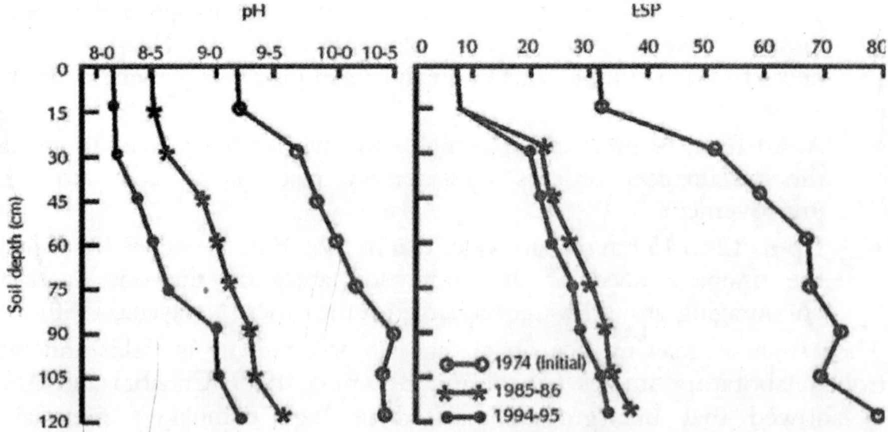

Fig. 4.8 Long-term effect of rice-wheat cropping on pH and ESP of a sodic soil at Karnal

Table 4.15 Cropping sequences recommended for sodic soils (Post reclamation phase)

Soil pH_2 (0-15 cm layer)	Years after reclamation	Crops	
		Kharif	***Rabi***
9.2 to 9.3	3 to 4	Cotton, sorghum, pearl millet, rape seed	Mustard
8.9 to 9.0	5 to 8	Sugarcane, groundnut, pigeon pea, soybean	*Berseem*, sunflower, safflower, gram, pea, lentil
8.5 to 8.8	After 8 to 10 years	Any kind of crops including vegetables, flowers etc.	

Improved Agronomic Practices

Plant population

Germination is generally inhibited in sodic soils due to formation of a hard crust or other adverse physico-chemical properties. High mortality rates especially during early stages further depresses the plant population. This together with poor tillering reduces the yield of the crops. Increasing the plant population either by reducing the plant-to-plant and/or row-to-row space in transplanted crops or gap filling or increasing the seed rate in seeded crops helps to improve crop stand in sodic soils (Chhabra, 1996). In general, seed rate should be increased by about 25% over and above the seed rates recommended for normal soils. In rice, increasing the number of seedlings per hill also helps to achieve higher yields (Table 4.16). In general 4-6 seedlings should be planted per hill as against the practice of 2 seedlings per hill in normal soils (Dargan *et al.*, 1982). In a field experiment on a saline-sodic soil in China, grain yield significantly increased from 1 to 9 seedlings per hill, and then levelled off (Wang *et al.*, 2010). Cultivation of 9 seedlings per hill is recommended, which is significantly more than the recommended cultivation practice of 3-5 seedlings per hill in normal soils.

Table 4.16 Grain yield (t ha^{-1}) of various varieties of rice as affected by plant population

Varieties	Plants per hill			Mean of varieties
	2	4	6	
IR-8-68	4.07	4.68	5.79	4.83
IR-8	2.75	3.89	4.80	3.81
Jhona 349	2.78	3.51	3.33	3.21
Jhona 351	2.31	2.92	3.42	2.88
Jaya	1.52	3.30	3.68	2.83
B-370	2.16	2.47	2.72	2.94
B-217	1.60	2.05	2.66	2.11
Mean	2.46	3.25	3.76	

Puddling

Puddling in lowland rice cultivation is performed mainly for the following reasons:

- To reduce the infiltration rate to conserve water
- To control weeds

In sodic soils, infiltration rate is naturally low and weed problem is also not encountered at least during the first 2-3 years. Therefore, puddling can be avoided in these soils unless the soil is too hard to transplant the rice seedlings (Table 4.17). It helps to reduce the cost, save energy and reduce environmental pollution.

Table 4.17 Effect of puddling versus no puddling on rice yield

Treatment	Grain yield (t ha^{-1})
Puddling	1.25
No puddling	2.15
CD at 5%	7.8

Variety Jhona348; Source: Dargan (1979)

Growth stage and age of seedlings

Rice has been categorized as only a semi-tolerant crop for sodic conditions if crop is raised from seeds and not from transplanted seedlings. Since the tolerance of crops varies from one stage to another, age of rice seedlings can make a difference in the yield during first few years of reclamation. Somewhat older rice seedlings, 35 to 40 days of age, are preferred in sodic soils instead of the usually recommended 30 day old seedlings (Table 4.18). Still better results have been obtained when rice seedlings are prepared in normal soils as it provides healthy seedlings for transplanting. Similarly, to establish tree species, planting older seedlings grown in normal soil has proved beneficial.

Table 4.18 Effect of age of seedlings on grain yield of rice

Age of seedling (days)	Yield (t ha^{-1})
24	4.94
28	5.33
32	5.31
36	5.65
40	5.47

Source: Dargan and Gaul (1974)

Green Manuring

Application of green manure enhances organic matter (OM) content of the soil, increases partial pressure of CO_2, lowers pH, enhances solubility of native $CaCO_3$ and adds considerable amount of plant nutrients to the soil. *Sesbania*, a green manuring (GM) crop, being tolerant to high ESP and water logging is an ideal crop. It is grown during summer months (May-June), a lean period in the rice- wheat cropping sequence. As such, it fits well into this cropping pattern. Normally, 45-50 days old *Sesbania* crop, when it has attained a height of 1.5 to 1.8 m, is incorporated into the soil. Longer growth period renders the plant hardy to make its decomposition difficult. A fifty days old *Sesbania aculeata* is able to supply about 4.2 t ha^{-1} dry matter and 90 kg N, accumulated mostly through biological N fixation, 11 kg P and 90 kg K. Having a narrow C:N ratio (about 25), it decomposes very easily and quickly. Ghai *et al.* (1988) reported that a decomposition period of 5-days for *Sesbania* species suffices in a reclaimed sodic soil as compared to a period of 10-15 days normally practised in non-sodic soils. However, the advantages of green manuring in sodic soils are more when it is decomposed for one week under submerged conditions prior to transplanting of rice (Swarup, 1988). The standing water in rice fields hastens the decay without any harm to rice crop. Allowing more time for decomposition of a succulent crop like *Sesbania* results in greater loss of N through volatilization and N recovery by the crop is less. Green manuring ensures steady supply of N to the rice crop, since about 50% of the organic nitrogen contained in the green manure is converted into the readily available ammonical form within 4 to 6 days and the rest in another 10 to 20 days of incorporation. Turning over the crop for green manuring, increased the yield of the following rice crop significantly (Singh *et al.*, 1991) with a saving of 60 kg N and 11 kg P ha^{-1} (Yaduvanshi, 2001). In another study, the yield increase in rice crop due to green manuring has been equivalent to about 80 kg of N applied through chemical fertilizers. Long-term field experiment in sodic soils at CSSRI, Karnal showed that the rice and wheat yield can be maintained even when only 50% of the recommended dose of NPK is used in an integrated manner with FYM or *Sesbania* green manuring (Yaduvanshi, 2003; Table 4.19).

Table 4.19 Effect of different treatments on yield and N:P:K soil status of rice-wheat cropping system in reclaimed sodic soils after 12 years

Treatments	Yield (t ha^{-1})		Soil available macro-nutrients ($kg\ ha^{-1}$)		
	Wheat (2004-05)	Rice (2005)	N	P	K
Control	1.97	2.23	91	7.8	160
$N_{120}\ P_0\ K_0$	3.64	3.78	144	6.2	158
$N_{120}\ P_{26}K_0$	4.25	4.65	168	14.8	155
$N_{120}\ P_{26}K_{42}$	4.46	4.92	169	15.8	275
$N_{60}\ P_{13}\ K_{21}$	3.87	4.08	132	11.4	232
$N_{60}\ P_{13}\ K_{21}$ + GM	4.66	6.07	175	18.2	285
$N_{120}\ P_{26}K_{42}$ + GM	5.02	5.62	201	20.8	301
$N_{60}\ P_{13}\ K_{21}$ + FYM	4.22	5.26	176	19.1	291
$N_{120}\ P_{26}K_{42}$ + FYM	5.17	5.54	198	21.5	298
$N_{180}\ P_{39}\ K_{63}$	5.12	5.64	197	24.7	315
CD (P=0.05)	0.49	0.54	9.7	0.98	11.7

NH_3 volatilization is another pathway through which N is lost from the soil-plant system. Ammonium fertilizers are particularly liable to volatilization loss.

The ESP has a marked effect on the amount of added N volatilized, increasing with increasing ESP although quantum may vary with soil type. The green manuring helps to reduce the volatilization losses in sodic soils. The loss of N as NH_3 volatilization with green manuring combined with urea was 13.4% as compared to 19.55% when urea was applied alone (Table 4.20).

Table 4.20 Ammonia losses from integrated nutrient management system in rice fields in a semi-reclaimed sodic soil

Treatment combination	Urea N loss per split			Total urea N lost	% Urea N loss	pH
	1st	2nd	3rd			
Control (No fertilizer application)	1.23	-	-	1.23	-	8.56
N_{120}	8.49	8.21	6.76	23.46	19.55	8.49
N_{120} P_{22}	8.28	7.35	6.70	22.33	18.61	8.48
N_{120} P_{22} K_{42}	8.14	7.24	6.65	21.75	18.13	8.45
N_{120} P_{22} K_{42} + GM	5.82	5.20	5.06	16.08	13.40	8.10
N_{120} P_{22} K_{42} + FYM	6.73	5.74	5.28	17.75	14.79	8.15
N_{180} P_{39} K_{63}	12.12	10.60	9.48	32.20	17.89	8.49
Mean	8.26	7.39	6.66			
CD(P=0.05) (i)Treatments	0.51	0.91	1.19			
(ii) Stage of Urea application	0.32					

Source: Yaduvanshi (2001)

Continuous cropping

Since most sodic soils have sufficient amount of salts, leaving the land fallow for a long time results in upward movement of salts. Once the reclamation process begins, the land should be cropped continuously without long gaps between the crops, it may result in resodification of the reclaimed lands.

Soil Management/Land Forming

Land forming is a general term that includes wide ranging activities such as laser grading, land planing and activities that involve moving soil from one area to another. The land forming methods vary with the aim for which land forming is practised. To produce an even surface of a desired grade helps to make irrigation more efficient. For rain water management, probably more efforts are needed depending upon the average annual rainfall and likely intensities. It may include raised and sunken beds, construction of on-farm ponds etc.

Salts move from wet to dry areas, is well known. As such, land forming has been extensively used to manage saline lands. Ridge and furrow method of cultivation is largely used as a management option to manage saline soils. But unlike soluble salts, considerable variations in the exchangeable sodium status of soils are not likely to occur through land forming alone during a crop season. Therefore, only few studies have been made to explore the usefulness of land forming in sodic soils.

The distribution of ESP in the soil profile also governs the kind of land forming practice that needs to be adopted. In a soil profile, if topsoil has less ESP than the bottom soil, even land grading need to be practised cautiously to ensure that cuts do not expose the high ESP subsoil. Top soiling, the process by which cut topsoil is set aside during land forming and replaced on heavily cut areas, is practised under such situations. If the top soil has high ESP than subsoil, deep tillage with inversion can be adopted to manage such soils.

In a trial on sugar beet on highly sodic soil, sowing the seeds on one side of the ridge halfway between the top and bottom opposite to the sun direction gave significantly higher yield than in flat method (Table 4.21). Since the soil was saline-sodic in nature (pH_2 10.5 and EC_e 6.36 dS m^{-1}), the benefit might have occurred due to low salinity rather than managed sodicity.

Table 4.21 Effect of method of sowing on sugar beet yield

Treatment	Plant survival (%)	Yield (t ha^{-1})
Flat sowing	37.5	12.8
Ridge sowing	71.6	19.4
Alternate ridge sowing	67.8	15.8
Transplanted on flat surface	58.2	13.4
Transplanted on ridges	84.3	13.7

Raised and sunken bed technology has been found useful in a rain fed sodic Vertisols where rain is insufficient to meet the water requirement of the rice crop but causes water stagnation in upland crops. The intervention improves surface drainage and supplements limited rainwater to cultivate rice in a limited area. The chemical amendment is applied as per the GR calculated for the soil. Any upland crop such as maize or cotton is grown on beds (raised beds) and rice is grown in furrows (sunken beds) (Fig. 4.9). The 1:1 raised to sunken beds ratio is found appropriate although it may change with rainfall and other agro-climatic conditions. This technology is now becoming quite popular even in normal lands of sub-humid and humid regions, where surface stagnation is a perpetual problem (Khandkar *et al.,* 2013).

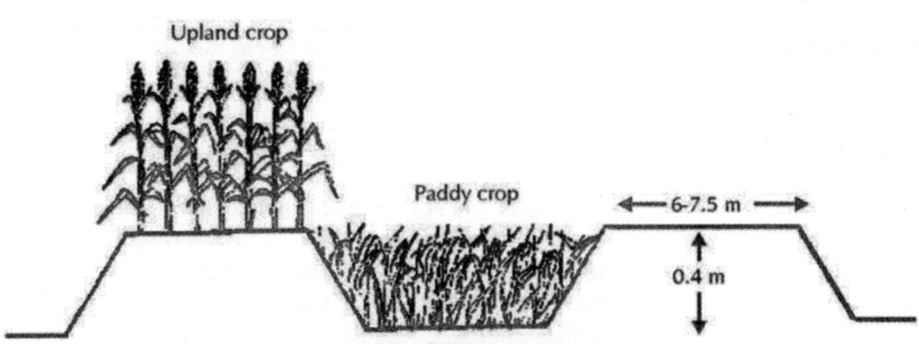

Fig. 4.9 A model to reclaim rain fed black cotton sodic soils (Vertisols)

Water Management

Irrigation water management in sodic soils poses peculiar problems due to clogging action of dispersed soil particles and low stability of the soil aggregates, which affects the water and air permeability. Because of the restricted entry of water, depleted soil moisture is not recharged fully during irrigation. Moreover, the amount of available water held by the soil between field capacity (0.1 bar) and wilting point (15 bars) goes on reducing with increasing ESP (Acharya and Abrol, 1991). In other words, the total water storage capacity of the soil decreases with increase in exchangeable sodium. Moreover, in early stages of reclamation, bulk of the roots is confined to surface 30 cm layer, imposing additional limitations as it restricts the moisture extraction from deeper layers following irrigation. All these constraints adversely impact the ability of a reclaimed sodic soil to supply adequate amount of water to the growing plants. As the depth of irrigation water and interval between two irrigations depends upon the evaporative demand, nature of plant response to soil water stress, depth of root development and the water storage and transmission characteristics of the soil, light and frequent irrigation regime is preferred in sodic soils. The total amount of irrigation water remaining the same; the water should be applied frequently in small quantities. Heavy irrigations can cause temporary water logging and oxygen stress in the root zone, which would affect the crop growth/yield adversely. Laser land levelling helps to achieve shallow and frequent irrigation regime in surface irrigation methods.

During the post-irrigation period also, the water transmission from the lower to the upper layers is severely restricted as a result of low hydraulic conductivity of the soil. Acharya and Abrol (1975) observed that with increasing ESP of the surface soil, the water uptake by *Brassica juncea* was reduced drastically (Table 4.22). Not only that, the contribution of lower layers was low and practically none when the ESP was high. Thus, even if the water content is same, the available water is much less in a alkali than a normal soil (Fig. 4.10).

Table 4.22 Water extraction by *Brassica juncea* from different depths as related to the ESP of the surface soils

Water extraction	ESP of surface 15 cm soil				
	4	11	16	23	18
Extraction rate (cm d^{-1})	0.23	0.24	0.20	0.18	0.14
Soil layer depth (cm)	Per cent of total extraction				
0-15	46.4	40.0	67.5	75.0	78.0
15-30	30.0	31.1	28.0	25.0	22.4
30-45	16.3	18.7	4.5	0	0
45-60	7.3	6.2	0	0	0

*Mean of 17 days extraction period

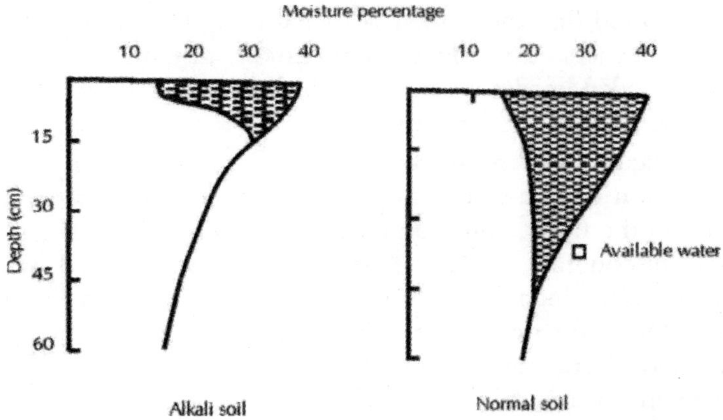

Fig. 4.10 The variation of available water in a alkali and a normal soil

Time of first irrigation to wheat

Sharma *et al.* (1985) reported that compared to 21 days in a normal soil, the first irrigation to wheat should be given at 30 days after sowing in a sodic soil (ESP = 28 to 50). The maximum yield of wheat was recorded when 5 irrigations were given at different crop growth stages in rice-wheat cropping system (Table 4.23). Maximum water use efficiency of 6.5 kg grain mm^{-1} of water was recorded when irrigation was given at 1.0 IW/CPE ratio (Ratio of irrigation water to cumulative pan evaporation) than the other treatments (Sharma *et al.*, 1990). Sharma and Singh (1992) reported that maximum yield of mustard under rice-mustard cropping system in sodic soils is obtained when water stress is avoided at rosette and siliquae formation stages.

Table 4.23 Yield and water use efficiency of wheat as influenced by different treatments in a sodic soil

Treatments	Grain yield (t ha^{-1})	% yield compared to yield with 5 irrigations	Water use efficiency (kg ha^{-1}-mm^{-1})
Rain fed	4.4	14.8	2.65
One (CRI)	11.8	39.7	5.38
Two (CRI+F)	17.2	58.0	6.24
Three (CRI+J+M)	21.6	72.8	6.57
Three (CRI+T+M)	25.0	84.6	7.53
Three (CRI+T+F)	21.9	74.0	6.70
Four (CRI+T+J+M)	27.7	93.5	7.23
Four (CRI+J+F+M)	26.6	88.5	6.96
Five (CRI+T+J+M+D)	29.6	100	6.78
CD at 5 %	1.53	-	0.12

CRI = crown root initiation, T = Tillering, J = Jointing, F = Flowering, B = Booting, M = Milk formation and D = Dough stage

From practical standpoint, Narayana (1980) reported that in tube well irrigated areas where tube well discharge is around 10 L s^{-1}, the borders for rice-wheat cultivation, should be 60-70 m in length and 6-8 m in width. Similarly,

Tyagi (1978) reported that the graded border with a grade of 0.1-0.2% perform well than level borders. The highest application, distribution and water use efficiencies in a graded border is achieved with a cut-off ratio of 70% except for the first irrigation where cut off ratio should be 85%.

A major difficulty with pressurized irrigation techniques in sodic soils has been the cultivation of rice-wheat cropping sequence. Since rice crop required standing water in the fields, adoption of pressurized irrigation such as sprinkler irrigation was discounted. With the spread of direct seeded rice in semi-reclaimed sodic soils, which does not require standing water in the fields, utility of sprinkler irrigation for rice-wheat cropping has been established (Singh *et al.*, 2015, Table 4.24). Water saving in wheat was recorded as 17.1 % with mini-sprinkler than with surface irrigation method.

Table 4.24 Effect of surface and mini-sprinkler on hybrid rice (Arize 6129) yield, irrigation water requirement, water productivity and saving of water and energy

Particulars	Conventional rice transplanting	DSR without wheat residue/ wheat sowing in zero tillage with 100% rice straw mulch	DSR without wheat residue/ wheat sowing in zero tillage with 100% rice straw mulch	DSR with wheat residue incorporation/ wheat sowing in zero tillage with 100% rice straw mulch
Mode of irrigation	**Surface**	**Surface**	**Mini-sprinkler**	**Mini-sprinkler**
Irrigation criteria	1 day after disappearance of water	After small cracks appear	CPE (Alternate day)	CPE (Alternate day)
Grain yield (t ha^{-1})	7.30	7.10	6.70	6.80
Total crop productivity(t ha^{-1})	14.58	13.90	13.45	13.67
Total irrigation water (ha-cm)	112.5	80.0	62.1	62.1
Crop water productivity (kg m^{-3})	1.30	1.75	2.17	2.20
Grain water productivity (kg m^{-3})	0.65	0.89	1.08	1.10
Irrigation water saving (%)	-	29.2	45.1	45.1
Electricity saving (%)	-	28.9	11.7	11.7
NUE (kg kg^{-1} nitrogen)	48.7	47.3	60.9	61.8

Rainfall received 389.7 mm and pan evaporation 565.2 mm during June to September 2014. CPE is the cumulative pan evaporation criteria used for irrigation through mini-sprinkler system, NUE is the nitrogen use efficiency, DSR is direct seeded rice

Nutrient Management

High pH, high exchangeable sodium, high $CaCO_3$, negligible to low organic matter content and inhospitable soil physical environment of sodic soils affect the transformation and availability of native and applied fertilizer nutrients (Gupta and Abrol, 1990). Sodic soils, in general, are deficient in Ca, N and Zn, but are rich in available P and K. Following practices are recommended for efficient nutrients management in sodic soils:

- Because of low available N status and slow transformation of organic and inorganic-N in salt affected soils, added N exhibits poor use efficiency and as a result crops grown in these soils generally respond to higher levels of N than those raised in normal but otherwise similar soil and climatic conditions (Dargan and Gaul, 1974). Therefore, general recommendation has emerged that apply 20 to 25% more nitrogen to rice-wheat in the initial years than recommended for normal soils.
- Rice and wheat plants use nitrogen most efficiently when it is applied at the maximum tillering stage. Rice plants use N around the panicle initiation/jointing stage also. Therefore, split application of N for wheat (1/2 at sowing, remaining 1/2 in two splits at tillering (21 days) and 42 days after sowing and for rice (half at transplanting + 1/4 at tillering + 1/4 at panicle initiation) resulted in maximum efficiency (Dargan *et al.*, 1973, 1975). Other field experiments have shown that maximum yields of rice and wheat were obtained when N was applied in 3 equal splits (Table 4.25), as basal and at 3 and 6 weeks after transplanting/sowing and foliar application of N (3% solution of urea).
- Practise green manuring with *sesbania* every year or at least once in two years to improve organic matter status and to overcome nitrogen deficiency problem.
- The source of nitrogen may have its own influence on growth and yield. Ammonium sulphate proved to be a better source of N for rice and wheat in terms of grain yield, as compared to urea and calcium ammonium nitrate. This was attributed to the beneficial effect of residual acidity of ammonium sulphate. Singh and Sharma (1984) observed that when optimum dose of gypsum is used, source of nitrogen does not make much difference on the rice and wheat yields. Otherwise also, farmers are often attracted towards urea because of the cost differential between the two kinds of fertilizers.
- Since sodic soils are generally rich in P and K; the application of P can be withheld for three to four years. After this, apply P to only rice crop preferably on soil test basis. Singh (1998) reported that rice crop responds to phosphorus when the Olsen's P of the surface soil falls below 7.5 kg ha^{-1}. The wheat yields do not show decline even at this critical level for the next few years. Application of P to either one or both crops significantly increased the grain yield of rice only after 5 years of cropping and of wheat after 11 years. The application of K can be avoided for about 8-10 years.
- Application of Zn plays an ameliorative role in sodic soils by selective absorption enhancing Ca^{++}/Na^{+} and K^{+}/Na^{+} ratios in plants. Field experiments on standardization of Zn requirement of crops showed

that application of 10 kg $ZnSO_4$ ha^{-1} to each crop of rice and wheat or 20 kg $ZnSO_4$ ha^{-1} to rice increased the grain yield significantly over no Zn (Singh and Abrol, 1986). Thus, apply 10 to 20 kg $ZnSO_4$ ha^{-1} for initial two to three years both to rice and wheat crops. Later on, depending upon soil test values, zinc may be applied to rice only.

- Symptoms of manganese (Mn) deficiency have recently been reported in wheat crop grown on reclaimed calcareous and light textured sodic soils. To overcome the deficiency, $MnSO_4$ @ 10 kg ha^{-1} should be applied through soil or 0.5% solution may be sprayed when such symptoms are noticed.

Table 4.25 Effect of splits of N on rice and wheat yields

Method of application		Grain yield (t ha^{-1})			
Rice	Wheat	Rice		Wheat	
		1981	1982	1981-82	1982-83
0+1/2+1/4+1/4	0+1/2+1/4+1/4	5.45	5.46	3.95	3.74
1/4+1/4+1/4+1/4	1/4+1/4+1/4+1/4	6.62	6.22	4.09	4.76
1/3+1/3+1/3	1/3+1/3+1/3	5.69	5.59	3.65	3.93
1/2+1/4+1/4	1/2+1/4+1/4	5.91	5.36	3.72	3.93
CD at P =0.05		0.22	0.31	NS	0.29

Treatments for rice respectively are: Basal+5 days+20 days+40 days, Basal +20 days+40 days + 60 days, Basal +20 days+40 days, same as before

Treatments for wheat respectively are: Basal+1st irrigation+ 2nd irrigation + 3rd irrigation; same as before; Basal+1st irrigation+ 2nd irrigation; same as before

Integrated nutrient management

Disproportionate and inappropriate continuous application of chemical fertilizers has lowered the fertility status and caused nutrient hunger in soils. Crop yields have either stabilized or even declined over the years. Integrated nutrient management comprising of a good blend of organic and chemical fertilizers helps to sustain crop production in reclaimed lands. Besides, farmyard and green manure increase organic carbon and also increase the availability of plant nutrients. FYM also helps to meet a part of the zinc requirement of the crops. Evidences exist to show that application of farmyard manure and/or green manure helps to build-up nitrogen status of the soils and over the years it increases the yields of crops. Experiments in gypsum amended sodic soils revealed that when 50 days old *dhaincha* crop is turned into the soil, it resulted in an increased yield of rice and wheat by 1.48 and 0.67 t ha^{-1} respectively.

Yield of rice and wheat significantly increased with integrated use of green manuring or FYM at 100% recommended dose of chemical fertilizers (120 N, 26 kg P and 42 kg K ha^{-1}) as compared to 100% recommended inorganic fertilizers alone (Table 4.26). The yields in integrated nutrient management treatments were nearly equal to the yield obtained with N_{180} P_{39} K_{36} applied solely through chemical fertilizers.

Table 4.26 Grain yield of rice-wheat on sodic soil with or without integrated use of fertilizers

Treatment	Grain yield (t ha^{-1})									
	1994-95		1995-96		1996-97		1997-98		1998-1999	
	Rice	Wheat	Rice	Wheat	Rice	Wheat	Rice	Wheat	Rice	Wheat
$N_{120}P_{26}K_{42}$	5.57	5.00	5.50	5.16	5.52	5.05	5.57	4.96	5.45	4.87
$N_{120}P_{26}K_{42}$ + GM	6.25	5.30	6.35	5.28	6.98	5.29	6.94	5.57	6.65	5.41
$N_{120}P_{26}K_{42}$ + FYM	6.12	5.15	6.15	5.10	6.56	5.14	6.67	5.25	6.59	5.52
CD (P=0.05)	0.52	0.52	0.58	0.51	0.56	0.40	0.56	0.50	0.33	0.53

Rainwater Management

Rainwater is the most precious source of water in the management of salt affected soils. Rainwater management plays a very significant role in conserving good quality water so as to reduce dependence on ground water for irrigation. Three tier rainwater management technology developed at CSSRI, Karnal can be implemented as it also ensures timely and adequate surface drainage besides providing other benefits. The technology consists of the following three components:

- Rainwater is allowed to accumulate and remain on the cropland till such time and extent that it does not adversely affect the crop (Fig. 4.11 left). It significantly reduces the water required to cultivate rice and also minimizes the drainage requirement.
- Excess water from the fields is led to the dugout ponds or natural depressions of sufficient capacity (Fig. 4.11 right). The stored water is reused during dry spells or for irrigating winter season crops. Later on it could also be developed as a multi-enterprise farm with aquaculture. It helps improve the livelihood security and provides high and regular income to the farmers.
- Rainwater in excess of these two compartments of storage is led out of the area to the natural drainage system through tertiary and main drains or is used for ground water recharge (Fig. 4.4).

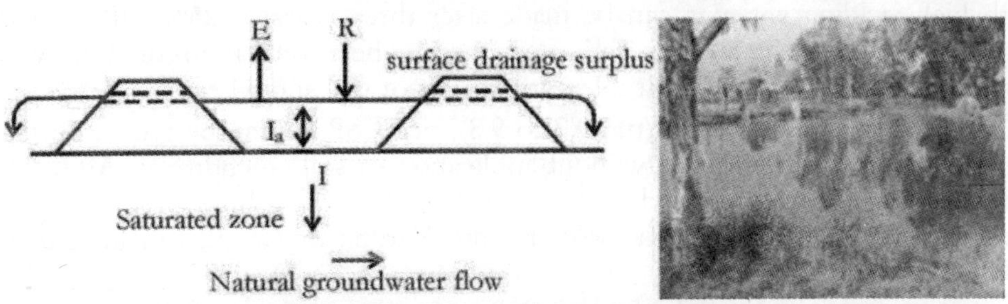

Fig. 4.11 A view of the provision for rainwater drainage in rice fields (left) and excess rainwater stored in on-farm pond (right)

PHYTOREMEDIATION

Synergy of the chemical amendments and the salt tolerant crop varieties can be utilized to phytoremediate a sodic soil. The actual strategy depends upon the initial ESP of the soils. For highly sodic soils ($pH_2 > 10.0$), growing salt tolerant grasses is attractive proposition (See Chapter 5 also). While Karnal grass proliferates under high ESP and initiates the reclamation process, para grass is equally good alternate to reclaim sodic lands. The results revealed that after growing Karnal and para grasses for 1 year, satisfactory yield of rice can be obtained. After 2 years, even good yield of wheat is possible. During these years, good yield of Karnal and para grasses is also obtained for use as fodder (Table 4.27). Although slow and less efficient, it can be adopted by resource poor farmers who cannot afford high investments on reclamation program.

The other approach is to cultivate CSR10 variety of rice during *kharif* and sugar beet during the *rabi* season (rice-sugar beet cropping sequence) for three years with small doses of an amendment or even without it. The cropping of the land initiates a self reclamation process through dissolution of native $CaCO_3$ by biological action of plant roots. Within three years, the soil pH comes down from 10.0 to 9.3. In the fourth year, any other high yielding variety of rice in *kharif* and wheat (KRL 1-4) or raya CS 52 can be grown successfully.

Table 4.27 Yield of Karnal and para grasses grown in sodic soil without any amendment

Grass	Year 1	Year 2	Year 3
Karnal	26.9	28.4	25.4
Para	21.9	29.3	31.5

The part data reproduced from Kumar and Abrol (1986)

Even salt tolerant varieties of rice, wheat and mustard can be grown with a small dose of about 50% of the recommended dose of the amendment. The data in Table 4.28 reveal that salt tolerant varieties gave higher yields at low doses of gypsum while normal high yielding varieties yielded higher at higher doses of amendments. The long-term results revealed that salt tolerant rice varieties can be grown up to two years with 25% GR. Later on these varieties can be replaced by other high yielding varieties. The replacement of salt tolerant wheat varieties with high yielding varieties can be made after three years in 25% GR treated plots. The economic analysis indicated that highest B:C ratio of 1.70 was obtained with 25% GR and salt tolerant varieties of rice and wheat.

In moderately sodic soils (pH 9.2 to 9.8), rice CSR 10 can be grown during *kharif* season as the first crop without application of soil amendment. After the rice crop, the soil pH comes down to a level so that wheat KRL 1–4 or Indian mustard varieties such as Pusa bold, Kranti, Varuna or CS 52 can be grown during the *rabi* season.

Table 4.28 Effect of gypsum levels and varieties on grain yield (t ha^{-1}) of rice and wheat

Gypsum levels (% GR)	Rice			Wheat		
	CSR 13	Pant 4	Mean	KRL 19	PBW 343	Mean
0	1.03	0.64	0.83	0.64	0.50	0.57
15	3.48	2.93	3.20	1.90	1.62	1.76
25	4.71	4.56	4.63	3.03	3.18	3.15
50	4.97	5.02	4.96	3.46	3.67	3.51
Mean	3.54	3.29		2.28	2.22	

RESODIFICATION OF RECLAIMED SODIC LANDS

With the initiation of land reclamation process and because of the roots' activities especially rice, the pH of the soil both at the macro and micro levels decreases sharply. These processes help in self reclamation of the soils both at top as well as down through the profile. As such, gypsum is applied only once during the first year with no further addition of gypsum in later years. On the contrary, many farmers have reported resodification of the reclaimed soils. In Uttar Pradesh, large areas have been reported to have lost their productivity because of the resodification. Extensive surveys have revealed that such a situation has arisen mainly due to mismanagement of the reclaimed lands or due to the addition of Na from outside sources. On the basis of these surveys, following reasons are attributed to the resodification:

- Entry of runoff from unreclaimed sodic lands
- Use of high RSC water for irrigation
- Shallow water table more so of high residual sodium carbonate
- Fallowing of the reclaimed land for prolonged periods.

It has emerged that in Haryana, use of high RSC water and in Uttar Pradesh shallow water table and leaving the land fallow are major factors causing resodification.

5

ALTERNATIVE LAND USE

Current pace of land reclamation is not commensurate with the land degradation rate. As a result projections have been made that the salt affected areas will continue to increase. Studies have been made to understand the limitations of the land reclamation programs as well as to suggest other practices that might help to increase the land reclamation rates. One of the major finding has been that a sizeable area of sodic lands is categorized as village community lands, government lands reserved for specific purposes, lands lying abandoned along the canals, railway tracks and roads and absentee lands. Reclamation of such areas for growing annual crops like rice and wheat may pose difficulties because of common property rights. One or a combination of the following three measures can be adopted for the reclamation and greening of these lands.

- Appropriate choice of crops including fodder crops, grasses, forestry and tree species
- Appropriate choice of cropping and agro forestry systems
- Other crops and cultivars such as medicinal and aromatic plants, economically important halophytes aquaculture and bio-fuel crops

Cultivation of trees and grasses on such lands appears promising options to produce food, fuel and forage on one hand and to increase forest cover, moderate climate change and other issues of environmental concerns on the other. During last four decades, few viable alternate management options/ technologies have been developed which can be implemented in different agro-eco regions of the country albeit some location specific modifications. The package of practices includes identification of suitable grass, forest and fruit trees, medicinal/aromatic plants, halophytes, planting techniques, filling mixture composition and selection of a suitable silvi-pasture/agro-forestry model. Studies have also been conducted to assess the efficiencies of these alternate practices in soil amelioration, their social acceptance, benefit cost ratio and environmental impacts. All these issues form the subject matter of discussion in this chapter

FORAGE GRASSES AND CROPS

Currently, India is facing a net deficit of 61.1% green fodder, 21.9% dry crop residues and 64% feeds. Due to ever-increasing population pressure of human beings, arable lands are mainly used for food and cash crops, leaving little scope of diversifying good-quality arable lands for fodder production, unless milk production becomes remunerative to the farmer as compared to other crops. Therefore, wastelands especially salt affected soils present an opportunity for cultivating fodder crops.

Distinct sodicity tolerant grasses have been identified through ecological studies on undisturbed sites in Karnal and Kurukshetra districts of Haryana. The grass species identified are: *Diplachne fusca, Sporobolus diander, Sporobolus marginatus, Chloris barbata, Kochia indica, Desmostachya bipinata* and *Cynodon dactylon.* The occurrence of these grasses in native sodic soils shows their potential for cultivating these grasses in these soils possibly without or with limited use of amendments and/or fertilizer and manure.

Grasses without Amendment and Fertilizers

The native grass, *Displachne fusca* (Australian grass) or now well known in India as Karnal grass with two cultivated grass species *Panicum antidotale* (Blue panic) and *Brachiaria mutica* (Para grass) were tried in a sodic soil. These were planted alone as well as in a mixture on a soil with average pH_2 of 9.8 and ESP of 71 in the 0-15 cm layer. The para grass yielded the maximum green fodder followed by para grass + Karnal grass mixture (Table 5.1). Though it was anticipated that the combination might give higher yield because of the creeping nature of the para grass and erect nature of Karnal grass but it didnot happen probably because of limited nutritional and soil factors. The other sodicity tolerant grasses are: *Sporobolus diander (ren ki ghas)* and *Cynodon dactylon (doob)* among the native grasses and *Chloris gayana (Rhodes grass)* among the cultivated grasses. In a pot house trial, comparative evaluation of 5 grass species at various ESP levels revealed that performance of Rhodes grass was the best at all ESP levels (Table 5.2).

Table 5.1 Forage yield of three grasses in monoculture and mixed planting under sodic conditions (pH_2 = 9.8, ESP 71)

Grass species	Green fodder yield (t ha^{-1})			
	1st cut	2nd cut	3rd cut	Total
Para	16.3	18.6	13.1	48.0
Blue panic	11.6	14.8	5.4	31.8
Karnal	9.4	11.9	4.0	25.3
Para + Blue panic	10.6	13.8	8.2	32.3
Para grass + Karnal	15.4	18.9	13.0	47.3
Blue panic + Karnal	11.3	15.8	5.1	32.2

Source: Chandra (1979)

Grasses with Amendments and Fertilizers

Several sodicity tolerant grasses can be cultivated on barren sodic soils with pH_2 as high as 10.6 albeit with slightly more efforts and inputs. Major operations are: land levelling and *bunding* of fields to conserve rain water, application of amendments and assured irrigation water for leaching. The latter two processes will help to partially neutralize excess exchangeable sodium. The effect of gypsum application on a sodic soil with pH_2 10.6 on the performance of 5 perennial forage grasses is summarized in Table 5.3. Blue panic and *Panicum leavifolium* had significantly lower forage yield than the Rhodes and Karnal grasses. Rhodes grass gave the highest green forage yield, which was significantly higher than all other grasses. The green forage yield of *Panicum laevifolium* and Blue panic improved with each increment of the gypsum level. In general, there was an appreciable improvement in the forage yield of all the grasses during second (1980) as compared to first year (1979).

Table 5.2 Green yield (g $plot^{-1}$) of 5 grasses at different ESP levels

Grass	ESP					Mean
	5	25	64	78	85	
Marval	307	281	198	178	173	228
Rhodes	488	415	409	380	268	392
Bermuda	283	271	248	241	202	249
Pangola	176	142	113	76	49	111
Karnal	270	210	214	208	206	222
Mean	305	264	236	216	180	80

CD at 5%, grasses or ESP = 23.5, Interaction = 5.14. Source: Kumar *et al.* (1980)

Table 5.3 Green forage yield (t ha^{-1}) of 5 grasses as affected by gypsum application

Grass	Year	Gypsum level (t ha^{-1})			Mean
		0	5.2	10.4	
Karnal	1979	18.8	19.9	22.2	20.2
	1980	51.6	48.9	47.9	49.5
Panicum laevifolium	1979	2.5	9.1	11.8	7.8
	1980	13.9	57.8	67.2	46.2
Blue panic	1979	3.68	9.0	11.0	7.9
	1980	1.5	50.2	60.3	42.3
Rhodes	1979	12.5	27.1	24.2	21.3
	1980	70.7	82.5	86.0	79.3
Coastal Bermuda	1979	4.0	6.9	8.0	6.3
	1980	28.3	33.4	34.0	32.0
CD (p = 0.05)	1979	2.6			
Grasses	1980	10.5			
Gypsum levels	1979	2.2			
	1980	2.6			

Source: Kumar and Abrol (1983)

Later observations in the experiment cited in the foregoing paragraph revealed that the yield of Karnal grass declined progressively after the second year in plots where gypsum was applied. It is attributed to land reclamation. The yield of the Rhodes grass also declined with time after the second year but at a much slower rate than the Karnal grass. Coastal Bermuda did not show much variability in its yield, after second year. *Panicum laevifolium* and Blue panic had almost similar growth pattern throughout with maximum production in the third year. The Karnal grass, Rhoades grass, Bermuda grass and para grass are therefore, the four promising grasses for large scale utilization of sodic lands.

Characteristics of some Sodicity Tolerant Grasses

Karnal grass (*Diplachne fusca* Linn. P. Beauv. Or *Leptochloa fusca*)

Karnal grass also known as Australian *ghas* and Kallar *ghas* is most tolerant to sodic conditions. In Uttar Pradesh (India), this grass is locally known as *Chhoti dandar* or *Narri*. It comes on its own at relatively lower elevations in sodic fields where monsoon rain water stands for comparatively longer duration (Fig. 5.1). It is perennial, erect type and grows as high as 75 to 100 cm. The grass can be propagated by root slips or stem cuttings or by seeds although it is difficult to collect its seeds. Once the grass is established, replanting is not required as the fallen seeds germinate in due course of time. The grass yields 40-50 t ha^{-1} green matter at ESP levels as high as 60 with minimal reductions even at the higher ESPs. Irrigation is necessary after planting and even subsequently except during monsoon season in semiarid regions. Drainage requirement is minimal as it has well developed aerenchyma in the roots (Qureshi and Barrett-Lennard, 1998). Rain water can be stored in *bunded* fields for as much as 8 days without losing the yield. It is capable to fix nitrogen in the root zone. Karnal grass is suitable to feed to the cattle as it contains almost all the elements and sufficient salts for the animal. It is very effective for bio reclamation of sodic lands. Other uses of this grass are: substrate for growing mushroom, production of methane (bio-gas) and production of ethanol (Mailk *et al.*, 1986).

Fig. 5.1 A view of *Leptochloa fusca* (Karnal grass) growth under natural sodic environment

Rhodes grass (*Chloris gayana* Kunth.)

The grass is an erect type and grows as high as 1 to 1.5 m. Initially, it gets affected by sodic conditions but once established, it grows well. It requires less number of irrigations since the grass is able to tolerate aridity and high temperature. Rhodes grass can be propagated by root slips or by seeds. In sodic soil, propagation by root slips is preferred because surface crust formed at the soil surface hinders seed germination. It is planted in the month of July without any amendment but immediately after planting irrigation is necessary. In the first year it yields 15-20 t ha^{-1} green matter which increases to 65-75 t ha^{-1} in the second year. One of the disadvantages of this grass is that it cannot tolerate water stagnation commonly occurring in sodic soils especially during monsoon season. Therefore, to get best results provision of good surface drainage must be made.

Bermuda grass (*Cynodon dactylon* Linn. Pers)

This grass is commonly known as *doob* in India or *Khabbal grass* in Pakistan and is commonly grown in pastures and soil conservation programs (Qureshi and Barrett-Lennard, 1998). Bermuda grass is moderately tolerant to sodicity. It is an early volunteer during reclamation process (It can tolerate ESP up to 50, but at 80 ESP its yield decreases to about 60% at an ESP of 80 compared to its yield at ESP 50 or less, being around 50 t ha^{-1}. It is a perennial grass, with creeping habit and rooting from many nodes. The grass is propagated by seeds as well as by root slips. In sodic soils, root slip method is preferred. During summer months, irrigation at an interval of 10 to 15 days is needed. Similar to Rhodes grass, Bermuda grass also requires a well laid out surface drainage system since it is also sensitive to water stagnation. Yield reduction to the extent of 50% occurs if water stands in the grassed land for 8 days. The grass is not only good for bio reclamation but also helps to reduce soil erosion, which is quite common in sodic soils being dispersed in nature.

Para grass (*Brachiaria mutica* (Forsk) Stapf.)

This grass is suited to medium sodic conditions and even grows well in standing water. The grass is propagated by root-slips or stem cuttings. Frequent irrigations are needed compared to other grasses, but drainage requirement in the monsoon season is minimal. The grass yields up to 50 t ha^{-1} annually at an ESP of 50.

Relative Tolerance

In general grasses are more tolerant to salts than most other crops. The grasses such as Karnal grass, Rhodes grass, Gatton panic and Bermuda grass can be grown without application of gypsum even up to pH 10.4. The relative tolerance of grasses and forage crops to different levels of soil sodicity is given in Table 5.4. Since the crops cultivated in sodic lands must also be tolerant to water logging especially the surface stagnation, the relative tolerance of few grasses to surface stagnation reveals that yield of para and Karnal grasses increased with increasing duration of stagnation (Table 5.5). Apparently, these two grasses are ideal as drainage requirement is minimal.

Table 5.4 Promising forage grasses/crops for sodic soils

pH_2	Grass/forage crop (Botanical name)
Grasses	
10.4	Karnal grass (*Leptochloa fusca*), Rhodes grass (*Chloris gayana*), Gatton panic (*Panicum maximum*), Bermuda grass (*Cynodon dactylon*)
10.0	Para grass (*Brachiaria mutica*)
9.6	Blue panic (*Panicum antidotale*)
9.2	Hybrid Napier (*Pennisetum purpureum*), *Setaria grass* (*Setaria anceps*)
8.8	Anjan grass (*Cenchrus ciliaris*), Guinea grass (*Panicum maximum*) Deenanath grass (*Pennisetum pedicellatum*)
Forage crops (Rainy season, *kharif*)	
9.6	Sorghum (*Sorghum bicolor*), Sudan grass (*Sorghum sudanence*) *Dhaincha* (*Sesbania aculeate*)
9.2	Teosinte (*Euchlaena americanum*)
8.8	Cowpea (*Vigna unguiculata*), Guar (*Cyamopsis tetragonalob*)
Forage crops (Winter season, *rabi*)	
9.6	Oat (*Avena sativa*), Barley (*Hordeum vulgare*), *Shaftal*/Persian clover (*Trifolium resupinatum*)
9.2	*Berseem* (*Trifolium alexandrinum*)

The pH values at which yield reductions is 50% or less

Table 5.5 Effect of surface stagnation on forage yield* of grasses (g pot^{-1})

Treatment	Para	Karnal	Coastal Bermuda	Bermuda	Rhodes
Control (no flooding)	596	273	383	384	530
Flooding for 2 days	717	311	336	297	552
Flooding for 4 days	775	330	312	279	494
Flooding for 8 days	806	351	194	219	466

*Average of 1978 and 1979, Source: (Kumar and Abrol, 1983)

Phytoremediation

The potential of halophytes in bioremediation has been well illustrated by Qadir and Oster (2004). Since grasses have the potential to grow well under sodic condition without or minimal application of amendment, studies have been conducted to assess their phytoremediation potential. Qadir and Oster (2004) compared the results of 14 experiments where sodic soils were either reclaimed with gypsum application or through vegetative reclamation. Results were slightly in favour of chemical treatments (62% sodicity decrease versus 52%) but in the bioremediation treatments sodicity was reduced throughout the whole root zone whereas gypsum was effective only in the layer where it was applied. Furthermore, the plant action improved soil structure and formed macropores enhancing air and water infiltration. A study conducted at Karnal (India) revealed that Karnal and Para grasses have the great potential to phytoremediate a sodic soil for the following reasons:

- The grasses can be cultivated in highly sodic lands without application of amendments

- The grasses are tolerant to water logging (surface stagnation), which is a common phenomenon in sodic lands
- The lands are reclaimed within 2-3 years (Table 5.6). After 5 years any forage crop can be cultivated on the phytoremediated sodic soil.
- The grasses have high yield and can be used as fodder for the animals.

Table 5.6 Crop yield during and after phytoremediation

Treatment	1979-80		1980-81		1981-82		1982-83		1983-84		*Kharif* 1984	
	R	W	R	W	R	W	R	W	R	W	Maize	Teosinte
R-W (C)	0	0	1.0	0	4.6	0.6	5.7	2.7	5.1	3.8	27.7	37.8
R-W (G)	3.7	2.6	4.5	2.3	5.8	3.4	6.5	3.6	4.8	4.6	34.5	43.6
R-W (P1)	-	-	3.8	0.1	5.8	1.6	6.8	3.3	5.4	4.7	30.4	37.1
R-W (P2)	-	-	-	-	5.3	2.6	5.9	3.4	5.1	4.9	29.8	40.2
R-W (P3)	-	-	-	-	-	-	5.6	3.0	5.9	4.3	26.7	37.2
R-W (K1)	-	-	4.1	0.3	5.4	2.1	6.6	3.6	5.4	4.6	28.3	46.0
R-W (K2)	-	-	-	-	6.1	3.4	5.8	3.8	5.2	5.2	28.9	44.7
R-W (K3)	-	-	-	-	-	-	5.8	3.2	5.5	4.1	23.6	41.7
Cd at P=0.05	-	-	-	-	-	-	NS	0.5	0.5	0.6	4.5	6.0

Source: Kumar and Abrol (1986). R, W, C, G, P and K stands for rice, wheat, control, gypsum, para grass and Karnal grass. 1,2 and 3 indicate para or Karnal grasses were grown for 1, 2 and 3 years

Forage Crops

Forage crop can be grown even in the first year of reclamation following the rice crop. The yields of wheat, *berseem*, *shaftal* and rice (unhusked) under different rotations reveal that inclusion of *berseem* and *shaftal* increased the yield of succeeding rice crop significantly, over the rice-wheat rotation (Table 5.7). Although yield of the fodder crop in the first year is not good, yet overall the rice fodder rotation helped to reclaim the soils better than rice-wheat (Sharma and Mehta, 1981). It is illustrated through higher infiltration rates of sodic lands under different crop rotations. Infiltration rates measured in the field after the harvest of wheat, *berseem* and *shaftal* crops revealed that water infiltrated under rice-*berseem*, rice-*shaftal* and rice-wheat-*dhaincha* rotations was higher than rice-wheat rotation. The basic intake rates of water for rice-wheat. rice-wheat *dhaincha*, rice-*berseem* and rice-*shaftal* were 13.15,17 and 16 mm d^{-1} respectively, whereas the basic intake rate of original sodic soil was only 4 mm d^{-1} (Fig 5.2, Sharma and Mehta, 1981).

Table 5.7 Yield (t ha^{-1}) of different crops under different crop rotation on farmers' field

Rotation	*Rabi*		*Kharif*	
	1979-80	**1980-81**	**1980**	**1981**
Rice-wheat	1.7	1.9	5.1	6.0
Rice-wheat-*dhaincha*	1.7	2.1	6.1	6.1
Rice-*berseem*	18.0	52.8	6.8	7.2
Rice-*shaftal*	24.0	55.8	7.0	7.1
CD)P=0.05)	-	-	0.10	0.65

Reclamation began from *Kharif* 1979

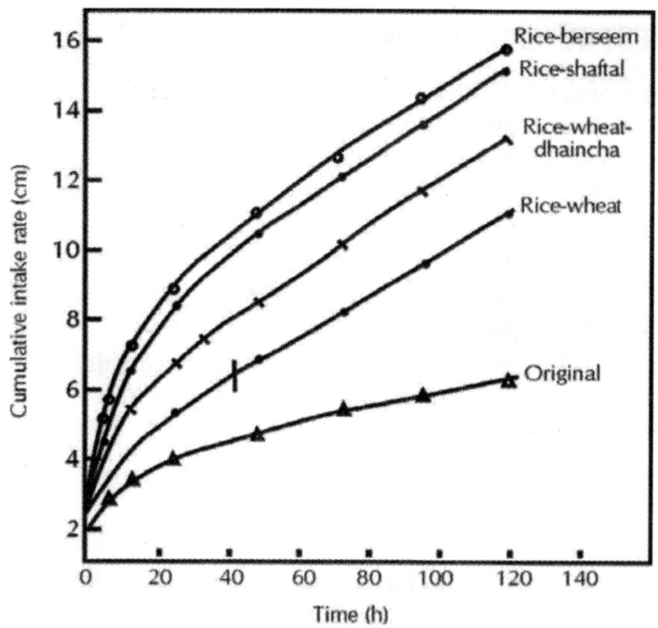

Fig. 5.2 Cumulative infiltration rate following various cropping rotations

Forage crops, such as *berseem*, *shaftal*, oat, sorghum, pearl millet, Napier pearl millet hybrid can be grown in semi-reclaimed sodic lands i.e. 1-2 years after reclamation (Table 5.8). Tyagi *et al.* (1996) also reported that shifting from rice-wheat to rice-forage pattern with addition of animal component in the second year of reclamation provides better livelihood security than the crop based cropping pattern alone. This practice is quite suited to farmers with small parcels of sodic lands (Tyagi *et al.*, 1996). The forage crops can also be grown using sodic water with appropriate neutralization of the residual sodium carbonate. Under sodic water irrigation (EC_{iw}=1.6 dS m^{-1}, RSC=8.6 meq L^{-1}), sodium absorption ratio (SAR) = 9.0 (mmol L^{-1})$^{0.5}$, application of gypsum at 2 t ha^{-1} improved the green forage yield and its quality. Sole crop of Persian clover (95.7 t ha^{-1}) gave significantly higher total green forage yield than oat (45.4 t ha^{-1}) (Kumar *et al.*, 2006).

Table 5.8 Production potential of different forage crops in sodic soils

Crops	Varieties	Green fodder yield (t ha^{-1})
Sorghum	MP chari, PC -6, PC -9, UP Chari-2, UP Chari-3	35-40
Pearl millet	Giant bajra, Raj bajra Chari-2	35-45
Napier peral millet hybrid	IGFRI-6	80-100
Oat	Kent, JHO-822	35-45
Berseem	Mescavi, Bundel *berseem* -3, Wardan	60-80
Shaftal	SH-48	50-70

Source: Sunil Kumar *et al.* (2012)

AGROFORESTRY ON SODIC LANDS

Sodic soils not amenable to cropping can be usefully exploited through growing trees. Besides meeting the wood requirement such an approach offers an opportunity to increase the forest cover of the country, which presently is far below the minimum requirement of one third of the total land surface. Afforestation in these barren lands would also contribute to the maintenance of healthy agro-ecological system and averting the environment hazards. Village community (*Panchayat*) lands and sodic lands with undulating topography offer enormous scope for this purpose.

The reclamation process for agro-forestry or horticultural plantations differs significantly than the package discussed for arable cropping. The two kinds of commonly adopted packages are:

- Reclamation for arable cropping and switchover from arable to horticultural plantations
- Spot reclamation for agro-forestry and/or horticultural plantations

While the first component has been highlighted in Chapter 3, only the second kind of reclamation process is discussed in this chapter. Irrespective of the package, any successful plantation should take into account the following:

- Good survival percentage and rapid early growth of saplings
- Choice of a right species matching the planting sites
- Adequate site preparation
- Application of fertilizer, if necessary
- Irrigation during initial stages
- Weed control and protection against insects/pests, disease, etc.
- Protection from animals.

For successful tree plantation in sodic environment, proper method of site preparation is of utmost importance. Some principles of good site preparation are:

- Production of a loose soil amenable to rapid proliferation of the root system

- Modification of root environment to retain and supply moisture to plants
- The addition of chemical amendments only around the plant than in the entire area
- Application of nutrients like nitrogen, zinc, etc
- Perforation of hard pan in the sub-soil if any.

Depending upon the sodicity, topography and resources, one of the following techniques can be successfully used for tree plantation.

- Pit method
- Auger hole method
- Combination of pit and auger hole methods
- Ridge plantation
- Trench plantation

Pit Method

Depending upon the type of soil and the depth of the *kankar* layer, pits of different sizes are used to plant trees. If necessary, pits are dug to a depth of about 1 m, so that the *kankar* layer if any is broken (Fig. 5.3). The soil of the pits is either replaced by a good soil from nearby place or amended suitably by gypsum or other amendments. FYM and basal dose of nutrients is also added at this stage. The deep pits having good soil or soil amended with addition of gypsum and FYM are found to give better results (Yadav, 1980).

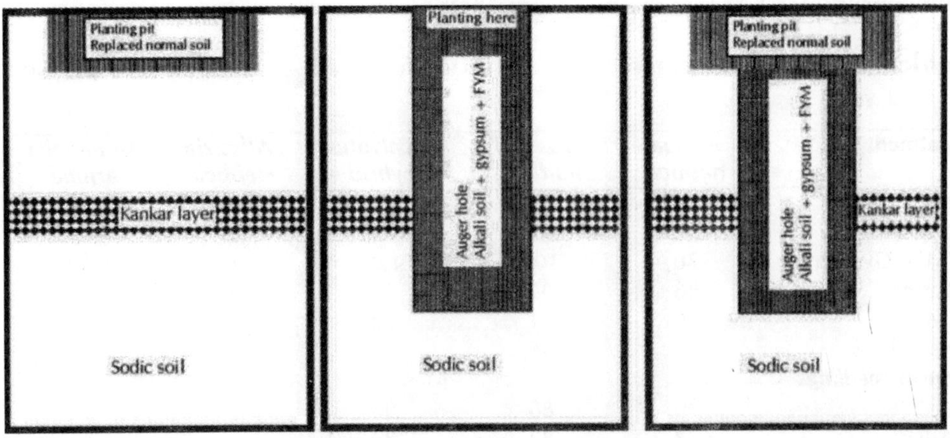

Fig. 5.3 Planting methods in sodic soils

Eucalyptus terticornis, Prosopis juliflora and *Acacia nilotica* were planted in pits 60 cm x 60 cm x 60 cm filled with good soil. *Eucalyptus hybrid, Acacia nilotica* and *Albizzia lebbek* were also planted in similar pits after treating the original soil of each pit with 5 kg gypsum + 20 kg FYM. The survival percentage in the case of *Eucalyptus* hybrid and *Acacia nilotica* was 60%, while it was only 16% in *Albizzia lebbek*. Replacement of the original soil with good soil or the original soil treated with gypsum yielded almost similar results. *Prosopis juliflora* performed better compared to *Acacia nilotica* (Singh and Singh, 1993).

The pits (90 cm *90 cm * 90 cm) dug in sodic lands (pH_2 10.0) before the onset of rainy season were used to plant trees after the treatments: (1) original soil (O.S) (2) O.S. + FYM @ 25 kg pit^{-1} (3) O.S. + gypsum @ 50% GR (3.5 kg pit^{-1}) (4) O.S. + gypsum @ 50% GR + FYM and (5) replacement of original soil by normal soil. Two types of seedlings of each species viz. small and tall were used for transplanting. The small seedlings represented those raised in the polyethylene bags and the tall seedlings represented those, which were first raised in polyethylene bags and subsequently allowed to grow in a temporary nursery bed. An additional treatment of fertilizer application (N and P) was also superimposed. The survival percentage of the planted seedlings was the lowest in the control (Yadav *et al.*, 1972). Though the treatment of FYM gave the higher survival percentage than the control, application of gypsum resulted in more than 90% survival in case of *Albizzia lebbek*, *Eucalyptus* hybrid, and *Prosopis juliflora*. Gypsum + FYM proved as good as the soil replacement and were at par. *Eucalyptus* hybrid and *Prosopis juliflora* attained maximum height as compared to remaining species (Singh *et al.*, 1993). In the untreated soil, performance of *Prosopis juliflora* and *Acacia nilotica* was better than other three species. The height increment was greater in case of plants raised from the tall seedlings as compared to those from the small seedlings. The fertilizer application increased the height, the magnitude being maximum in case of *Eucalyptus* hybrid (Table 5.9). It was concluded that the species like *Eucalyptus* hybrid, *Prosopis juliflora* and *Acacia nilotica* can be grown successfully by treating the soil of the planting pit with gypsum + FYM (Singh and Dagar, 2005).

Table 5.9 Height increment (cm) of different forest species in a sodic soil (1 year)

Treatment	*Acacia nilotica*	*Prosopis juliflora*	*Eucalyptus hybrid*	*Albizzia lebbek*	*Terminalia arjuna*
Soils					
Normal	107	100	119	87	98
FYM + Gypsum	76	102	120	71	95
Gypsum	88	92	91	62	85
FYM	76	81	62	46	68
Original	40	46	33	15	25
Type of seedlings					
Tall	90	88	93	58	77
Small	77	82	83	55	72
Fertilizer					
Fertilized	91	94	103	57	80
Un-fertilized	95	76	74	56	69

Auger Hole Technique

This technique of tree plantation in sodic soils has been devised keeping in view that a hard *kankar* layer is always encountered at about 1 to 1.5 m depth. Thus, to pierce this layer, an auger hole is made from 1.2 m to 1.8 m depth. The diameter of the auger hole is about 15 cm (Gill and Abrol, 1985). A manually or mechanically operated auger is used (Fig. 5.4). Auger holes are refilled with either the normal soil from adjoining areas or properly amending the original sodic soil before planting the saplings. This technique is cheaper than the pit method as it involves reclamation of a very small percentage of the soil and aims at training the roots to grow deeper into the soil where relatively better soil environment exist. This technique was first described by Agarwal and Gupta (1968) but has recently been adopted on large scale in tree plantation at the Central Soil Salinity Research Institute and Forest Department of Haryana. Initially deep auger holes (150 cm) have a distinct advantage over the pit method or a shallow auger hole (90 cm) in which *kankar* layer is not pierced through (Table 5.10).

Table 5.10 Mean biomass production after one year of planting in shallow and deep auger holes on oven dry basis (g)

Tree species	Flat surface plantation				Ridge plantation			
	With grass		Without grass		With grass		Without grass	
	Shallow	Deep	Shallow	Deep	Shallow	Deep	Shallow	Deep
Eucalyptus	160	860	125	667	342	897	314	816
Acacia	80	697	90	540	212	925	186	857
Parkinsonia	124	714	120	585	265	878	226	757

Source: Grewal and Abrol (1983)

Fig. 5.4 A view of the tractor operated auger for making auger holes (left) and a view of tree plantations (right)

Mixing of 2-3 kg gypsum and 8 kg FYM with the original sodic soil per post hole of 15 cm diameter and 150 cm depth resulted in good survival and growth of selected species as with 12 kg gypsum and 24 kg FYM in conventional pit method of planting (Table 5.11). *Acacia nilotica* (Babul), *Prosopis juliflora* (Mesquite), *Casuarina equisetifolia* (Jor-tor) and *Sesbania* (*Dhaincha*) have shown promising results when planted in a highly sodic soils using auger hole technique. The moderately good species are *Albizzia labbek (Siris), Azadirachta indica (Neem)* and *Eucalyptus tereticornis.* However, it was observed that later on the growth of *Eucalyptus* plants gets restricted. Visual symptoms like yellowing or chlorosis indicated possibility of nutritional disorders.

Table 5.11 Survival (%) and growth parameters of *Eucalyptus tereticornis* and *Acacia nilotica* after 7 years of planting in a sodic soil

Post hole filling mixture O. S. +G+FYM (kg)	Size of the post hole (cm x cm)	*Eucalyptus* Survival (%)	*Eucalyptus* Height (cm)	*Eucalyptus* DSH (mm)	*Acacia* Survival (%)	*Acacia* Height (cm)	*Acacia* DSH (mm)
3+4	10x120	100	822	138	100	659	140
3+6	10x180	100	884	158	100	666	155
3+8	15x120	88	838	144	100	680	153
3+8	15x180	100	902	163	100	733	164
12+24	90x90	100	921	167	100	741	177
Mean		98	873	154	100	696	158

O. S. = Original soil (pH_2 10.3-10.5); G = gypsum, FYM= farm yard manure.
DSH = Diameter at stump height; Source: Gill and Abrol (1986)

Combination of Pit and Auger Hole Methods

This technique uses a combination of pits of 60 cm x 60 cm x 60 cm dimensions with auger holes of 20-25 cm diameter and 140-160 cm depth made in the center of these pits that pierces through the *Kankar* pan. These pits cum auger holes are filled back with a mixture of either the normal soil or the original soil + 8 to 10 kg gypsum + 10 kg FYM + 20 kg river sand. The planting is done in the center of the pit over the auger hole.

Amongst the 3 methods, auger hole technique has the following advantages:

- The cost of plantation is less, as small amount of amendment is required
- The process of planting can be mechanized such that planting can be speeded up
- The *kankar* layer can be pierced to facilitate root penetration to greater depths.

While auger hole technique has several plus points, it has some limitations as well. This technique is useful only when carried out mechanically. Making auger holes manually is a tedious task and requires large labour force. It is also observed that the initial advantage gained with this method in the first 2-3 years is evened out in later years. This was particularly observed in the case of *Eucalyptus* plantations. It seems that roots proliferation in this case is vertical and the lower soil layers which are deficient in the nutrients are unable to support the fast growth of this tree. Moreover, in the vicinity around the auger hole, the pH remains higher than the auger hole and therefore, root proliferation in the horizontal direction is restricted.

Ridge Plantation Technique

Runoff potential of sodic lands is quite high. The run-off from these barren lands results in flooding of low-lying areas downstream. Thus, a technique to conserve rainwater within the planted area has been devised. In this technique, ridges of different sizes, shapes and spaces are made depending upon the location specific situation of the planting sites. Plant saplings are planted on the ridges while trenches are used to conserve rainwater. The biomass of the plants was much higher in the ridge than the flat method of planting (Table 5.10). The biomass was higher in deep than shallow auger holes in the case of both the methods. The cost of ridge making and the overall economics would determine the application of this technology by the forest department.

Trench Plantation Technique

In this technique, trenches of different sizes and shapes are opened either manually or mechanically. Trenches are refilled by adding suitable amendment in the soil. The plants are then planted in these trenches. It was observed that mesquite can be successfully established with this technique. (Singh and Abrol, 1986; Singh *et al.*, 1988). Mesquite performed better in the case of no grass compared to with grass when Karnal grass was grown in association with mesquite (Table 5.12 and Fig. 5.5). The technique is inferior to the pit or the auger hole method but is cheaper and easier to apply for large scale plantations.

Table 5.12 Effect of treatments and planting techniques on survival, height, DSH and DBH of mesquite 2 years after plantation

Treatment	Survival (%)	Height (cm)	DSH (mm)	DBH (mm)
With grass	92	294	38.6	13.1
Without grass	92	370	48.0	18.3
CD (p=0.05)	Ns	62	6.9	4.9
Filling mixture				
Original soil	63	153	18.0	5.8
Gypsum	88	323	38.8	13.9
Gypsum + Rice husk	100	394	55.0	21.1
Gypsum + FYM	100	412	54.0	18.9
CD (p=0.05)	24	67	16.8	7.9
Planting technique				
Pit	100	323	47.5	17.2
Auger	100	358	46.5	17.3
Trench	88	349	38.8	13.9
CD (p=0.05)	ns	ns	7.0	ns

Original soil pH_2 = 10.4-10.6 , DBH = Diameter at breast height. Source: Singh *et al.* (1988)

Cultural Practices for Growing Trees

Although the success of tree plantation in sodic lands depends to a great extent on the planting technique, soil preparation, selection of suitable species and use of amendments, cultural practices can help to improve the performance of tree species under such an adverse environment.

Planting time

Generally, there are two planting seasons for woody species-February to March and July to September. The latter is preferred under rain fed or scarce irrigation situations. Under assured irrigation facilities and if the trees are tolerant to cold at the initial establishment stage, the February-March season is preferred. Deciduous trees like *shisham* and poplar should always be planted in this season.

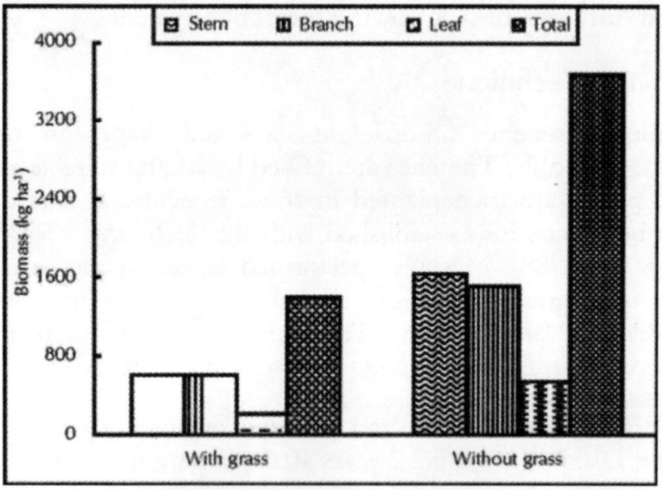

Fig. 5.5 Oven dry biomass of mesquite 720 days after planting

Irrigation

Irrigation especially during the initial establishment stage of the woody species is critical. Fortnightly irrigations for the first three months followed by monthly irrigations up to two years results in significance increase in survival, growth and bio-mass production of some forest tree species (Fig. 5.6). Under normal monsoon rainfall conditions, irrigation may be withdrawn after two years.

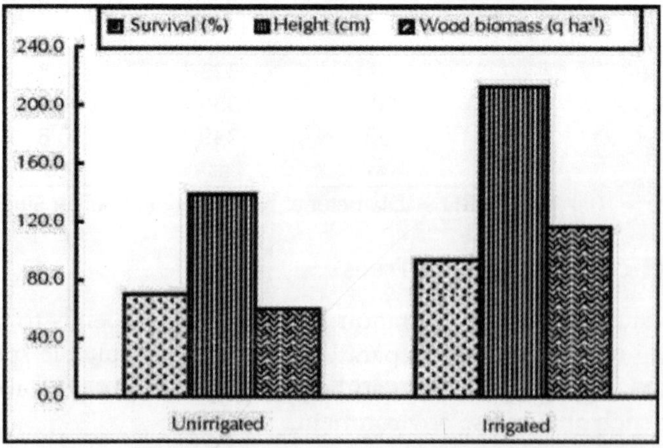

Fig. 5.6 Effect of irrigation on mesquite plants

Plant spacing

If the primary purpose of plantation is firewood production or energy plantation then 2 m x 1 m or 2 m x 2 m spacing should be adopted. In sodic soils, closer planting covers the risk of high mortality. In agro-forestry, the minimum spacing between rows and plant should be kept at 4 m and 2 m, respectively (Fig. 5.7). If the trees are spreading type like *Prosopis, Acacia, Dalbergia* etc. the spacing should not be less than 5 m x 3 m. Further, if the purpose is both firewood and timber production, high density plantation may be preferred. After about 3 or 4 years, 50% of the plants may be harvested for firewood while remaining 50% may be kept for timber production.

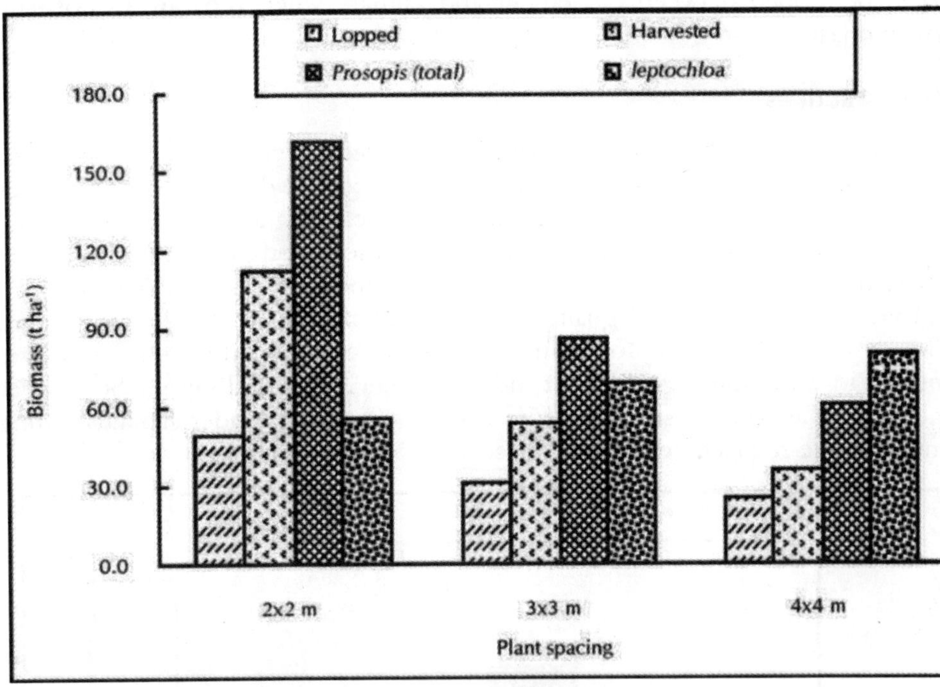

Fig. 5.7 Effect of spacing on *Prosopis* and *Leptochloa* grass

Nutrient application

Sodic soils are generally deficient in N, Ca and Zn. Although N application can be avoided for N-fixing species, application of N to the non-nitrogen fixing trees significantly improves the growth. A basal dose of N (50 g urea $plant^{-1}$) should be mixed in the soil mixture used to fill the auger-holes. Zinc sulphate at the rate of 10 g $auger\text{-}hole^{-1}$ may also be applied at the time of planting. Later, 50 gram urea should be applied in two equal splits every year, one during the monsoon season and the second in February-March. Application of phosphorus and potash is not required as these are present in sufficient quantities. Application of farm-yard manure does help in meeting the iron needs of the woody species.

Pruning/Lopping

Pruning refers to the removal of unwanted side branches of trees which are photosynthetically less active. It is practised to give proper shape to the trees and to enhance the growth of vegetation underneath and inter crops. The biomass received during lopping can be used as firewood. Lopping of *Prosopis juliflora* at 16 and 40 months after planting (at 2 m x 2 m spacing) yielded about 13 t of woody bio-mass ha^{-1} and increased the green forage yield of inter-planted Karnal grass (Fig. 5.8 Singh, 1987). At the first lopping, all the side branches are removed up to one meter height and at the second lopping 2 m stem should be kept free of unwanted side branches. Most of the trees of the tropical regions should be lopped between November and January when they are in dormant growth phase.

Other practices

Tolerance of plants to salinity is mainly influenced by climate, particularly the abundance or lack of rainfall to leach salts, soil type and drainage characteristics within the root zone which influence the ease of leaching and salt accumulation. Other factors include rootstock or variety, irrigation method (surface or flood, overhead sprinkler, drip), stage of plant growth and irrigation management. To improve upon survival and quality, salt tolerant rootstock should be planted, followed by *in-situ* grafting for desired characters. Some of the fruit species are sensitive to water logging and frost such as pomegranate and sapota. So, proper drainage in the rainy season and protection of the young tender saplings during winter is a pre-requisite for the overall success of the plantations.

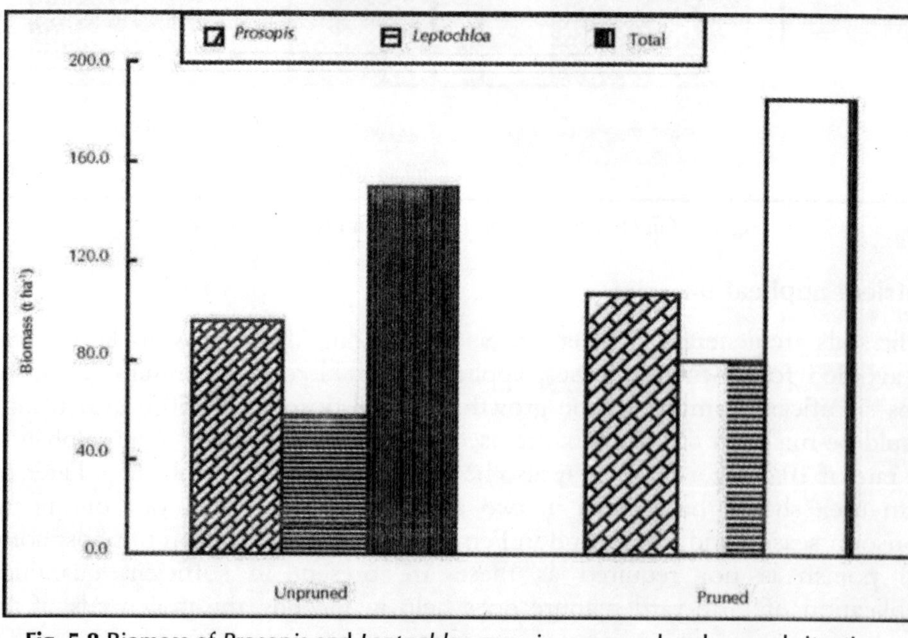

Fig. 5.8 Biomass of *Prosopis* and *Leptochloa* grass in unpruned and pruned situations

Suitable Tree Species for Sodic Soils

Certain forest species tolerant to soil sodicity have been identified. Some species like *Capparis aphylla, Caparis horrida, Salvadora oleoides* and *Zizyphus* sp. occur naturally under moderate sodic conditions (Yadav, 1978, Sharma and Bhargava, 1978). Species like *Acacia nilotica* (Babul), *Prosopis juliflora* (mesquite), *Butea monosperma (Dhak)*, *Tamarix articulata* and *Albizzia lebbek* (Siris) can grow within certain limits in sodic soils (Khan and Yadav, 1962). Among fast growing exotic species, *Leucaena leucocephala,* (Varieties K8 and Fiji) has shown good growth in moderate to heavy sodic soils (Lohani, 1979). *Acacia nilotica* and *Prosopis spicegera* were found to be promising in Punjab. On the basis of available data, some species have been arranged tentatively in descending order of their tolerance to soil sodicity as: *Prosopis juliflora, Acacia nilotica, Eucalyptus* hybrid, *Zizyphus* species, *Capparis decidua, Capparis horrida, Pongamia pinnata, Butea monosperma, Casuarina equisetifolia, Acacia aurtiduliformis, Phyllanthus emblica, Terminalia arjuna, Azadirachta indica, Dalbergia sissoo, Tamarindus indica, Albizzia procera, Albizzia lebbek, Syzygium cumini, Leucaena leucocephala, Populus deltiodes, Ailanthus excelsa* (Singh, *et al.*, 1992). The most promising forest species identified for planting on highly sodic soils are *Pahari Kiker (Prosopis juliflora), kikar (Acacia nilotica), Pharans (Tamarix articulata).* The relative tolerance of different species to soil sodicity is reported in Table 5.13 (Singh *et al.*, 1997a) The most promising species for planting in soils having pH_2 more than 10.0 included: *Pahari Kiker* (*Prosopis juliflora*), *babul* (*Acacia nilotica*), *Frans* (*Tamarix articulata*) and Australian pine (*Casuarina equisetifolia*). The performance of four tree species on highly sodic is shown in Table 5.14.

Table 5.13 Relative tolerance of forest tree species to soil pH/sodicity

Highly tolerant pH_2 > 10.0	**Tolerant pH_2 9.5 -10.0**	**Moderately tolerant pH_2 9.0 – 9.5**	**Sensitive pH_2 < 9.0**
Pahari Kiker *(Prosopis juliflora)*, Kiker *(Acacia nilotica)*, harans *(Tamarix articulata)*, Jor-tor *(Casuarina equisetifolia)*	Jangal jalebi *(Pithecello – bium dulce)*, Safeda *(Eucalyptus tereticornis)*, Arjuna tree *(Terminalia arjuna)*, Imly *(Tamarindus indica)*, Papri *(Pongamia pinnata)*, Khair *(Acacia catechu)*, Dhak *(Butea monosperma)*, Lasura (Cordia rothii), Siris *(Albizzia lebbeck)*, Parkinsonia aculeata	Kigelia *(Kigelia pinnata)*, Shisham *(Dalbergia sissoo)*, Sieamea *(Cassia siamea)*, Neem *(Azadirachta indica)*, Siliver oak *(Grevillea robustra)*, Dhaincha *(Sesbania sesban)*, Bakain *(Melia azedarach)*, Susbabul *(Leucaena leucocephala)*, Saunjana *(Moringa olifera)*, Bamboo *(Morus alba)*, Siris *(Albizia procera)*, Acacia auriculiformis, Acacia senegal	Poplar *(Populus deltoides)*, Teak *(Tectona grandis)*, Ficus virena, Bombax ceiba, Hardwickia binnata, Thespesia populnea, Bauhinia variegata, Lagerstro-emia speciosa, Cadrella toona

Table 5.14 Growth performance and biomass production of salt tolerant trees in a soil of pH_2 10.2 after 10 years of planting

Growth parameters	Trees			
	Prosopis juliflora	***Acacia nilotica***	***Casuarina equisetifolia***	***Eucalyptus tereticornis***
Height (m)	12.9	11.6	14.5	14.9
DSH (cm)	15.9	15.4	15.6	13.6
DBH (cm)	12.5	13.6	12.0	11.0
Bole weight (kg $tree^{-1}$)	112.6	85.4	84.2	65.6
Branches + Leaves weight (kg $tree^{-1}$)	43.2	43.8	28.4	23.5

DSH : Diameter at stump height (5 cm above the ground level);
DBH : Diameter at breast height (137 cm above the ground level)

Agro-forestry for Moderately Sodic or Reclaimed Lands

Most promising trees for agro-forestry on reclaimed sodic lands are*: Populus deltoides* (*Poplar*), *Eucalyptus tereticornis* (*Safeda*), *Acacia nilotica* (*Kikar*) and *Tectona grandis* (*Teak*). The suitable intercrops during establishment stage are rice-*berseem* and rice-wheat. From third year onwards, as tree canopies develop, shade loving species such as *berseem*, guinea grass, turmeric and colcasia perform well. Poplar based agro-forestry is more remunerative because of its faster growth and higher wood price in the market compared to *safeda* and *kikar* based systems. The intercrops have been found to aid the growth of the trees.

Six years results of a study on comparative evaluation of agriculture, forestry and agro-forestry practices in terms of productivity, sustainability and economics in a moderately sodic soil are summarized as follows:

- Poplar and *Eucalyptus* growth was better when grown with intercrops, whereas *acacia* grew better without crops.
- Yields of almost all the intercrops were higher with *Eucalyptus* while minimum with *Acacia.*
- Effect of trees on the fertility build-up in the soil was in the order *Acacia* > poplar >*Eucalyptus* > sole crops.
- The benefit-cost ratio was highest in the case of poplar based and minimum in the case of *Acacia* based system.

Agro-forestry Systems for Commercial Farms

Viewing agro-forestry as a viable proposition, many progressive farmers have adopted trees and inter-crops in the very first year of land reclamation. In this set-up, the trees are planted with standardized auger-hole technique and the inter-space between the trees is reclaimed with amendments like gypsum or pyrites. For good returns and speedy reclamation, rice crop is taken as the first crop as in the case of normal land reclamation. Singh *et al.* (1997b) evaluated this agro-forestry system on a moderately sodic soil ($pH_2 < 9.2$) for 6 years. Three commercial trees viz. Poplar (*Populus deltoides*), Eucalyptus (*E. tereticornis*) and babul (*Acacia nilotica*) were planted either sole or in association with rice-wheat, rice-*berseem*, pigeon pea/sorghum- mustard rotations. Results after 5 years showed that inter-crops of *berseem*, rice, wheat and mustard can be successfully grown along with these trees during initial 3 years (Table 5.15). Later on these crops can be replaced with shade-loving crops such as turmeric. These inter-crops help *Populus* and *Eucalyptus* to grow faster but adversely affect the growth of low water demanding trees like *Acacia*.

Table 5.15 Total grain (G), fodder (F) and rhizome (R) yield from various crop rotations in 5 years

Crop rotation	Yield (t ha $^{-1}$)			
	Eucalyptus	***Acacia***	***Poplar***	**Open**
Rice-wheat/cowpea-*berseem*				
Rice (G)	14.4	12.5	11.8	21.6
Berseem (F)	239.7	212.7	234.6	389.1
Cowpea (F)	18.0	4.6	2.5	45.0
Rice-wheat/ Guinea grass-oat				
Rice (G)	13.4	11.8	10.3	21.0
Wheat (G)	9.0	8.1	8.4	16.0
Guinea grass (F)	18.1	12.3	2.5	30.0
Oats (F)	23.8	24.0	25.6	42.0
Pigeon pea-mustard/turmeric				
Pigeon pea (G)	0.7	0.8	0.2	0.7
Mustard (G)	2.3	1.8	2.0	4.0
Sorghum (G)	18.3	8.6	22.8	50.0
Turmeric (R)	22.3	5.9	8.3	22.1

Source: Singh *et al.* (1997)

Phytoremediation by Trees

In addition to bio-mass production, growing of trees on sodic soils results in their amelioration by improving physical, chemical and biological properties of the soil. The incorporation of organic matter through leaf litter and other foliage results in favourable changes in the physico-chemical properties of the soil (Table 5.16). Within 20 years, the pH of the soil reduced from the original 10.3 to around 9.0 and organic carbon increased around four fold except in *Eucalyptus tereticornis*. The reclamation was higher in *Prosopis juliflora* compared to other tree species. Growing leguminous trees like *Prosopis, Acacia* and *Casuarina* ameliorates these soils at a much faster rate than non-leguminous trees, because of the build-up of organic matter and the recycling of important nutrients. It has also been proved that highly sodic soils reclaimed through tree plantations can be successfully used for growing all kinds of agricultural crops after removing the trees as revealed by the yield in pots filled with the soil reclaimed under the trees (Table 5.16).

Table 5.16 Ameliorative effect of tree plantation on a sodic land

Species	Original soil		After 20 years		Yield (g pot^{-1})	
	pH_2	O.C. (%)	pH_2	O.C. (%)	Wheat	Oat
Eucalyptus tereticornis	10.3	0.12	9.18	0.33	32	43
Acacia nilotica	10.3	0.12	9.03	0.55	56	62
Albizia lebbek	10.3	0.12	8.67	0.47	45	53
Terminalia arjuna	10.3	0.12	8.15	0.58	44	46
Prosopis juliflora	10.3	0.12	8.03	0.58	62	88
Cropped land					13	24

Source: Singh and Dagar (1998)

Fuel Wood-cum-Forage Production for Phytoremediation

In this system, salt tolerant trees like *Pahari Kikar* (*Prosopis juliflora*) and Karnal grass (*Laptochloa fusca*) are grown together without any amendment application. In addition to firewood and forage production, the system reclaims sodic soils to the extent that normal crops such as oats, *senji, berseem* and *shaftal* etc. can be grown after 5-6 years. Large area of salt affected village community lands has been reclaimed by following this bio-amelioration approach (Singh *et al.*, 1993).

HORTICULTURE IN SODIC SOILS

Earlier, it was considered highly difficult to raise fruit trees in highly sodic soils of $pH_2 > 10.0$. But recent experiments at ND University of Agriculture and Technology, Faizabad and CSSRI, Karnal with some salt tolerant species of fruits have shown that there are about half a dozen fruit plants which can be grown in these soils by adopting proper site preparation techniques and by using higher dose of organic and inorganic amendments. The fruit species which can be established successfully in sodic soils are: *ber* (*Zizyphus mauritiana*) guava (*Psydium guajava*), pomegranate (*Punica granatum*), *amla* (*Emblica officinalis*), *karaunda* (*Carrisa carandus*), *imly* (*Tamarindus indica*) and *jamun* (*Syzygium cuminii*). The fruit species, which have shown high degree of salt tolerance and were found promising at the ND University of Technology at Faizabad are reported in Table 5.17.

Table 5.17 Suitable fruit species and their relative tolerance

Fruit species	Tolerance		Suitable varieties
	pH	ESP	
Aonla- Emblica officinalis	9.5	40	Chakaiya, NA-7, NA-6
Guava-*Psidium guajava*	9.2	30	Sardar (L-49)
Ber- Ziziphus mauritiana	9.5	45	Banarasi, Karaka, Ponda
Bael- Aegle marmelos	9.0	30	Narendra- 5, Narendra-9
Mahua- Madhuca latifolia	9.2	-	Narendra selection
Grape- *Vitis vinifera*	9.0	30	Beauty seedless, Perlette
Karaunda-Carrisa carandus	9.2	35	Narendra selection

Source: Singh et al. (1994)

Planting Technique and Cultural Practices

The planting techniques are similar to the one used for forest plantations yet the best one is a combination of pit and auger hole. It consists of making pits of 60 cm x 60 cm x 60 cm or other suitable dimensions manually. In the center of each pit, an auger hole of 20-25 cm diameter and 140-160 cm depth is made using tractor mounted augers. These pits cum auger holes are filled back with a mixture of original soil + 8 to 10 kg gypsum + 10 kg FYM + 20 kg river sand. Comparative performance of eight fruit trees planted by two different techniques after 10 years of planting is given in Table 5.18. The average fruit yield of *Aonla*, *Karaunda*, Guava and *Ber* from 6th year onwards in a sodic soil of pH_2 10.0 is given in Table 5.19 (Singh *et al.*, 2008). The financial analysis revealed that the B:C (benefit to cost) ratio for *Aonla*, *Karaunda* and guava was 2.79, 1.62 and1.60, respectively.

Table 5.18 Comparative growth of eight fruit species in a sodic soil

Fruit species	Planting technique			
	Pit (90 cm x 90 cm x 90 cm)		Pit-cum-auger hole	
	Survival (%)	Height (m)	Survival (%)	Height (m)
Jamun (*Syzygium cumini*)	100	4.82	100	5.05
Imli (*Tamarindus indica*)	96	3.45	96	3.54
Karaunda (*Carissa carandus*)	96	1.65	100	1.96
Ber (*Zizyphus mauritiana*)	96	3.03	100	4.16
Guava (*Psydium guajava*)	91	3.03	100	3.15
Aonla (*Emblica officinalis*)	88	2.61	96	4.17
Mango (*Mangifera indica*)	30	1.33	55	2.12
Pomegranate (*Punica granatum*)	20	1.55	45	2.12
Mean	77	2.68	86	3.28

Table 5.19 Yield and economic gains from promising fruit trees after 6 years of plantation

Years after planting	Fruit species							
	Aonla		*Karaunda*		Guava		*Ber*	
	Y	I	Y	I	Y	I	Y	I
6	2.2	9.8	1.4	5.7	1.5	5.6	1.2	2.2
7	2.4	14.7	1.7	7.4	1.8	7.0	1.4	3.0
8	2.7	19.6	2.0	9.5	1.9	8.3	1.5	3.7
9	3.2	26.9	2.1	10.6	2.1	11.0	2.1	6.3
10	-	-	2.5	14.0	2.3	12.4	2.2	7.8

Y : Yield, I : Income in thousand rupees ha^{-1}

To improve upon survival and quality, sodicity tolerant rootstock is planted followed by *in-situ* grafting for the desired characters. For fruit species sensitive to water logging and frost, proper drainage in the rainy season and protection of the young tender saplings during winter are made for the overall success of the horticulture plantations. The data presented in Table 5.20 reveal that pomegranate and *bael* require good drainage if cultivated in such soils. A source of water is a pre-requisite to apply regular irrigations as well as to off-set frost damage. For better growth of the plants, apply 25% higher dose of nitrogen and FYM every year than recommended for normal soils. Grow *Sesbania* during rainy season as a green manure crop to improve soil fertility.

Table 5.20 Effect of prolonged water stagnation on fruit trees recorded during October each year

Fruit trees	Plant mortality (%)	
	1994	1995
Punica granatum (Pomegranate)	25	75
Achras zepota (Sapota)	10	Fully damaged by frost
Tamarindus indica (Imli)	2	0
Carrisa carandus (Karaunda)	12	22
Ziziphus maurtiana (Ber)	6	0
Phoenix dactylifera (date palm)	12	6
Syzygium cuminii (Jamun)	0	0
Psidium guajava (Guava)	9	9
Aegle marmelos (Bael)	50	42
Emblica officinalis (Aonla)	16	12

Source: Singh and Dagar (2005)

Multi-storey mixed plantations in which multiple combinations of fruit species are used for plantation was tested at Faizabad (Table 5.21). In this model, *aonla*, guava and *ber* were planted at a spacing of 8x8 m and *karaunda* and *phalsa* in between the rows at 4x4 m spacing. Initially the soil had high pH (9.5-10.6) and ESP (76-80). The pits used to plant the trees (1x1x1 m for *aonla*, guava and *ber*) and (0.6x0.6x0.6 m for *karaunda* and *phalsa*) were filled with a mixture of normal soil + BHC (100g) + FYM (25 kg). Data on growth in terms of trunk girth of plants of 1-10 years age indicate fast growth (Table 5.22). *Ber* appears to have excellent growth after 6th year and onwards under both *aonla* + *Ber* and guava + *ber* + *phalsa* models. *Aonla* recorded higher girth than the guava under *aonla* + guava + *karaunda* model.

Table 5.21 Species combination, variety and plant density of the experimental set-up

Species combination	Variety	Type of plant	Plant density ha^{-1}
Aonla +	Chakaiya + NA-7	Budded	144
Guava +	L-49 (sardar)	Budded	100
Karaunda	Local selection	Seedling	100
Aonla +	Chakaiya + NA-7	Budded	144
Ber	Umran	Budded	100
Guava +	L-49 (sardar)	Budded	144
Ber +	Gola + kaithli	Budded	100
Phalsa	Local selection	Seedling	100

Table 5.22 Growth performance of fruit species under mixed plantation

Combination	Average trunk girth * (cm) in successive years									
	1	2	3	4	5	6	7	8	9	10
Aonla+	15.3	24.8	29.7	38.8	45.2	48.5	52.0	57.8	60.1	62.0
Guava+	8.4	12.8	19.5	24.5	33.0	40.2	46.0	50.2	54.1	56.2
Karaunda	2.2	4.5	8.8	10.2	11.6	12.2	-	-	-	-
Aonla+	14.6	22.5	28.4	38.2	42.1	45.8	50.2	54.3	56.0	57.1
Ber	13.0	22.5	26.8	36.6	39.3	49.8	58.5	62.8	71.2	75.2
Guava+	8.8	13.7	20.2	23.6	28.5	39.5	48.4	56.5	61.2	64.0
Ber+	10.5	20.7	25.7	29.8	42.1	56.3	59.2	61.0	62.4	63.4
Phalsa	-	-	-	-	-	-	-	-	-	-

* Average of 20 plants, Source: Singh (1996)

Budded plants of guava and *ber* and seedlings of *karaunda* and *phalsa* started bearing from the 3rd year. This suggests that these plants are more precocious in bearing than *aonla* plants. *Karaunda* and *phalsa* lost productivity after 5th year. It might be a result of the shading effect of the other trees. Total productivity of guava + *ber* +*phalsa* was the highest till 6th year followed by *aonla*+guava+ *karaunda* and *aonla* + *ber* (Table 5.23). Higher productivity of *ber*+guava + *phalsa* was again observed in 9th and 10th years as compared to other combinations. Changes in some soil properties after 5th and 10th year of plantation revealed that guava + *ber* + *phalsa* model was the most effective in improving the soil properties (Table 5.24).

Table 5.23 Productivity of multi-storey mixed plantation

Combination	Average yield * (t ha $^{-1}$) in successive years							
	3	4	5	6	7	8	9	10
Aonla+	-	-	2.5	4.5	5.0	5.8	6.3	7.5
Guava+	1.3	2.2	3.2	4.0	5.2	5.5	6.0	6.6
Karaunda	0.4	0.8	1.2	0.2	-	-	-	-
Total	1.7	3.0	6.9	8.7	10.2	11.3	12.3	14.1
Aonla+	-	-	2.3	4.2	4.8	5.1	5.5	6.0
Ber	0.7	2.0	2.8	4.0	4.5	4.8	5.0	5.5
Total	0.7	2.0	5.1	8.2	9.3	9.9	10.5	11.5
Guava+	1.2	2.6	3.9	5.2	5.5	6.0	7.2	8.0
Ber+	0.6	2.0	3.0	4.5	4.7	5.2	6.0	6.5
Phalsa	0.3	0.6	0.8	0.3	-	-	-	-
Total	2.1	5.2	7.7	10.0	10.2	11.2	13.2	14.5

*Average of 20 plants, Source: Singh (1996)

Table 5.24 Physico-chemical properties of soil (0-45 cm) as affected by mixed plantation

Combination	Years after plantation	pH2	ECe (dS m^{-1})	ESP	Hydraulic conductivity (cm h^{-1})	Organic carbon (%)
Aonla+	5	9.8	15.7	67	0.12	0.17
Guava+ *Karaunda*	10	9.2	13.4	60	0.17	0.21
Aonla+	5	9.8	13.3	65	0.18	0.16
Ber	10	9.4	12.0	57	0.23	0.22
Guava+	5	9.2	10.9	59	0.24	0.25
Ber+	10	9.0	8.3	47	0.38	0.30
Phalsa						
Control (barren land)	5	10.5	21.2	80	0.08	0.13
	10	10.6	18.3	76	0.09	0.15

Source: Singh (1996)

Other Models

A number of models have been tried to make horticultural plantations economically viable alternative so as to get high returns during initial years of plantation. Following are some models that have the potential of adoption.

Agri-horti system

In this system agricultural crops are grown in association with fruit species. In a study conducted at Faizabad, mustard and sunflower were grown in association with guava and *aonla*. Mustard yield in both the years is more than sunflower (Table 5.25). Total productivity of mustard and sunflower is relatively higher with guava than with *aonla*. However, it seems that cultivation of these oil seed crops can be one of the options to utilize inter-space between guava and *aonla* orchards during the first few years. It should provide additional income to the farmers during initial years when the fruit plants do not provide any income.

Table 5.25 Productivity of oil seeds as inter-crop

System	Variety	Productivity (t ha^{-1})	
		2nd year	3rd year
Guava+	Sardar	-	-
Mustard+	Baruna	0.14	0.18
Sunflower	Modern	0.10	0.13
Aonla+	Chakaiya	-	-
Mustard+	Baruna	0.11	0.16
Sunflower	Modern	0.08	0.11

Silvi-horti system

In this system, forest species are planted in association with fruit species. Some horticultural based agro-forestry models have been introduced by ND University of Agriculture and Technology, Faizabad (Singh, 1996). In the silvi-horti-pasture model, the growth rate of guava + *Eucalyptus* + *subabul* was faster and production was higher in terms of fruits, fodder and fuel wood (Table 5.26). Reduction in fruit yield particularly in guava and *aonla* was noticed in the 7th year. It could be due to the suppression in growth and fruiting by the forest species. Hence the forest species need to be removed from the model at this time.

Table 5.26 Productivity of silvi-horti system

System	Product	Average productivity (t ha^{-1}) in successive years						
		4	5	6	7	8	9	10
Guava+	Fruit	1.8	2.7	3.0	2.5	3.3	3.5	4.0
Eucalyptus+	Wood	-	-	-	10.5	-	-	-
Subabul	Fodder+wood	0.7	1.2	1.4	1.0 (2.0)	-	-	-
Aonla+	Fruit	-	2.2	4.1	3.6	4.0	4.3	4.8
Eucalyptus+	Wood	-	-	-	9.0	-	-	-
Subabul	Fodder+wood	0.6	1.0	1.2	1.0 (2.1)	-	-	-
Bael+	Fruit	-	0.9	1.8	2.2	3.0	3.1	3.5
Eucalyptus+	Wood	-	-	-	7.8	-	-	-
Phalsa	Fodder+wood	0.1 (0.2)	0.2 (0.3)	0.2 (0.4)	0.1 (0.4)	-	-	-

Wood-dry, Fodder-green, Values in parenthesis are for wood

Data recorded on physico-chemical properties of soil in relation to silvi-horti systems reveal a clear-cut improvement in hydraulic conductivity and increase in organic carbon content of the soil (Table 5.27). Corresponding reduction in pH, electrical conductivity and exchangeable sodium are also noticed in all the treatments compared to the control. Guava + *Eucalyptus* + *subabul* were the most effective system in improving the soil properties (Table 5.27).

Table 5.27 Changes in soil properties under silvi-horti systems after 7 years (0-45 cm soil depth)

Systems	Hydraulic conductivity (cm h^{-1})	Organic carbon (%)	pH_2	EC_e (dS m^{-1})	ESP
Guava+*Eucalyptus*+*subabul*	0.28	0.26	9.2	10.0	49
Aonla+*Eucalyptus*+*subabul*	0.26	0.20	9.8	13.3	53
Bael+ *Eucalyptus*+*phalsa*	0.26	0.21	9.7	12.8	55
Control (barren land)	0.09	0.16	10.4	18.3	76

Oleri-horti system

In this system vegetable crops are grown in association with horticultural plants, which may consists of single or mixed plantation. Experimental results of a single specie plantation revealed that total productivity of the system was highest when cauliflower + onion were grown in association with guava plants (Table 5.28). It was followed by cabbage + okra, tomato + bottle gourd and chillies + bitter gourd. The net returns were also highest with cauliflower + onion followed by cabbage + okra in the 4th year. In the mixed plantation, productivity of bottle gourd + cabbage was higher than the tomato + spinach when grown in association with *aonla* + guava (Table 5.29).

Table 5.28 Productivity of guava based vegetable cropping system

System	Varieties	Spacing	Productivity (t ha^{-1})		Net returns in 4th year (Rs.)
			2nd year	4th year	
Guava	Sardar	8x8 m	-	2.4	18,632
Cauliflower	Pusa Synthetic	60x40 cm	5.3	6.6	
Onion	Pusa Red	15x10 cm	8.5	9.0	
Guava	Sardar	8x8 m	-	2.6	16,800
Cabbage	Golden Acre	60x45 cm	4.8	5.2	
Okra	Parbhani Kranti	30x15 cm	5.6	5.8	
Guava	Sardar	8x8 m	-	2.5	10,450
Tomato	Pusa Ruby	60x45 cm	4.7	5.6	
Bottle gourd	P. S. Prolofic long	2x1.5 m	4.0	4.6	
Guava	Sardar	8x8 m	-	2.2	5,837
Chillies	Pusa Sadabahar	45x30 cm	2.0	2.5	
Bitter gourd	Faizabadi	2x0.6 m	1.0	1.2	

Table 5.29 Productivity of vegetables as inter-crops in mixed plantation (*aonla* + guava)

System	Variety	Spacing	Average yield (t ha^{-1})	
			1st year	2nd year
Bottle gourd	P.S. Prolific long	2x1.5 m	2.6	3.8
Cabbage	Golden acre	60x45 cm	3.0	4.0
Tomato	Pusa Ruby	60x45 cm	2.3	2.8
Spinach	All green	Broad cast	2.0	2.6

Note: In both the systems aonla + guava were grown at a spacing of 8x8 m

Silvi-oleri System

In this system vegetable crops are grown in association with forest plantations. Studies at Faizabad, revealed that the productivity of brinjal + bottle gourd is maximum when grown in association with teak plantation (Table 5.30). It is followed by tomato + onion, cauliflower + okra and bitter gourd + okra. The net returns per unit area were the highest in the case of cauliflower + okra in the 5th year of cultivation.

Table 5.30 Productivity of vegetables as inter-crops under teak plantation

System	Variety	Spacing	Average yield (t ha^{-1})			Net returns in 5th year (Rs. ha^{-1})
			1st	2nd	5th	
Brinjal	Pant Samrat	60x45 cm	7.8	9.5	9.8	9,750
Bottle gourd	P. S. Prolofic long	2x1.5 m	4.0	4.2	4.5	
Cauliflower	Pusa Synthetic	60x40 cm	5.2	5.4	5.8	21,508
Okra	Parbhani Kranti	30x15 cm	4.3	5.0	5.2	
Tomato	Pusa Ruby	60x45 cm	4.0	4.6	5.0	16,580
Onion	Pusa Red	15x10 cm	7.6	8.0	8.3	
Bitter gourd	Faizabadi	2x0.6 m	0.8	1.2	1.5	16,232
Okra	Parbhani Kranti	30x15 cm	4.0	4.5	5.0	

Note: In each system teak plants were planted at a spacing of 4x4 m

Relative Tolerance

The relative tolerance of fruit trees as assessed by Singh *et al.* (1994) is reported in Table 5.31. The most promising fruit species are: *Aonla* (*Emblica officinalis*), *Karaunda* (*Carissa carandus*) and Guava (*Psydium guajava*).

Table 5.31 Relative tolerance of tree species to soil alkalinity/sodicity/pH

Average pH_2 (0-120 cm)	Fruit trees
More than 10.0	-
9.5 to 10.0	*Achras japota* (sapota) *Carissa carandus* (Karaunda) *Psidium guajava* (Guava) *Emblica officinalis*(Aonla) *Zizyphus maurtiana* (Ber) *Syzgium cumini(Jamun)* *Phoenix dectylifera* (Khajoor) *Aegle marmelos* (Bael)
9.0 – 9.5	*Punica granatum* (Pomegranate)
< 9.0	*Prunus persica* (Peach) *Pyrus communis* (Pear) *Vitis vinifera* (Grapes) *Mangifera indica* (Mango)

Source: Singh *et al.* (1997)

MEDICINAL AND AROMATIC CROPS

Medicinal and aromatic plants are in great demand to meet the national requirement and for export. Since these crops are minor in nature, it is not always possible to allocate fertile lands to these crops, which are mostly used to produce food grains. Wastelands especially the lands affected by sodicity and salinity problems where profitable returns are not possible from agricultural crops, could be utilized for the cultivation of these high-value crops with marginal inputs. The threshold limits of pH and ESP for some of the medicinal and aromatic plants (Table 5.32) indicate that most of the aromatic grasses (except citronella) can withstand sodicity levels given by a soil pH_2 of 9.5 or an ESP of 55. Among the medicinal crops, rye for ergot can be grown successfully in soils having pH_2 up to 10.0 (Patra and Singh, 1998). Detailed studies on ameliorative potential of these crops for different kinds of sodic soils and their efficacy to withstand different stresses are required for their cultivation under specific situations. These crops have the potential to grow in inter-spaces between fruit plants and rows. For example; if pH_2 is less than 9.5, crops of *Babuna* (*Matricaria camomila*) and *Isabgol* (*Plantigo ovata*) can be grown in between fruit trees.

Table 5.32 Safe limits (threshold) of sodicity/alkalinity for some important aromatic and medicinal crops

Crop	pH	ESP
Aromatic		
Palmarosa	9.5	55
Lemongrass	9.0	50
Vetiver	9.5	55
Citronella	8.5	25
Jamrosa	10.0	55
Medicinal		
Periwinkle	9.5	-
German Chamomile	9.5	-
Rye for ergot	10.0	-

Source: Patra and Singh (1998)

A study on cultivation of German chamomile (*Chamomilla recutita, Asteraceae*) indicates that the plant height, number of flowers and dry matter yield were significantly lower in sodic than in normal (non-sodic) soil. However, application of gypsum and pyrites to sodic soils significantly increased the plant height, number of flowers and dry matter yield. The gypsum was more effective than the pyrites in increasing the growth and yield of chamomile in sodic soil (Sukhmal Chand *et al.*, 2005). Experiments carried out at CIMAP research farm confirm the prospects of cultivation of some of the aromatics crops on sodic lands. *Palmorosa (Cymbopogonmartini), Lemon grass (Cymbopogon flexuosus) and Vetiver (Vetiveria zizanioides)* among aromatic grasses and also the *German chamomile (Matricariachamomilla)* are the promising crops (Kudesia *et al.,* 2009). In a recent paper Dagar *et al.* (2015) reported that liquorice (*Glycyrrhiza glabra* Linn.) also known as Mulahatti, can be successfully grown on sodic soils. Besides getting (2.4–6.1 t ha^{-1} forage per annum), a root biomass of 6.0–7.9 t ha^{-1} can be obtained in three years of growth fetching about Rs 6.0 to 8.0 lakhs ha^{-1} with a net profit of. Rs 2–2.65 lakhs per annum per ha at or around high pH of 9.7-9.9 (ESP 58-62). Besides, the sodic soil can be reclaimed substantially in terms of reduced pH and exchangeable sodium percentage.

Relative Tolerance

Parveen Kumar *et al.* (2015) reviewed and documented the relative tolerance of various medicinal and aromatic plants to abiotic stress due to soil and water salinity and soil sodicity, which is reproduced as Table 5.33.

Table 5.33 Salinity and sodicity tolerance limits of important medicinal and aromatic plants

Plant species	Salinity tolerance ($EC\ dS\ m^{-1}$)		Sodicity tolerance	
Medicinal plants	**Soil (EC_e)**	**Irrigation water**	**Soil (pH_s)**	**ESP**
German Chamomile (*Matricaria chamomilla* L.).	10-12	8-10	9.5	–
Blonde psyllium (*Plantago ovata* Forsk.)	7-8	11-12	9.5	–
Periwinkle (*Catharanthus roseus*)	6-8	8-10	10.0	–
Rye for ergot (*Claviceps purpurea*)	10-12	12-15	9.6	–
E. henbane (*Hyoscyamus muticus*).	7-8	8-10	8.9	–
Wormwood (*Artemisia* spp.)	8-10	8-10	8.4	–
Licorice (*Glycyrrhiza glabra*)	6	10	-	-
Dill (*Anethum graveolens*)	8-9	6-8	8.6	–
Malabar nut (*Adathoda vasica*)	–	8-10	–	–
Aloe (*Aloe vera*)	8-10	10-12	8.4	–
Kair (*Capparis decidua*)	18-20	15-18	8.6	–
Euphorbia (*Euphorbia antisyphilitica*)	12-14	10-12	8.8	–
Jatropha (*Jatropha curcas* Linn.)	6-8	6-8	8.7	–
Holy basil (*Ocimum sanctum* Linn.)	–	8-10	8.6	–
Indian senna (*Cassia angustifolia*)	9-10	10-12	8.7	–
Garden cress (*Lepidium sativum*)	10-12	15-17	8.6	–
Asparagus (*Asparagus racemosus*)	7-8	10-12	8.2	–
Winter cherry (*Withania somnifera*)	8-10	10-12	8.4	–
Mint (*Mentha citrata*)	4-5	4-5	8.6	–
Aromatic Plants				
Palma rosa (*Cymbopogon martinii*)	9-12	14-16	9.5	55
Lemon grass (*Cymbopogon flexuosus*)	8-10	7-8	9.0	50
Citronella (*Cymbopogon nardus*)	4-5	5-6	8.5	25
Jamrosa (*Cymbopogon khasans*)	10-12	9-10	10.0	45
Vetiver (*Vetiveria zizanioides* Linn.)	9-10	10-11	9.5	55
Marigold (*Calendula officinalis*)	4-5	5-6	8.8	40

Note: pH as an indicator to define the degree of sodic lands holds good only for alluvial soils of the Indo-Gangetic plains

HALOPHYTES

About 2,600 halophytic species are known to exist while only few have been extensively studied for their potential in agriculture as biological resources having economic potential as sources of oils, flavours, gums, resins, oils pharmaceuticals and fibers (Galvani, 2007) or for environmental protection and conservation of ecosystems (Barnes and Baylor, 2007, Zhang *et al.,* 2007). The relative growth of halophytes vis-a-vis other plant types in relation to salinity/sodicity of the growth media is shown in Fig. 5.9 (Hasanuzzaman *et al.* 2014). It shows that potential yield of halophytes is at much higher salinity/sodicity in the growth media mainly because halophytes have salt excluding, salt excreting or salt accumulating mechanisms that blunt the adverse effect of salts. The success of halophytes cultivation as irrigated crops depends upon some basic conditions that must be met:

- High yield potential
- Low irrigation requirement than the conventional crops
- Able to grow with poor quality waters
- The products must be able to replace the conventional crop products
- These must fit in within the existing agricultural infrastructure.

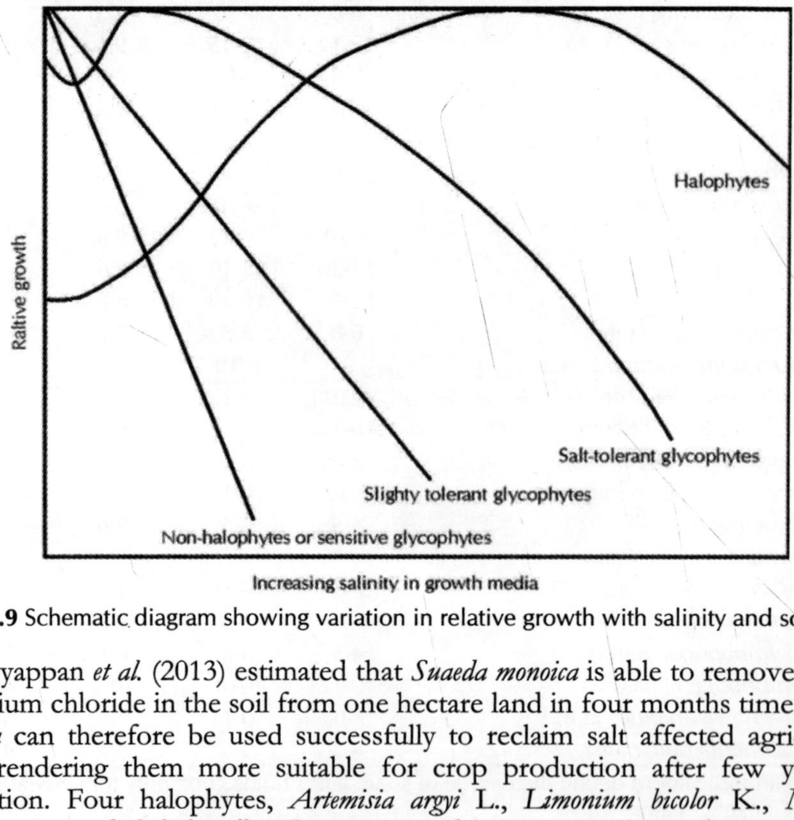

Fig. 5.9 Schematic diagram showing variation in relative growth with salinity and sodicity

Ayyappan *et al.* (2013) estimated that *Suaeda monoica* is able to remove 453.55 kg sodium chloride in the soil from one hectare land in four months time. *Suaeda monoica* can therefore be used successfully to reclaim salt affected agricultural lands rendering them more suitable for crop production after few years of cultivation. Four halophytes, *Artemisia argyi* L., *Limonium bicolor* K., *Melilotus suaveolens* L., and *Salsola collina* P. were tested in pot experiment by using three salt-affected soils in Tianjin. The ESP in two soils was high and could be reduced from 66.2 to less than half in Dagang soil and from 51.4 to less than 30 in Xiqing soil by the cultivation of four halophytes (Sakai *et al.*, 2012). A major advantage of halophyte cultivation is that even worthless lands and waters can be used to grow dedicated halophytes, able to produce some useful yield, albeit at a lower level than in good arable lands. The economic framework in this situation can be totally different. Moreover, there is a great scope to improve halophytes into new salt-resistant crops or used as a source of genes to be introduced into conventional crop species that in general show decline in their economical production with increasing salinity or sodicity. Similarly, bio-fuel production could be another option to utilize sodic lands. For example, in the USA a large amount of waste land is being used for biodiesel plant production. A major advantage in cultivation of halophytes is that it does not depend on fresh water. Considering all these issues, it appears that there is a great need for in depth studies on the potential use of halophytes in saline agriculture and their use and incorporation in the consumers' diet.

FISH PONDS IN SODIC LANDS

Sodic soils are characterized by low infiltration rate. As a result water infiltration and percolation through these soils is quite low. As such, farm ponds constructed in these soils provide opportunities to practise aquaculture without the fear of high seepage losses (Fig. 5.10). These ponds can be located in the lowest portion of the farm while the remaining upland portion can be put under agricultural cropping. These ponds also serve as drainage structure for the cropped area. The field studies have shown that ponds constructed in highly sodic soils have no impact on the fish production. The drainage water quality is quite good except during the first rain storm of the monsoon season and has no deleterious effect on the growth of the fishes. Alternative arrangement for filling the pond in the post monsoon season has to be made to ensure fish farming in semiarid regions. The experience at CSSRI, Karnal and even in adjoining areas in farmer's fields has been quite encouraging for such a set-up. In a study conducted at N.D. University of Agriculture and Technology, Faizabad, survival of fishes was around 50% with grass carps recording the highest survival (59%) followed by common carps (56%). A total productivity of 0.637 t was obtained from 0.4 ha area meaning thereby that the productivity could be around 1.6 t ha^{-1}.

Fig. 5.10 A semi-reclaimed sodic lands converted to a fish pond

Another way to reduce seepage losses in aqua ponds constructed on normal land has been to use Na_2CO_3 and $NaHCO_3$ salts at the bottom to increase ESP of the bottom soil. This practice also had no deleterious effect on fish production and can be used where water is limited.

6

MANAGEMENT OF SODIC WATER

Amongst various causes of formation of sodic soils, one of the causes is continuous use of sodic water for irrigation of crops. ESP of the soils increases with each irrigation and if the practice is continued, it renders the soil sodic resulting in inhibited germination, reduced infiltration rate of rain and irrigation water and show other characters often associated with sodic soil. The productivity losses get accentuated with each year finally leading to the failure of the crop. This chapter describes the current status of the theory of the mechanisms of sodicity build-up to understand the potential of sodic water to turn a soil sodic. Reclamation and management practices for sodic water have been described through collation of the latest information available on the subject. The chapter also discusses the management options for irrigation with such quality of water. While many practices that are similar to the ones adopted to manage sodic soils are briefly discussed (Details covered in Chapter 4), others are discussed in details. These practices either alone or in combination will help to avoid formation of sodic soils or at least minimize their adverse impacts on soils and crop productivity.

CHARACTERIZATION OF WATER QUALITY

Quality of water has always been important to the mankind yet it assumed serious dimension only in recent times because of increased pollution of surface and ground water resources. Knowledge of irrigation water quality therefore, is critical to understanding management options for long-term sustainable productivity in arid, semiarid and coastal regions as well as in other ecologies where ground water is polluted. Some of the known information illustrates the impact of water quality on soil resource or crop productivity as follows:

- Water of even low electrical conductivity, as low as 0.3 dS m^{-1} contributes as much as 8,460 t day^{-1} of salt at the expected inflow of 510 m^3 s^{-1} or 4,400 ha m day^{-1} of water as is being introduced in Rajasthan through the Indira Gandhi *Nahar Pariyojana.* Thus, without proper water management, salts may accumulate to render soil saline/sodic.
- In many situations, irrigation water quality has impacted crop productivity more than other constraints such as soil fertility and weed infestation.

- Irrigation water quality has been found to alter fertilizer needs, irrigation application methods to be used and other management options necessary to maintain long-term productivity of lands.

The concentration and kind of constituents present determine water quality and management options needed to reclaim and manage the water. Currently, irrigation water quality is characterized on the basis of one or more of the following characteristics of the water.

- **Salinity hazard** - Total soluble salt content (Electrical Conductivity, EC).
- **Sodium hazard** - Relative proportion of divalent to monovalent cations mainly the sodium, calcium and magnesium expressed in terms of Sodium Adsorption Ratio (SAR).
- **Sodic hazard** - Carbonate and bicarbonate- (Residual Sodium Carbonate (RSC); Adjusted Sodium Adsorption ratio- (Adj. SAR) and Adjusted R_{Na}.
- **Specific ions**- Chloride (Cl), sulphate (SO_4^{2-}), boron (B), nitrate-nitrogen (NO_3-N) and heavy metals.
- **Other potential contaminants** - Contaminants such as BOD, COD, heavy metals and pathogenic contaminants.

Elaborate studies have been conducted and volumes of literature are available on the characterization, laboratory investigations, management and reclamation of saline/sodic waters. Technologies have also been developed and extensively applied for these kinds of water especially in northwest India. Relatively speaking, the understanding on saline water has been far better than sodic water. It appears that there are strong reservations on the characterization of water using SAR as it may lead to erroneous conclusions. Similar reservations are being expressed for RSC as well. While in the international literature concept of SAR has slowly replaced the RSC concept, yet in the Indian context RSC is still widely used to characterize sodic water because of its simplicity in use as well as its practical applications under field situations. Although continuous use of sodic water is known to make the soil sodic, yet the complexity in understanding dissolution-precipitation and adsorption-desorption makes it difficult to predict the anticipated problems with the continuous use of sodic water. Moreover, role of magnesium in sodic water has not been well understood. Even today, not much headway has been made in describing its role while calculating SAR or RSC. It is because of such complexities, it has not been possible to relate the sodic water quality parameters with the observed ESP upon its long-term use.

EXTENT AND DISTRIBUTION OF POOR QUALITY WATER

No systematic attempts have been made so far in the country to arrive at the estimates of saline/sodic ground water resources. The major monitoring agencies such as Central Ground Water Board (CGWB) and the State Ground water Departments monitor the water quality mainly from the drinking point of view. The development of saline ground water for irrigation purpose being insignificant, so far not much information on the distribution of ground water quality is known. Yet based on whatsoever information was available, a first approximation map of ground water quality for irrigation on a 1:6 million scale was prepared by Gupta *et al.*, (1994). Four legends were used in mapping i.e. good water (EC_{iw} < 2 and SAR < 10), saline water (EC_{iw} > 2 dS m^{-1} and SAR < 10), high-SAR saline water (EC_{iw} > 4.0 and SAR > 10) and sodic water (EC_{iw} variable, SAR variable and RSC > 2.5) mainly based on the characterization reported in Table 6.1 (Fig. 6.1). High salinity ground water is mostly encountered in arid parts of northwest states like Rajasthan, Haryana and Punjab (Fig. 6.1). About 30-84 % of the ground water aquifers especially in the states of Rajasthan, Haryana and Uttar Pradesh are considered unfit for irrigation, as their use is likely to build-up of salinity or sodicity in the soils (Table 6.2), the most affected state being Rajasthan followed by Haryana. Of the poor quality water in the four states, the sodic water constitutes about 37% of the total poor quality resource while another 43% might pose problems of sodicity on sustained use (Table 6.3). CGWB (1997) approximated that leaving aside minor patches, the total area underlain with the saline ground water (EC >4 dS m^{-1}) is 193,438 km^2 with annual replenishable recharge of 11,765 million m^3 Yr^{-1} (Table 6.4). However, this does not differentiate between the kinds of water available in a particular state or region.

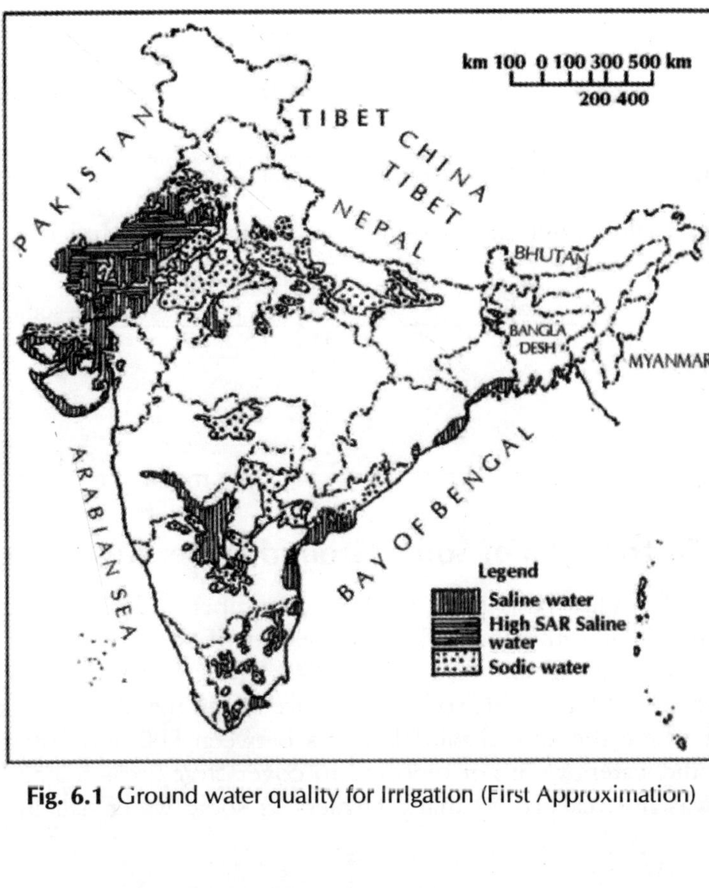

Fig. 6.1 Ground water quality for Irrigation (First Approximation)

Table 6.1 Grouping of poor quality water

Water quality	EC_{iw} (dS m^{-1})	SAR_{iw} (mmol $L^{-1})^{1/2}$	RSC (meq L^{-1})
A. Good	< 2	< 10	< 2.5
B. Saline			
i. Marginally saline	2-4	< 10	< 2.5
ii. Saline	> 4	< 10	< 2.5
iii. High-SAR saline	> 4	> 10	< 2.5
C. Sodic* water			
i. Marginally sodic	< 4	< 10	2.5 – 4.0
ii. Sodic	< 4	< 10	> 4.0
iii. High-SAR sodic	variable	> 10	> 4.0
D. Toxic Water	The toxic water has variable salinity, SAR and RSC but has excess of specific ions such as nitrate, boron, fluoride, chloride, sodium, silica or heavy metals such as selenium, cadmium, lead and arsenic etc.		

Source: Minhas and Gupta (1992), Toxic waters added by Gupta (2011). *In literature it is also referred as alkali water.

Table 6.2 Distribution of ground water in different states based on ground water EC

State	Ground water quality (%)		
	Good	Marginal	Poor
Haryana	37	8	55
Punjab	59	22	19
Uttar Pradesh	37	20	43
Gujarat	70	20	10
Rajasthan	16	16	68
Karnataka	64	10	25

EC < 2 dS m^{-1} (Good), 2-6 dS m^{-1} (Marginal) and > 6 dS m^{-1} (Poor)

Table 6.3 Poor quality water distribution (%) in some states of India

States	Categories		
	Saline	Sodic	Saline-sodic
Punjab	22	54	24
Haryana	24	30	46
Rajasthan	16	35	49
Gujarat	20	28	52
Average	20	37	43

Source: Manchanda *et al.* (1989a, b)

State Wise Distribution of Sodic Ground Water Resources

A number of studies have been made and numbers of good review have recently appeared that summarizes the distribution of sodic waters in various states of India (Minhas and Gupta, 1992, CGWB, 1997, Minhas *et al.*, 1998 and Gupta and Gupta, 2014 a,b). The sodic waters are found mainly in the semi-arid parts of India where the annual rainfall varies between 500 and 700 mm. Till recent past, sodic waters were not reported to cover large areas mainly because the problem was not realized by many farmers as sodic water usually has low salinity. Gupta and Gupta (1987) categorized problems associated with salinity of water that reveal that SAR is generally associated with high and RSC with low salinity waters with some exception of course (Table 6.5). Sharma *et al.* (1975 a,b) also observed that RSC problem is more pronounced at low salinity compared to high salinity of waters (Table 6.6 and Table 6.7). Wherever, high sodicity was associated with relatively high salinity, the problem has generally been characterized as that of salinity. In recent times, the spread of sodic waters has been widely reported in various parts of the country.

Table 6.4 Poor quality ground water resources in states

State	Utilizable ground water resource for irrigation in net terms (M ha-m yr^{-1})	Net draft (M ha-m yr^{-1})	Level of ground water development (%)	Use of poor quality water (M ha-m yr^{-1})	Area underlain by saline ground water (EC > 4 dS m^{-1}) (km^2)
Punjab	1.47	1.67	98	0.68	3,058
Haryana	0.86	0.72	76	0.47	11,438
Uttar Pradesh	6.31	2.98	42	1.42	1,362
Rajasthan	0.95	0.77	73	0.65	141,036
Bihar	2.06	0.82	36	NA	NA
West Bengal	1.77	0.63	32	NA	NA
Delhi	0.01	0.01	120	NA	140
Gujarat	1.56	0.85	49	0.26	24,300
Karnataka	1.24	0.45	33	0.17	8,804
Tamil Nadu	2.02	1.40	63	NA	3,300
Madhya Pradesh	2.66	0.73	25	0.20	NA
Maharashtra	2.29	0.88	35	NA	NA
Andhra Pradesh	2.70	0.78	26	0.25	NA
Total (India)	32.63	13.50	37		193,438

Table 6.5 Associated problems in relation to EC of the ground water

EC (dS m^{-1})	Degree of problem		
	SAR	RSC	B
<1.5	-	-	-
1.5-3.0	+	+ +	-
3.0-5.0	+ +	+	-
5.0-10.0	+ + +	-	+
>10.0	+ + + +	-	+ +

- no , + low, + + medium, + + +, high and + + + + very high problem
Source: Gupta (1984)

Table 6.6 Distribution of water samples in different EC and RSC ranges (Jind district)

EC range (dS m^{-1})	Number of samples	% water with RSC (meq L^{-1})		
		< 2.0	2.0 – 5.0	> 5.0
< 1.0	41	61	34	5
1.0 – 2.0	39	41	23	36
2.0 – 4.0	45	82	11	7
4.0 – 6.0	24	–	–	–
6.0 – 8.0	11	–	–	–

Source: Sharma et *al.* (1975a)

Table 6.7 Distribution of water samples in different EC and RSC range (Karnal including Kurukshetra districts)

EC (dS m^{-1})	No. of samples	Average RSC (meq L^{-1})
< 1.0	154	2.2
1.0 – 2.0	47	5.3
2.0 – 4.0	16	6.5
4.0 – 6.0	2	12.5
> 6.0	3	0.0

Source: Sharma *et al.* (1975b)

Studies in the state of Haryana have revealed that sodic water problem is increasing in areas where water table is going down and farmers are installing submersible tube wells. In Nissang block of district Karnal, out of a total of 57% water being salty, the percentage of sodic waters is 47% (Anonymous, 2007). On the basis of 1224 samples collected from all over the Bhiwani district covering seven blocks it was found that 16.3% waters were of good quality, 66.8% saline and 16.9% sodic in nature. In sodic water, sodic class recorded the minimum number of samples (1.4%) followed by highly sodic (2.4%) and marginally sodic (13.1%) classes. Besides waters of high sodicity are found in Ballabgarh, Sohna, Faridabad, Gurgaon, Mohindergarh, Narnaul, Bawal, Pataudi (Mohindergarh), Dadri (Bhiwani), Salhawas (Rohtak), Ganaur (Sonepat), Rajound, Narwana (Jind), Karnal, Nilokheri, Assandh, Madlauda, Pundir, Guhla and Kaithal.

In Punjab low to medium sodic water constitute about 47% area of the state. The bicarbonate water having high residual sodicity covers nearly 25% of the total area and poses a serious threat to sustained agricultural production. Water of high residual sodicity are found in Khara Majha (Amritsar), southern parts of Ludhiana, Kharar (Ropar), Zira and Dharamkot (Moga), Phul block (Mansa), Moga, Baghapurana and Nihalsingh Wala (Moga), Malerkotla and Sangrur (Sangrur). The distribution of saline and sodic water in an area of 5.475 lakh ha in three districts namely Faridkot, Bhatinda and Sangrur indicated that sodicity problem is more pronounced and both salinity and sodicity problems increased with depth (Table 6.8).

Table 6.8 Ground water quality distribution in southwest Punjab

Depth	Area (%)					
	Electrical conductivity (dS m^{-1})			Residual sodium carbonate (meq L^{-1})		
	< 2.0 Fresh	2.0-4.0 Marginal	> 4.0 Saline	< 2.5 Fresh	2.5-5.0 Marginal	> 5.0 Sodic
10	84.0	15.5	0.5	58.1	18.8	23.1
15	80.6	18.2	1.3	45.2	24.3	31.1
20	80.7	17.3	2.2	33.9	26.5	39.6
25	79.5	17.9	2.6	33.1	28.1	38.8
35	77.2	19.7	3.1	31.1	28.5	40.4

Source : Gupta *et al.* (1995)

Out of total 2,296 samples, 7% of samples in Madhya Pradesh are of sodic nature (Anonymous, 2007). The percentage is as high as 26% in Datia and 18% in Gwailor. Nearly 4% of the 773 samples of water are of sodic nature in the state of Maharashtra.

The quality of ground water in north-eastern and eastern parts of Gujarat is predominantly bicarbonate type. In Gujarat, 84% of ground water have EC_{iw} > 3 dS m^{-1}. Waters of medium to high residual sodicity are found in the districts of Rajkot, Jamnagar, Bhavnagar and parts of Surendranagar and Kutch.

Ground water of major parts of Uttar Pradesh is primarily bicarbonate type. Dixit (1971) reported that 50% of the waters in the districts of Agra, Aligarh, Etah, Mainpuri and Mathura have salinity < 3.0 dS m^{-1} and RSC > 2.5 meq L^{-1}. In Mathura district, about 24% waters also have RSC > 2.5 meq L^{-1} having SAR < 10. Waters having SAR > 10 in these districts are about 10%. In Iglas and Gonda blocks of Aligarh district, waters of low saline sodic nature (EC < 3 dS m^{-1}, SAR < 10 and RSC > 5 meq L^{-1}) constitute nearly 50% of ground water.

Areas of north-eastern and southern parts of Rajasthan (Jaipur, Kota, Udaipur where rainfall is >500 mm) have residual sodicity in the ground water in significant amounts. Ground water in the Nagaur, Sikar and Jhunjhunu districts also have high residual sodicity. Although only 13.3% of ground water of Bikaner is sodic in nature yet the percentage is much higher in Jaisalmer and Churu districts (Table 6.9). In Bhartapur district, the sodic water problem was recorded in Nadvai tehsil where 20-27% water have high sodic problem followed by Nagar with 18% and Deeg, Kama, Pahadi and Veir with 12-14%.

Table 6.9 High SAR saline and sodic water in 3 districts of Rajasthan

District	High SAR Saline	Marginally sodic	Sodic	Highly sodic
Jaisalmer	25.5	10.6	0.5	29.4
Churu	16.8	7.1	0.4	20.5
Bikaner	4.8	7.3	3.1	2.9

In general, ground water in Andhra Pradesh (including Telangana) is HCO_3, HCO_3-Cl, Cl-HCO_3 type. Waters with high sodicity hazards having RSC >5 meq L^{-1} and SAR >15 are found to cover 10-20% of samples in Prakasam, Krishna and West Godawari districts (Fig. 6.2, Anonymous, 2009). About 25% water of Anantapur district has RSC >5 meq L^{-1}. Ground water of Kapileswarapuram block in East Godawari and Nhenda block in Guntur district is high in residual sodicity (RSC >5 meq L^{-1}).

In Odisha, except for the coastal belt, ground water in contact with river alluvium is mostly Ca-Mg-HCO_3 type with EC < 3 dS m^{-1}. In coastal regions ground water quality is highly variable from fresh (Na-HCO_3 type) to residual sodicity to highly saline in nature.

Fig. 6.2 Ground water quality of Andhra Pradesh including Telangana

Out of the 1,287 samples from various parts of Tamil Nadu, 8% had the problem of RSC more than 2.0 meq L^{-1} (Fig. 6.3). More than 10% of water had RSC problems (2.5 meq L^{-1} or more) in the districts of Cuddalore, Karur, Tiruchirapalli, Thanjavur and Thiruvarur, the last three districts being part of the Cauvery Delta Zone (Haroon *et al.*, 2009).

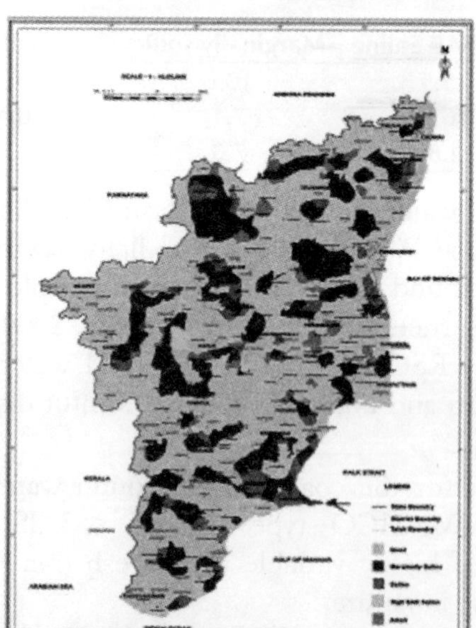

Fig. 6.3. Water quality map of Tamil Nadu

Most ground waters in Karnataka are bicarbonate type. The ground water of Chitradurg district having residual sodicity> 3.5 meq L^{-1} when used to irrigate some shallow and deep red and black soils made the soil sodic. Most ground waters of the district had a CO_3:HCO_3 ratio of 1:2. From sodicity point of view, the water of the Gadak taluka of the Gadag district had 53% of ground waters that have RSC values more than 2.5 meq L^{-1} (Table 6.10).

Table 6.10 Water quality ratings of ground waters of different taluka of Gadag district

Taluka	Sample size	Good	M. saline	Saline	High SAR saline	M. sodic	Sodic	Highly sodic
Gadag	156	31.41 (49)	21.15 (33)	8.33 (13)	5.13 (8)	12.18 (19)	8.33 (13)	13.46 (21)
Mundaragi	98	45.92 (45)	14.29 (14)	1.02 (1)	4.08 (4)	8.16 (8)	12.24 (12)	14.29 (14)

Source: Anonymous, 2009; M= marginal; Figures in the parenthesis represent number of samples

THEORY OF SAR/RSC

Sodium adsorption ratio (SAR) has been defined as the ratio of the sodium to square root of the calcium plus magnesium ions present in the water as follows:

$$SAR = \frac{Na}{(Ca + Mg)^{1/2}} (mmol\,L^{-1})^{1/2} \tag{6.1}$$

The concentration of the ions in eq. (6.1) is expressed in mmol L^{-1}. If the concentration of the ions in eq. (6.1) is expressed in meq L^{-1}, then eq. (6.1) is written as:

$$SAR = \frac{Na}{[(\frac{Ca + Mg}{2})]^{1/2}} (meq\ L^{-1})^{1/2} \tag{6.2}$$

An unusual aspect of the SAR is that Ca and Mg have been lumped together making it controversial. Considering various opposing arguments, scientists are veering around the view that Mg does not affect the soil as adversely as Na but is not as beneficial as Ca. Being intermediate of Na and Ca, its clubbing with Ca is questionable especially for Indian conditions where Mg/Ca ratios of 2-4 are commonly encountered even going up to 16 (Gupta and Gupta, 1987). AICRP guidelines stipulate that if Mg/Ca ratio is more than 3, some chemical amendments are needed to manage such kind of water (Minhas and Gupta, 1992). Thus, under Indian conditions where Mg/Ca ratios tend to increase with salinity of water, SAR may underestimate the sodium hazard of irrigation water. The only justification to club calcium with magnesium appears to be that in the past Ca and Mg were reported together and since the calcium plus magnesium is divided by 2, one need not worry much about the Mg until the Mg/Ca ratio remains around 1.0. With slight manipulation, eq. (6.2) can be written as:

$$SAR = \frac{Na}{\sqrt{Ca[(1+Mg/Ca)/2]^{1/2}}} (meq\ L^{-1})^{1/2} \quad (6.3)$$

Apparently, SAR is underestimated with increasing Mg/Ca ratios, even though high Mg/Ca ratios are known to cause dispersion and build-up of higher ESP in soils than waters with low Mg/Ca ratios (Yadav and Girdhar, 1981). For the same amount of Na and Ca in waters, SAR will be 0.63 times the SAR of the water for Mg/Ca ratio of 4 compared to Mg/Ca ratio of 1.0. Therefore, to avoid such underestimation, eq. (6.2) should be used with the following stipulations:

For non-calcareous soils, actual value of Mg/Ca may be taken in eq. (6.3) if it is less than 1 (Gupta, 2015). If it is more than 1, Mg/Ca in the water may be taken as 1.0 irrespective of its value such that:

$$SAR = \frac{Na}{\sqrt{Ca}} (meq\ L^{-1})^{1/2} \quad (6.4)$$

Gupta (1984) called this as SCAR (Sodium to Calcium Activity Ratio)

For calcareous soils, limit of Mg/Ca ratio may be increased to 2.0 such that actual values below 2 are used in eq. (6.3). For Mg/Ca greater than 2.0, eq. (6.3) is given as (Gupta, 2015):

$$SAR = \frac{0.82\,Na}{\sqrt{Ca}} (meq\ L^{-1})^{1/2} \quad (6.5)$$

It may be noted that contrary to their appearances, eq. (6.4) or eq. (6.5) do not neglect Mg but takes it equal to Ca or twice the Ca respectively. Eq. (6.4) has earlier been proposed by Gupta and Gupta (2003) albeit without this justification. Data in Table 6.11 reveal that SAR calculated with eq. (6.4) is more closely related to ESP of the soil than conventional SAR.

Table 6.11 Comparison of SAR (Eq. 6.2) and SCAR (Eq. 6.4) values of irrigation waters in relation to observed ESP of soils (Gupta, 1984)

Location	Irrigation water				Irrigated soils	
	EC (dS m^{-1})	Mg/Ca ratio	SAR (eq. 6.2) (meq L^{-1})$^{1/2}$	SCAR (meq L^{-1})$^{1/2}$	SAR_e (meq L^{-1})$^{1/2}$	ESP
Kaparda	10.8	4.4	28	59	31	60
Jelwa	5.9	8.1	23	36	28	38
Shikarpura	4.5	16.0	8	26	32	35

General guidelines on the use of SAR to characterize irrigation waters are given in Table 6.12 along with some comments on the management options. Several other Tables have also appeared in the literature making the issue quite complex because of wide differences in the guidelines. For example; tolerance of crops to SAR of the irrigation water for non-saline conditions given in Table 6.13 are quite high especially for moderately tolerant and tolerant crops. Since SAR in general is high in high EC waters, to have a non-saline environment with such high SAR values is not possible. Clearly, such Tables may have limited applications under Indian conditions. Until guidelines conforming to Indian conditions are available, guidelines given in Table 6.12 can be safely used.

Table 6.12 Some parameters/indices for rating ground water quality for irrigation (Ayers and Westcot, 1985, Eaton, 1950 and Wilcox, 1948).

Class	SSP (%)		SAR (meq L^{-1})$^{1/2}$	Sustainability for irrigation
			Values	Comment
I	<20	Excellent	< 10	Use on sodium sensitive crops such as avocados and on heavy textured soils# needs caution
II	20-40	Good	10-18	Amendments (such as gypsum, sulphitation press mud*, distillery spent wash) and leaching needed
III	40-80 (40-60)*	Fair (Permissible)	18-26	Generally unsuitable for continuous use
IV	>80 (60-80)	Poor (Doubtful)	> 26	Generally unsuitable for use
V	(>80)	Unsuitable		

*Classification in Parenthesis by Wilcox (1955); #added by the author based on Indian experience on heavy textured Vertisols.

Table 6.13 Tolerance of crops to SAR under non-saline environment

Tolerance	SAR of irrigation water	Crop	Condition
Very sensitive	2-8	Deciduous fruits, nuts, citrus, avocado	Leaf tip burn, leaf scorch
Sensitive	8-18	Beans	Stunted, soil structure favourable
Moderately tolerant	18-46	Clover, oats, rice, tall fescue	Stunted due to nutrition and soil structure
Tolerant	46-102	Wheat, barley, tomatoes, beets, tall wheat grass, crested grass, lucerne	Stunted due to poor soil structure

Source: Extracted from the Australian Water Quality Guidelines for Fresh & Marine Waters (ANZECC, 2000) http://www.mfe.govt.nz/publications/water/anzecc-water-quality-guide-02/revision-water-quality.html. Adopted from Pearson (1960)

USSL Diagram for Water Quality Characterization

High SAR adversely affects the crop yield but not in isolation from EC of water. Thus, Richards (1954) combined the effect of EC and SAR in a USSL diagram which has been frequently used and quoted to categorize irrigation waters. Some confusion has emerged on the use of this diagram. In this diagram, for a better quality rating of the water, water having high EC needs to have low SAR. Some users compare it with Rhoades diagram or FAO data set wherein high SAR water should have high EC to maintain the infiltration rate. This opposing contentions cause confusion. It should be noted that both these classifications are complementary and not contrary to each other. While the former is related to crop performance, latter is to maintain the soil's infiltration rate. Studies have proved that for the same EC of water, high SAR water will produce lesser yields than low SAR water (Table 6.15). Similar results have also been reported by Minhas and Gupta (1992). Notwithstanding other limitations, if any, USSL diagram can be safely used to have a first guess of EC and SAR hazard of poor quality irrigation waters, in specific situations.

Table 6.14 Effect of saline water (EC 8 dS m^{-1}) with varying SAR on yields of two cropping sequences (2003-07)

SAR	Sequence 1		Sequence 2	
	Pearl millet	Wheat	*Dhaincha*	Wheat
control	3.03	4.14	2.09	3.69
10	1.94	3.78	1.79	3.36
20	1.53	3.52	1.67	3.12
30	1.06	2.92	1.58	2.43
40	0.56	2.58	1.44	2.10

Source: Anonymous (2009)

Absolute Concentrations of Carbonate/Bicarbonate and Residual Sodium Carbonate

When water containing carbonates is applied, calcite is formed and Ca concentration in the soil solution goes a sea change. As such Ca does not counteract the negative effects of Na and problems related to high Na are exuberated. Such waters may cause severe problems if used for irrigation through overhead sprinklers resulting in precipitation of calcium carbonate forming a white scale on leaves and fruits and possibilities of clogging of the irrigation equipment. Absolute concentration of bicarbonate ions in amounts greater than 10.0 meq L^{-1} also affects plant growth in other ways such as: by affecting the soil structure and mineral nutrition and by reducing the availability of iron in many plants thereby causing iron chlorosis.

Two indices are mainly used to assess the carbonates hazard of irrigation water namely direct measurement of carbonate and bicarbonate (Table 6.14) and the residual sodium carbonate (RSC). Often the negative impact of absolute concentrations is ignored if RSC values are within the desired range, which may not always be true as shown later in this section.

In order to have an idea of the sodic hazard of irrigation water, concept of residual sodium carbonate (RSC) has been propounded, which is given as:

$$RSC = (CO_3 + HCO_3) - (Ca + Mg) \quad (6.6)$$

The basic premise is that it is not the absolute bicarbonate and carbonate concentrations that are important, but instead, the relative concentrations of bicarbonate and carbonate compared to concentrations of calcium and magnesium.

Table 6.15 Potential limitation of irrigation water due to carbonate level

Carbonates (meq L^{-1})	Potential limitation
<1.5	Generally safe for irrigation
1.5 - 8.5	Increasing problem
>8.5*	Severe problem

In literature a value of 7.5 meq L^{-1} is also given; Camberato (2001). http://www.scnla.com/Irrigation_Water_Quality.pdf

This index is also quite controversial. While it appears in few reports, in others it is not even mentioned. In some reports it has been included in initial versions but does not find mention in the revised editions. Argument is also made that this parameter only indicates whether the calcite is likely to be formed or not and has no diagnostic value. It is the ultimate SAR or Adj. R_{Na} likely to remain in the soil solution that governs the quality of irrigation water. In fact, Adj. R_{Na} does not consider RSC but the absolute values of HCO_3 and Ca present in the water without making any distinction for waters having RSC or no RSC. If RSC guidelines are any indications, very low values of RSC have been recommended for no or even for severe hazard. Most reports propose severe damage at RSC (meq L^{-1}) of 2.5 (Table 6.16). Under the Indian condition, this value is considered safe.

Table 6.16 Bicarbonate (HCO_3) hazard of irrigation water (meq L^{-1})

RSC	None	Slight to moderate	Severe
International	1.25	1.25-2.5	> 2.5
India	<2.5	2.5-4.0	>4.0

Irrespective of what is stated, the parameter is widely used in India. Nonetheless, it needs a fresh look especially the way RSC is calculated. Since the conditions required to precipitate Mg do not prevail in agricultural lands, it is suggested that Mg may be taken out from the RSC equation. In low salinity waters which do not contain significant amounts of carbonate ions, the eq. (6.6) can be written as below because HCO_3 ions do not precipitate Mg ions.

$$RSC = HCO_3 - Ca \qquad (6.7)$$

and called this as RSBC (Residual Sodium Bicarbonate) (Gupta, 1984).

Based on EC_{iw}, SAR and RSC, a Committee of Experts grouped the water into four main classes (Table 6.1). Saline and sodic water have further been sub-grouped into three classes each to grade the severity of the problems.

Eaton (1950) used two more terms i.e. soluble sodium percentage (SSP) and SSP (possible) to describe the hazard of water that contained high concentration of bicarbonate ions.

$$SSP = (Na * 100)/ (Ca + Mg + Na) \qquad (6.8)$$

$$SSP\ (possible) = (Na * 100)/ [(Ca + Mg + Na) - (CO_3 + HCO_3)] \qquad (6.9)$$

Such that deduction part in eq. (6.9) does not exceed (Ca + Mg). In essence it combines the SSP and residual sodium carbonate (RSC) as follows:

$$SSP\ (possible) = [(Na+K) * 100] / [(Na+K) - RSC] \qquad RSC < 0.0 \qquad (6.10)$$

For RSC > 0.0, SSP (Possible) =100%

Apparently, SSP (possible) is higher than SSP. It may be interesting to note that while RSC < 0.0 are of little concern, yet even the negative values of RSC are used in eq. (6.10). The guidelines to characterize the water on the basis of SSP are given in Table 6.12.

Wilcox (1955) proposed a diagram relating SSP and EC and rated the water quality as excellent to good, permissible to doubtful, doubtful to unsuitable and unsuitable (Table 6.12). Since there are no guidelines on SSP (possible), the guidelines for SSP can be used for SSP (possible) as well. The values of SSP or SSP (possible) are rarely used in India to assess the water quality although it may provide valuable information on the water quality.

In spite of many suggestions and proposals, no clear alternative solution has emerged although few new equations have been proposed to overcome the issue related to precipitation and dissolutions of Ca. For Example; Ayers and Westcot (1976) proposed the parameter adj. SAR. It is based on the postulation of the concept of Saturation Index (SI) as per the following equation (Langelier, 1936):

$$SI = pH_a\ (\text{Actual pH of water}) - pH_c\ (\text{calculated pH}) \qquad (6.11)$$

The positive values of SI indicate that $CaCO_3$ will precipitate while negative values indicate that $CaCO_3$ of the soil will dissolve in the water. The pH_c, a theoretical parameter is calculated pH of the irrigation water in contact with lime and in equilibrium with soil CO_2. It is given by the following equation:

$$pH_c = (pk_2 - pk_c) + p(Ca) + P(Alk) \qquad (6.12)$$

Since the values of pH_c above 8.4 indicate a tendency to dissolve lime from the soil and values below 8.4 indicate a tendency to precipitate lime from the applied water, Bower (1961) replaced pH_a by 8.4 in eq. (6.11) and p(Ca) by p(Ca+Mg) assuming that Ca and Mg behave in a similar fashion. In eq. (6.12) pk_2 and pk_c are the negative logarithms of the second dissociation constant of H_2CO_3 and solubility product of $CaCO_3$, respectively both corrected for ionic strength. Also p(Ca+Mg) and p (Alk) are the negative logarithm of the molal concentration of (Ca+Mg) and equivalent concentration of titratable bases (CO_3+HCO_3). pH_c is calculated using Table 6.17, which relate to the concentration values from the water analysis. The table values of ($pk_2 - pk_c$) are calculated using the sum of (Ca + Mg +Na) in meq L^{-1} in the concentration column 1, p(Ca+Mg) using the sum of (Ca + Mg) in meq L^{-1} and p (alk) using the sum of (CO_3 + HCO_3) in meq L^{-1}. These values are then used to calculate the pH_c (Ayers and Wescot, 1976) and Adj. SAR as per the following equation.

$$Adj.\ SAR = SAR\ (1 + 8.4 - pH_c) \qquad (6.13)$$

Such that pH_c in eq. (6.13) is calculated from eq. (6.12) with (Ca + Mg) as was proposed by Bower (1961). The tests on this equation proved that it over predicts the sodium hazard. For several waters, values of Adj. SAR are nearly double the values of SAR (Table 6.18, Minhas and Gupta, 1992) but can be up to 3 times the SAR values. Some workers attempted to use the eq. (6.13) taking Ca ignoring Mg to calculate p(Ca) instead of p(Ca +Mg). Although it resulted in lower values of adj. SAR, but that didn't resolve the overestimation problem (Table 6.18). The overestimation has also been recognized while developing guidelines for SAR and Adj. SAR in some countries, being 2-3 times higher for Adj. SAR than SAR (Table 6.19). Therefore, it is recommended that use of adj. SAR may be dispensed with except for studies related to its improvement.

While Rhoades (1968) agreed with the contention of Bower (1961) to replace Ca by (Ca+Mg), Suarez (1981) did not. He observed that magnesium carbonate does not precipitate. Magnesium precipitation in irrigated soils is only under agriculturally unusual conditions (Suarez, 1981). While carbonate completely precipitates the calcium, bicarbonate may precipitate calcium only partially under certain conditions increasing the proportion of the sodium in the soil solution but in no case precipitation approach completeness as per the following reaction.

$$Ca + 2HCO_3 = CaCO_3 + H_2O + CO_2 \qquad (6.14)$$

Table 6.17 Values* of pk_2-pk_c, p(Ca+Mg) and p(Alk) for calculating pH_c

Concentration	$pK_2 - pK_c$	P(Ca+Mg)	pAlk
0.05	2.0	4.6	4.3
0.10	2.0	4.3	4.6
0.15	2.0	4.1	3.8
0.20	2.0	4.0	3.7
0.30	2.0	3.8	3.5
0.50	2.1	3.6	3.3
0.75	2.1	3.4	3.1
1.00	2.1	3.3	3.0
1.25	2.1	3.2	2.9
1.50	2.1	3.1	2.8
2.00	2.2	3.0	2.7
2.50	2.2	2.9	2.6
3.00	2.2	2.8	2.5
4.00	2.2	2.7	2.4
5.00	2.2	2.6	2.3
6.00	2.2	2.5	2.2
8.00	2.3	2.4	2.1
10.00	2.3	2.3	2.0
12.50	2.3	2.2	1.9
15.00	2.3	2.1	1.8
20.00	2.4	2.0	1.7
30.00	2.4	1.8	1.5
50.00	2.5	1.6	1.3
80.00	2.5	1.4	1.1

* Ayers and Westcot (1976)

Table 6.18. SAR, Adj. SAR with and without modification

EC	SAR	Adj. SAR, p(Ca+Mg)	Adj. SAR, with p(Ca)
3.2	21.4	38.2 (1.8)*	ND
1.5	19.5	41.0 (2.1)	33.1
1.4	15.8	31.6 (2.0)	25.3
1.4	13.5	26.7 (2.0)	19.6
1.4	15.8	31.6 (2.0)	25.3
1.5	11.6	25.6 (2.2)	17.4
3.0	25.0	52.6 (2.1)	35.0
1.1	4.9	10.4 (2.1)	ND
1.4	15.8	33.0 (2.1)	27.2

ND not determined because Ca+Mg were reported together; Values in Parenthesis are ratios of Adj. SAR/SAR

To assess the dissolution and precipitation of calcium, therefore, Suarez (1981) proposed another equation and designated it as adjusted R_{Na}. It is given as follows:

$$\text{Adj.}\, R_{Na} = \frac{Na}{[(Ca_x + Mg)/2]^{1/2}} \tag{6.15}$$

Table 6.19. Guidelines for use of SAR and Adj. SAR of waters

Application	Anticipated problems				
	None	Increasing	Significant	High	Severe
SAR					
Most production systems	<1	1-2	2-4	4-5	>5
Adj. SAR					
Direct plant impact	<3	3-6	6-8	8-9	>9
Permeability of natural soil	<6	6-7	7-8	8-9	>9

Anonymous. Guide to Interpreting Irrigation Water Analysis. Spectrum Analytic Inc. Washington C.H. Ohio USA. 20 p. Comment: SAR values seem to be quite low and higher limits are used in India.

Here Ca_x is the value of Ca in the water that takes care of EC_{iw} and HCO_3/Ca ratio of the water and the estimated partial pressure of CO_2 in the surface few mm of the soil (Table 6.20). The Ca_x values represents Ca (meq L^{-1}) expected to remain in the soil solution at equilibrium. Interestingly, development of the index states that there is no precipitation of Mg. It only utilizes Ca, HCO_3 and EC of the water to calculate Ca_x. Minhas and Sharma (2006) stated that as on today, this index is most appropriate to calculate ESP development in soils. If it is so, then the role of Mg in various equations especially in RSC needs to be reassessed. Minhas and Gupta (1992) also mentioned that Adj. R_{Na} equation does not account for CO_3 in water although it is known that carbonates are more hazardous to soils than HCO_3. Currently, this issue can be resolved by

Table 6.20. Calcium concentration (Ca_x) expected to remain in near-surface soil-water following irrigation

HCO_3 / Ca	Salinity of applied water (EC_{iw}) (dS m^{-1})											
	0.1	**0.2**	**0.3**	**0.5**	**0.7**	**1.0**	**1.5**	**2.0**	**3.0**	**4.0**	**6.0**	**8.0**
0.05	13.20	13.61	13.92	14.40	14.79	15.26	15.91	16.43	17.28	17.97	19.07	19.94
0.10	8.31	8.57	8.77	9.07	9.31	9.62	10.02	10.35	10.89	11.32	12.01	12.56
0.15	6.34	6.54	6.69	6.92	7.11	7.34	7.65	7.90	8.31	8.64	9.17	9.58
0.20	5.24	5.40	5.52	5.71	5.87	6.06	6.31	6.52	6.86	7.13	7.57	7.91
0.25	4.51	4.65	4.76	4.92	5.06	5.22	5.44	5.62	5.91	6.15	6.52	6.82
0.30	4.00	4.12	4.21	4.36	4.48	4.62	4.82	4.98	5.24	5.44	5.77	6.04
0.35	3.61	3.72	3.80	3.94	4.04	4.17	4.35	4.49	4.72	4.91	5.21	5.45
0.40	3.30	3.40	3.48	3.60	3.70	3.82	3.98	4.11	4.32	4.49	4.77	4.98
0.45	3.05	3.14	3.22	3.33	3.42	3.53	3.68	3.80	4.00	4.15	4.41	4.61
0.50	2.84	2.93	3.00	3.10	3.19	3.29	3.43	3.54	3.72	3.87	4.11	4.30
0.75	2.17	2.24	2.29	2.37	2.43	2.51	2.62	2.70	2.84	2.95	3.14	3.28
1.00	1.79	1.85	1.89	1.96	2.01	2.09	2.16	2.23	2.35	2.44	2.59	2.71
1.25	1.54	1.59	1.63	1.68	1.73	1.78	1.86	1.92	2.02	2.10	2.23	2.33
1.50	1.37	1.41	1.44	1.49	1.53	1.58	1.65	1.70	1.79	1.86	1.97	2.07
1.75	1.23	1.27	1.30	1.35	1.38	1.43	1.49	1.54	1.62	1.68	1.78	1.86
2.00	1.13	1.16	1.19	1.23	1.26	1.31	1.36	1.40	1.48	1.54	1.63	1.70
2.25	1.04	1.08	1.10	1.14	1.17	1.21	1.26	1.30	1.37	1.42	1.51	1.58
2.50	0.97	1.00	1.02	1.06	1.09	1.12	1.17	1.21	1.27	1.32	1.40	1.47
3.00	0.85	0.89	0.91	0.94	0.96	1.00	1.04	1.07	1.13	1.17	1.24	1.30
3.50	0.78	0.80	0.82	0.85	0.87	0.90	0.94	0.97	1.02	1.06	1.12	1.17
4.00	0.71	0.73	0.75	0.78	0.80	0.82	0.86	0.88	0.93	0.97	1.03	1.07
4.50	0.66	0.68	0.69	0.72	0.74	0.76	0.79	0.82	0.86	0.90	0.95	0.99
5.00	0.61	0.63	0.65	0.67	0.69	0.71	0.74	0.76	0.80	0.83	0.88	0.93
7.00	0.49	0.50	0.52	0.53	0.55	0.57	0.59	0.61	0.64	0.67	0.71	0.74
10.00	0.39	0.40	0.41	0.42	0.43	0.45	0.47	0.48	0.51	0.53	0.56	0.58
20.00	0.24	0.25	0.26	0.26	0.27	0.28	0.29	0.30	0.32	0.33	0.35	0.37
30.00	0.18	0.19	0.20	0.20	0.21	0.21	0.22	0.23	0.24	0.25	0.27	0.28

Adapted from Suarez (1981); Assumes a soil source of calcium from lime ($CaCO_3$) or silicates; no precipitation of magnesium, and partial pressure of CO_2 near the soil surface (pCO_3) is 0.0007 atmosphere; Ca_x, is in meq L^{-1}.

taking CO_3 + HCO_3 together especially because in most cases these ions are reported together. The procedure to calculate various quality indices is illustrated in Example 1.

Example 1: Water analysis report of a given water sample is

Ca = 2.54 meq L^{-1} Mg = 1.34 meq L^{-1}

Na = 6.87 meq L^{-1} EC = 1.1 dS m^{-1}

CO_3 = 0.38 meq L^{-1} HCO_3 = 4.21 meq L^{-1}

Find RSC, SAR, Adj. SAR and adj. R_{Na}

$$RSC = (CO_3 + HCO_3) - (Ca + Mg) \text{ meq } L^{-1}$$

$$= (0.38+4.21) - (2.54+1.34) = 4.59 - 3.88 = 0.71 \text{ meq } L^{-1}$$

$$SAR = \frac{Na}{\sqrt{(Ca + Mg)/2}}$$ Where Na, Ca and Mg are in meq L^{-1}

$$= \frac{6.87}{\sqrt{(2.54 + 1.34)/2}} = 4.93$$

$$Adj.\,SAR = \frac{Na}{\sqrt{(Ca + Mg)/2}}[1 + (8.4 - pHc)]$$

Sum of Ca + Mg +Na = 2.54+1.34+6.87 = 10.75 meq L^{-1}

($pk_2 - pk_c$) obtained from the Table 6.17 using sum of Ca + Mg +Na = 10.75 meq L^{-1} is 2.3

Sum of Ca+Mg = 2.54+1.34 = 3.88 meq L^{-1}

p(Ca+Mg) obtained by using sum of Ca+Mg = 3.88 meq L^{-1} is 2.7

Sum of CO_3 + HCO_3 = 0.38 + 4.21 = 4.59 meq L^{-1}

p(alk) obtained by using the sum of CO_3 + HCO_3 = 4.59 meq L^{-1} is 2.3

pH_c = 2.3+2.7+2.3 = 7.3

Adj. SAR = 4.93 [1+(8.4-7.3) = 4.93[2.1] = 10.35

Note that the value of Adj, SAR is more than double the value of SAR

$$Adj.\,R_{Na} = \frac{Na}{\sqrt{(Ca_x + Mg)/2}}$$

EC is 1.1dS m^{-1}, HCO_3/Ca ratio 4.21/2.54 = 1.66

Ca_x value from Table 6.20 corresponding to the nearest values of HCO_3/Ca and EC_{iw} is 1.43

$$= \frac{6.87}{\sqrt{(1.43 + 1.34)/2}} = 6.87/1.18 = 5.82$$

Leaching Fraction and Water Sodicity Parameters

Leaching fraction is another important parameter that has been introduced in some of the equations. It has been observed that precipitation of $CaCO_3$ decreases with increasing LF. Thus, adj. SAR equation has been modified as below (Bower, 1961).

$$Adj.\,SAR = \left\{\frac{1}{LF}\right\}^{1/2} SAR_{iw}\,\{1 + (8.4 - pH_c)\} \quad (6.16)$$

Where LF is the leaching fraction. Rhoades (1968) further modified the equation as follows:

$$SAR_{dw} = \frac{Y^{1+2LF}}{(LF)^{1/2}} SAR_{iw} \{1 + (8.4 - pH_c)\} \quad (6.17)$$

Here Y is an empirically determined mineral weathering co-efficient usually equal to 0.65. Rhoades (1968) observed that while eq. (6.16) would predict the ESP of the soil surface, eq. (6.17) can be used to assess the ESP of the soil at the bottom of the root zone. Clearly, ESP at the soil surface is higher than at the bottom. But once it is observed that eq. (6.13) over predicts the sodium hazard, eq. (6.16) would also over predict the adverse effects as it is similar to eq. (6.13) at LF =1.

Minhas and Sharma (2006) on the other hand proposed that Adj. R_{Na} concept should be used to predict the sodic hazard of waters taking into account the dilution and dissolution processes commonly occurring under field conditions. As such, they modified eq. (6.15) by considering a concentration factor, CF and a simple dissolution factor, DF as follows:

$$CF = (Dlw/ET) + (Dlw/Drw)$$

$$DF = Drw / Dlw$$

Here, Dlw is the amount of sodic water applied (mm), Drw is the annual rainfall and ET is the annual crop evapotranspiration requirement. According to them eq. (6.15) then could be modified to:

$$\text{Adj. } R_{Na} \text{ (mod.)} = (CF/DF)^{1/2} * \text{Adj. } R_{Na}$$

$$\text{Adj. } R_{Na} \text{ (mod.)} = [1/(Drw/Diw)\ \{ET/(Drw+ET)\}^{1/2}] * \text{Adj. } R_{Na} \quad (6.18)$$

$$= A * \text{Adj. } R_{Na} \quad (6.19)$$

Values of A for the data reported by Minhas and Sharma (2006) are given in Table 6.21 for various cropping systems and climatic conditions. The values vary with rainfall, irrigation and ET depending upon the crop rotation. It may be noted that when the rainfall is in 60-70 cm range, factor A follows the order Rice-wheat > Maize/millet-wheat > cotton-wheat >fallow-wheat. But if rainfall is less as in two cases of fallow-wheat (30-40 cm), the factor in fallow -wheat is even higher than Maize/millet-wheat. All the data sets given in Table 6.21 are for the alluvial soils of Punjab and Haryana. Factors for other soils and climatic conditions need to be worked out.

In spite of the limitation as revealed through various developments in this section and tests failing to validate the universal application of even the modified equations, use of SAR given by eq. (6.1) or eq. (6.2) and RSC given by eq. (6.6) is still practised widely to characterize and arrive at management options for use of poor quality waters. However, there is a need to go deeper into the issues raised to arrive at more theoretically sound characterization scheme for poor quality sodic ground water.

Table 6.21 Values of A of eq. (6.19) for various agro-climatic and cropping conditions

Rainfall (D_{rw}) (cm)	Irrigation water (D_{iw}) (cm)	ET (cm)	Modification factor, A = $[1/(D_{rw}/D_{iw}) \{ET/(D_{rw}+ET)\}^{1/2}]$
71	38	40	0.89 (Fallow-wheat)
70	38	40	0.90 (Fallow-wheat)
35	40	42	1.55 (Fallow-wheat)
30	40	42	1.75 (Fallow-wheat)
72	53	105	0.96 (Cotton-wheat)
60	130	105	2.72 (Rice-wheat)
72	137	100	2.50 (Rice-wheat)
77	142	100	2.45 (Rice-wheat)
70	56	80	1.10 (Maize/millet-wheat)
72	62	80	1.19 (Maize/millet-wheat)

INFILTRATION/SOIL STRUCTURE PROBLEMS

High sodium concentrations besides affecting crop yield adversely impact the infiltration rate, hydraulic conductivity and structural stability of the soils. In some cases, rate may be reduced to such an extent that the crop may not be able to draw enough water. The hydraulic conductivity of the soil profile might become too low to ensure adequate transport of soil moisture and even result in impeded drainage. Other problems that may be encountered are crust formation, temporary saturation of the surface soil, high pH and increased potential for diseases, weeds, soil erosion, lack of oxygen and inadequate nutrient availability. To assess irrigation water response to likely changes in infiltration and or on soil structure, several Tables and diagrams have been proposed. In some only EC and SAR are considered. Since SAR does not account for precipitation of Ca in the soil solution, the second set also takes into account the anionic composition of water. It may be mentioned that irrigation waters in India are rarely examined for this hazard although it may be important consideration in deciding about the application of an amendment.

SAR and Soil Salinity

Values of EC that must be used with a given SAR to avoid adverse impact on infiltration rate and soil structure are given in Table 6.22. It may be noted that for extremely low salinity water, even low SAR may result in problems. Since it is difficult to specify universally applicable critical values of SAR and EC of irrigation water, a diagram has been devised by which the permeability hazard is assessed by ascertaining whether the SAR-EC combination lies to the left (problem likely) or to the right (no problem likely) of the threshold line (Fig. 6.4). Since SAR of the top soil is in near equilibrium with SAR of irrigation water when EC_e is equal to EC_{iw}, it can be concluded that the irrigation waters having SAR, 10, 20 and 30 should not have EC of water less than 1, 2 and 3 dS

m^{-1}, respectively (Gupta, 1990). The tabulated values reported by Ayers and Westcot (1985) are generally higher than these values (Table 6.22).

Table 6.22 Irrigation water quality criteria (EC, dS m^{-1}) for various SAR values of water

SAR	EC for various degrees of restriction on use		
	No	Slight to moderate	Severe
< 3	<0.7	0.7–0.2	>3
3-6	>1.2	1.2-0.2	<0.2
6-12	>1.9	1.9-0.5	<0.5
12-20	>2.9	2.9- 0.5	<0.5
20-40	>5	5 - 2.9	<2.9

Source: Ayers and Westcot (1985)

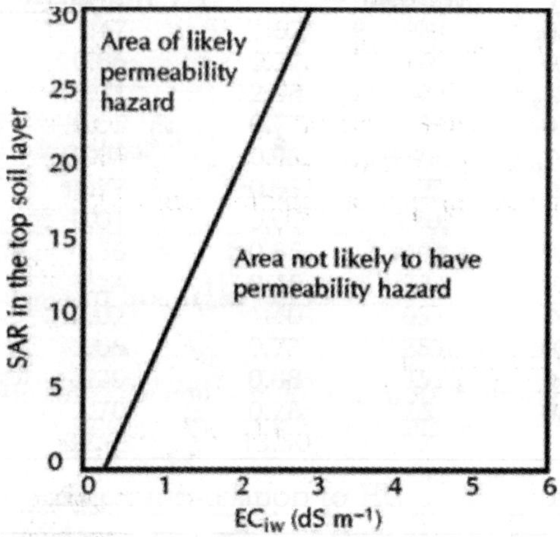

Fig. 6.4. Relative rate of water infiltration as affected by salinity and SAR (Rhoades, 1977)

To predict soil structural stability, a diagram showing relationship between SAR and EC has been proposed (Fig. 6.5). Water quality that falls to the right of the dashed line is unlikely to cause soil structural problems. Water quality that falls to the left of the solid line is likely to induce degradation of soil structure calling for corrective management (e.g. application of gypsum or some suitable amendment). Water that falls between the lines is of marginal quality and should be treated with caution depending upon the soil properties and rainfall.

Permeability Index

According to Doneen (1964), the permeability Index (PI) is calculated by using the following equation:

$$PI = 100 * (Na + \sqrt{HCO_3}) / (Na + Ca + Mg + K) \quad (6.20)$$

It may be noted that unlike other equation/diagrams, it takes into account the bicarbonate content of the water, although does not account for carbonates. Nonetheless, it may be possible to modify the equation by adding K and CO_3 in the numerator. The classification of the water is then made in the three groups shown in Fig. 6.6. Class 1 shows no problem while class II and Class III mean 25 and 75% reduction in permeability.

As an illustration, all these parameters for 7 water samples from Haryana have been calculated (Table 6.23). Some of these parameters have also been used in Fig. 6.5 and Fig. 6.6 to illustrate their use in identifying the problems. It has emerged that even a good quality water, if low in electrolyte concentration can pose permeability problems.

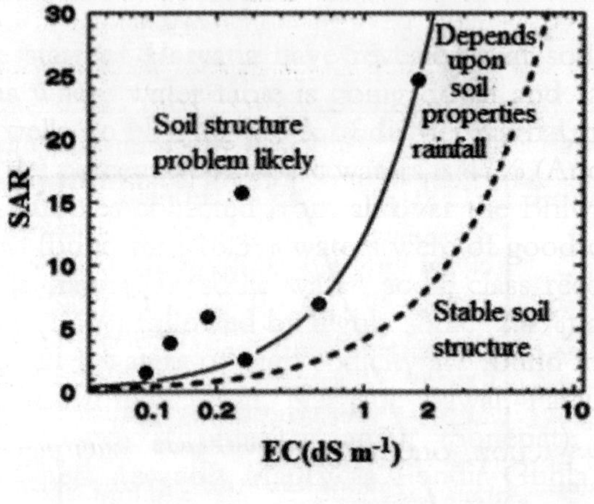

Fig. 6.5 Relationship between SAR and EC of irrigation water for predicting soil structural stability (Source: ANZECC, 2000)

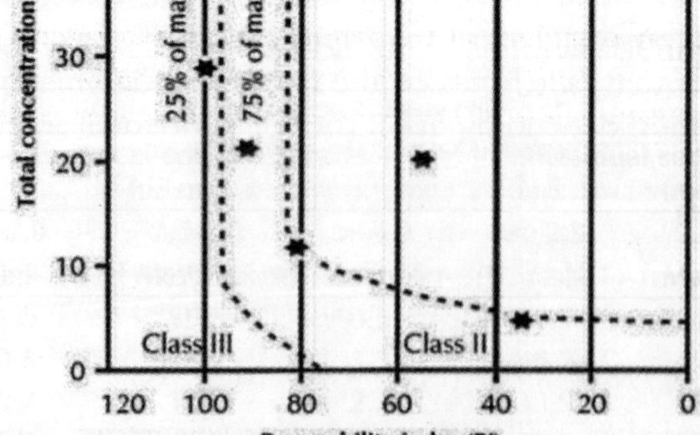

Fig. 6.6 Classification of irrigation water on the basis of permeability index

Table 6.23 Quality parameters of various kinds of water categorized in Table 6.1

Parameter	Good	Marginally saline	Saline	High SAR saline	Marginally sodic	Sodic	High SAR sodic
SAR (Eq. 4) $(meq\ L^{-1})^{1/2}$	1.6	2.9	8.2	24.8	3.9	7.2	15.7
SAR (Eq. 6) $(meq\ L^{-1})^{1/2}$	1.6	4.0	10.7	24.8	5.0	7.4	15.7
% change in SAR	-	39.8	29.7	-	27.6	3.1	-
Adj. SAR $(meq\ L^{-1})^{1/2}$	3.2	8.1	19.0	62.7	7.2	18.2	33.5
Adj. R_{Na} $(meq\ L^{-1})^{1/2}$	1.8	3.1	9.1	29.4	4.3	8.8	19.5
RSC $(meq\ L^{-1})$	0.4	1.5	Nil	Nil	2.8	7.2	6.8
RSC $(meq\ L^{-1})$	2.1	10.2	Nil	Nil	6.1	10.3	8.5
Mg/Ca ratio	0.8	3.2	2.4	0.6	2.2	1.15	0.8
Cl/SO_4 ratio	2.4	1.9	1.3	2.8	7.2	2.2	0.7
SSP	36.6	35.5	52.3	74.6	56.0	67.8	85.1
SSP possible	100	100	58.3	75.9	100	100	100
PI (%)	33(II)*	54 (I)	57 (I)	76 (I)	81 (II)	88 (II)	98 (III)

*values in parenthesis indicate the class as per Fig. (6.6)

PREDICTION OF SOIL ESP

Continuous use of sodic water results in build-up of soil ESP over time. To initiate any corrective action, one must be able to predict the ESP build-up. Many equations available in the literature relate ESP of the soil with SAR of the water. On the contrary, Indian experiences have revealed that if SAR of the water is high due to neutral salts, build-up of ESP is not to the extent that it could adversely affect the crops. Moreover, high SAR of the soil solution due to neutral salts gets reduced with leaching as is the case with saline water resulting in their characterization as high SAR saline (Table 6.1). Nonetheless, tests have revealed the usefulness of these approaches. More rigorous tests however, are needed to come to any firm conclusion under Indian conditions where monsoon climatic conditions prevail.

According to Richards (1954), the relationship between SAR of the water and soil ESP is give by the following equation:

$$ESP = \frac{100\,(-0.0126 + 0.01475 * SAR)}{1 + (-0.0126 + 0.01475 * SAR)} \tag{6.21}$$

Since SAR of the soil water could be 1 to 3 times more than the irrigation water, higher ESP build-up may be expected than given by eq. (6.21).

Bower and Maasland (1963) used eq. (6.16) to predict the sodium hazards of Punjab (Pakistan) ground water as follows:

$$ESP = 2\ SAR\ \{1 + (8.4 - pH_c)\} \tag{6.22}$$

Factor 2 has been obtained by assuming LF = 0.25 such that $1/(LF)^{1/2}$ is equal to 2.

Bajwa *et al.* (1983) reported that observed ESP is quite similar to SAR_{dw} (Rhoades, 1968) but the differences between observed ESP and predicted SAR increased with increase in $SAR_{adj.}$ of the irrigation water. Bajwa *et al.* (1992, 1998) compared different equations to predict the hazard of irrigation with sodic and saline-sodic waters in a cotton-wheat rotation. It was observed that modified Ayers and Westcot (1985) equation predicted the developed ESP for all sodic and saline-sodic (high RSC) waters but none of the models could predict the build-up of ESP for waters with high SAR and low RSC for which SAR_{dw} equation performed better.

In a recent paper, Minhas and Sharma (2006) used data of large number of experiments including the above referred studies to develop and test eq. (6.19). Following observations emerged:

1. Adj. R_{Na} concept (Ayers and Westcot, 1985) suitably modified to take care of concentration factor that is dependent upon applied sodic water, evapotranspiration and annual rainfall and dissolution process, which is further dependent upon rainfall and applied irrigation water is able to predict the hazards of sodic waters. Predicted ESP using the modified adj R_{Na} concept is given by ESP = 0.96 * adj R_{Na} (mod). Thus, it was made possible to establish one to one relationship between ESP and the sodicity parameter adj. R_{Na} (mod), which has eluded so far.
2. It was also observed that the threshold limit of ESP tolerance of paddy/millet is relatively less when ESP build-up accompanies the application of sodic water compared to when the ESP gets reduced from an initially high value as in sodic land reclamation following gypsum application (Table 6.24). It seems that in many cases adverse effects are more pronounced in case of sodic water irrigated lands than for lands under reclamation (Minhas and Sharma, 2006).

Table 6.24 Comparative tolerance of crops to soil ESP in sodic water irrigated crops and under sodic land reclamation

Crop	Threshold ESP	
	Sodic water irrigated land*	**Soil reclamation#**
Wheat	16.2 (1.86)	16.4 (2.1)
Paddy	20.1 (1.62)	24.4 (0.9)$
Millet/maize	6.1 (1.34)	13.6 (2.6)
Cotton	14.9 (1.26)	NA

*Minhas and Sharma (2006); # Gupta and Sharma (1990); Values in parenthesis are the slope of Maas and Hoffman (1977) equation; $ Values may not be used to extend beyond ESP_{50} i.e. ESP at which yield is 50% of the optimum; NA = Not available

Based on eq. (6.19) Minhas and Sharma (2006) predicted the adj. R_{Na} of the irrigation waters that could sustain the yield of various cropping sequences under different rainfall conditions (Table 6.25).

Table 6.25 Predicted values of adj. R_{Na} of irrigation water for sustaining yield under different cropping sequences

Cropping sequence	Permissible adj. R_{Na} for sustainable yields in rainfall zones (cm)		
	<40	40-60	>60
Fallow-wheat	16	21	27
Maize/millet-wheat	14	17	23
Paddy-wheat	6	9	14
Cotton-wheat	14	20	26

Source: Minhas and Sharma (2006)

HARMFUL EFFECTS OF SODIC WATER

When sodic waters containing higher proportions of sodium with predominance of carbonates and bicarbonates in relation to calcium and magnesium are used for irrigation, calcium gets immobilized due to its precipitation as calcium carbonate, resulting in excessive saturation of soil with sodium. It results in increased pH and exchangeable sodium percentage. On a long-term basis, it might lead to formation of sodic soils. The high ESP and pH adversely affect the soil-physical properties due to breakdown of soil structure as a result of swelling and dispersion of clay particles. Thus, fine textured soils remain dispersed and puddled when wet and hard when dry. Effects become apparent following rainfall or an irrigation event, which leads to stagnation of water and resultant aeration problems. These soils do not easily attain proper soil moisture conditions for cultivation. A thin crust is often formed at the soil surface, which acts as a barrier to water penetration and seedling emergence. As such, these soils below a few cm of the surface might remain almost dry and hard, though water may be stagnating at the surface. On drying, soil becomes very hard, affecting the proliferation of roots. Increase in soil pH reduces the availability of a number of plant nutrients like nitrogen, zinc, iron etc. It also results in decreased availability of calcium and magnesium and toxicity of sodium. Such changes also cause nutritional imbalance in the plants as uptake of some elements may be suppressed while for others it might be enhanced. Direct injury by burning of stems and leaves has also been observed. One or a combination of these factors adversely affects the yield of crops although in most cases a combination of these factors is responsible for yield decline.

To illustrate, what is stated in the previous paragraph, yield data in respect of wheat, pea, gram and lentil reveal increasing yield reduction with increasing RSC of water (Table 6.26). Apparently wheat is more tolerant than the other three crops tested. Similarly, data in respect of soybean and mustard presented in Table 6.27 reveal that yield of both the crops declined with increasing EC and RSC of water although soybean seems to be more sensitive to sodic water irrigation than mustard (Anonymous, 2009). While the yield reduction in the case of saline water is due to build-up of soil salinity, in the case of sodic water, it is because of the build-up of ESP.

Table 6.26 Crop yields as affected by irrigation with sodic water in a sandy loam soil at Agra

RSC (meq L^{-1})	Crop yield (t ha^{-1})			
	Wheat	Pea	Gram	Lentil
Control	4.04	1.33	0.76	1.38
05	3.98	1.22	0.69	1.25
10	3.87	0.93	0.50	1.11
15	3.43	0.86	0.34	1.05
20	3.04	-	-	-
CD 5%	0.2	0.34	0.22	0.15

Table 6.27 Effect of irrigation with saline and sodic water on the yield (t ha^{-1}) of soybean and mustard

Treatments	Soybean			Mustard			
	2004	2005	Mean	2003–04	2004–05	2005–06	Mean
Saline water (dS m^{-1})							
BAW	2.10	1.80	1.90	1.90	2.30	2.60	2.30
4	1.90	1.20	1.60	1.80	2.20	2.40	2.10
8	1.50	1.20	1.30	1.60	2.10	2.30	2.00
12	0.90	0.60	0.80	1.20	1.70	1.70	1.50
LSD ($p=0.05$)	0.40	0.20	0.30	0.40	0.30	0.30	0.30
Sodic water (RSC, meq L^{-1})							
BAW	2.10	1.80	1.90	2.40	2.30	2.60	2.40
5	1.20	1.10	1.20	2.30	2.20	2.50	2.30
10	0.70	0.70	0.70	2.30	2.20	2.30	2.20
15	0.50	0.50	0.50	2.00	2.00	1.90	2.00
20	0.30	0.40	0.40	1.90	1.70	1.60	1.70
LSD ($p=0.05$)	0.30	0.20	0.20	0.30	0.40	0.20	0.30

The changes in physico-chemical properties of soils irrigated with sodic water is manifested through changes in EC, pH and ESP of the soils, being controlled by the EC, SAR, RSC of the irrigation water and the amount of water applied to irrigate the crops. Higher the quantity of water applied more would be the pH and ESP of the soil. On an average, pH and ESP of the soil irrigated with RSC water at Agra was much higher than that of canal water (Table 6.28). In a trial, the average values of ESP over a period of two years in 0-30 cm depth were 13, 17, 19 and 25 at RSC 5, 10, 15 and 20 meq L^{-1} respectively (Fig. 6.7). As a result of high pH and ESP soils tend to be dispersed and infiltration and penetration rates are reduced (Table 6.29). Crust formed on the soil surface also inhibits the water infiltration. Thus, following an irrigation event or rainfall, water gets stagnated on the soil surface causing aeration problems (Fig. 6.8).

A major difference between the two kinds of stresses caused by saline and sodic irrigation water is that sodicity build-up happens slowly over the years and as such the adverse effect of sodic water might not appear in the first few years. As such continuous application of sodic water results in increasing yield reductions over the years because of increasing build-up of ESP. On the other hand, in saline water irrigated lands, salinity build-up occurs over the season with increasing number of irrigations. Besides, cyclic variation in salinity is observed in monsoon climatic conditions. On the contrary, once ESP build-up happens, it is not reduced to appreciable level due to rainfall, although steady state ESP might get delayed by few years with conjunctive use of sodic and fresh water including rainwater.

Table 6.28 Effect of RSC water on soil properties with potato-sunflower rotation (Average of 5 years)

Treatments	Depth (cm)	At sowing			At harvest					
		Potato			Potato			Sunflower		
		EC_e	pH	ESP	EC_e	pH	ESP	EC_e	pH	ESP
Canal Water	0-15	2.6	7.6	9.0	2.7	7.6	9.8	2.8	7.6	10.8
	15-30	2.5	7.6	9.7	2.6	7.5	10.7	2.7	7.6	11.3
RSC water	0-15	4.6	9.0	22.1	4.7	9.0	29.9	4.8	9.1	36.7
(15 meq L^{-1})	15-30	4.5	8.8	25.0	4.6	8.8	32.6	4.6	9.0	42.6

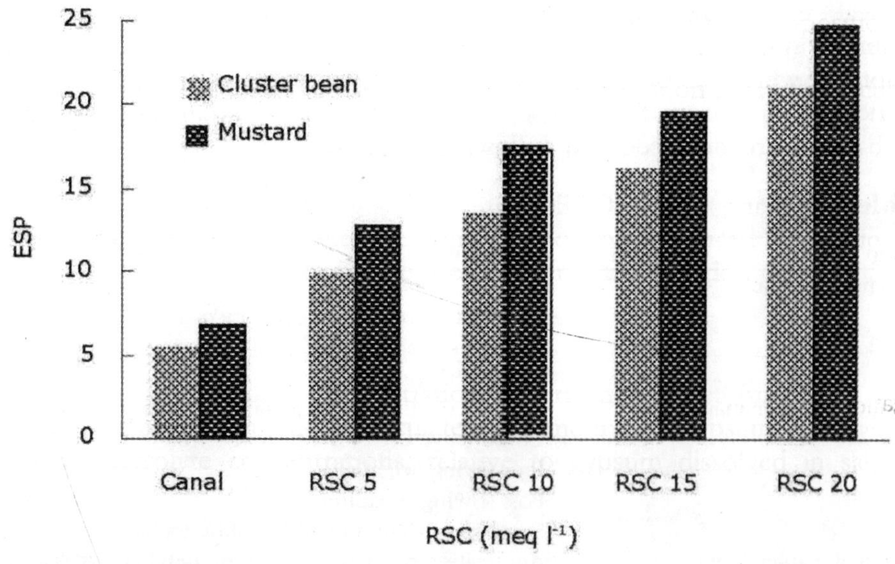

Fig. 6.7 ESP status after harvest of cluster bean and mustard (average of two years)

Table 6.29 Effect of gypsum and FYM treatments on infiltration rate of soil

Treatments	Infiltration (cm h^{-1})	Penetration rate
Sodic water	0.21	0.27
Sodic water with 100 % neutralization of RSC and 20 t ha^{-1} of FYM	0.29	0.41

Fig. 6.8 Water stagnation in a plot irrigated with sodic water for about a decade (AICRP, Hisar)

MANAGEMENT OF SODIC WATER IRRIGATED LANDS

Several options exist to manage sodic water for crop production (Table 6.30). Physical, economical and institutional factors determine which option or combination of options to use. Most of these strategies have been discussed in Chapter 4, which can be applied as such during the management of sodic water. A brief description of these and detailed discussions on strategies not included in Chapter 4 are included in the following sections.

Table 6.30 Management of sodic water

Group	Technology
Crop management	Crop selection
	Exploitation of varietal differences
	Growth stage sensitivity
	Agronomic practices
Irrigation water management	Shallow depth-high frequency irrigation
	Conjunctive use of multi-source waters
	Pre-sowing irrigation
	Post sowing irrigation
	Switchover to improved irrigation techniques
Chemical management	Application of chemical amendments
	Application of additional nutrients
Rainwater management	*In-situ* rainwater conservation
	Rainwater harvesting and reuse

Selection of Crops and Cropping Sequences

The inter and intra-genic variations in salt tolerance of plants can be exploited to select crops/cultivars that produce satisfactorily under a given root zone sodicity (Refer Chapter 4). Vegetable crops are usually known to be relatively less tolerant to sodic conditions. Yet there is a scope to select vegetable crops to suit root zone ESP (Table 6.31). In a recent study, Choudhary *et al.* (2009) reported that out of 3 high yielding wheat varieties PBW 343 fared significantly better than PBW 502 and PBW 550 with high RSC waters of 6.5 and 10 $mmol_c\ L^{-1}$ producing 90 and 72% relative yield compared to 85-87% and 58-64% for the other two varieties.

Table 6.31 Tolerance of vegetable crops to soil ESP

Sensitive crops (ESP < 20)	Semi-tolerant crops (ESP 20-40)	Tolerant crops (ESP > 40)
Pea, cowpea, cluster bean, ginger, turmeric	Tomato, garlic, okra, radish, carrot, cauliflower, chilli, onion, potatoes, ash gourd, coriander, Fenugreek, fennel	Brinjal, spinach, sugar beet

Source: Phogat *et al.* (2010)

Besides what has been stated, general guidelines to select crops and cropping sequences are: tolerant or semi-tolerant crops having low water requirement such as barley, wheat, mustard, pearl millet and sorghum should be grown while crops with high water requirement such as rice, sugarcane and *berseem* should be avoided. In low rainfall areas (average annual rainfall <400 mm) if good quality canal water is unavailable, it is advisable to keep the fields fallow during *kharif* season. During *rabi,* only tolerant and semi-tolerant crops like barley, wheat and mustard should be grown. In areas having rainfall > 400 mm $annum^{-1}$, sorghum-wheat, guar-wheat, pearl millet-wheat and cotton-wheat rotations can be practised, provided sowing of *kharif* crops is completed with rainwater or good quality canal water. Not more than 2 to 3 irrigation should be applied with sodic water during the *kharif* season. Sodic water should not be used to grow summer crops.

Improved Agronomic Practices

Besides what has been stated in the foregoing sections, improved agronomic or cultural practices could help to get better yields in sodic water irrigated lands. Some of the options are:

Pre-sowing irrigation

In most crops, from salt tolerance point of view, the period of germination and seedling emergence is the most critical stage. Failure of the crop at this stage scales up to poor crop stand. As a result, one ends up with considerable decrease in yield. By applying good quality water for pre-sowing irrigation, it is possible to reduce the risk of poor stand since germination is facilitated by fewer amounts of salts or low ESP in the root zone.

Other practices

- Since the seed mortality under saline/sodic water irrigated lands is more than the normal lands or lands irrigated with normal water, to have a good crop stand, apply at least 25% more seed rate in case of sodic water irrigated lands.

- If the fresh water is inadequate and is not available at the pre-sowing stage, crops could be sown with conserved moisture rather than after applying sodic water. A post-sowing irrigation with sodic water would help better germination of seeds.

Nutrient Management

To achieve optimum productivity, balanced use of essential nutrients is an important prerequisite. While it would be appropriate to apply nutrients as per regional recommendations for normal soils/waters or on soil test basis, some additional points should be considered.

- Application of 25% extra nitrogen is needed as compared to the normal conditions.
- Phosphorus, potassium and other limiting nutrients should be applied on soil test basis.
- Sodic waters high in bicarbonates decrease the availability of Zn, Mn, Fe and Cu. Significant decrease in DTPA extractable iron in soils irrigated with sodic water for 9 years was recorded by Bajwa *et al.* (1993). Application of Fe should be practised with use of amendments based on soil test reports. Similar to Fe, solubility and availability of Mn is also governed by pH and oxidation-reduction status of the soils after irrigation with sodic water. Presence of RSC in irrigation water decreases Mn availability to upland crops like wheat taken after low land rice. It is because under submerged conditions, Mn gets solubilized after conversion to reduced state and leached to deeper depths particularly in coarse textured soils. Since it also oxidizes readily under upland conditions, its soil application is ineffective. Repeated foliar application of $MnSO_4$ would be beneficial in overcoming its deficiency. Zinc sulphate @ 25 kg ha^{-1} should be added, particularly to winter crops. The application of Zn @ 25 and 50 kg ha^{-1} resulted in 11 and 19.6% increase in yield, respectively, as compared to control in a sodic water (RSC 11.0 meq L^{-1}) irrigated land (Table 6.32).

Table 6.32 Grain yield (t ha^{-1}) of wheat crop irrigated with sodic water in relation to Zn and gypsum application

Gypsum levels	Levels of Zn (kg ha^{-1})			Mean
	Control	25	50	
G_0*	4.53	6.53	8.73	6.60
G_{25}	15.90	18.50	20.00	18.13
G_{50}	23.70	26.23	27.80	25.91
G_{75}	28.05	30.10	31.90	30.01
G_{100}	29.70	31.80	33.50	31.66
Mean	20.38	22.63	24.39	
CD (5%)	Gypsum level (G) : 1.22	Levels of zinc : 0.94	G x D : NS	

*G_0, G_{25}, G_{50} and G_{100} means no gypsum and 25, 50, 75 and 100% neutralization of RSC through typsum

- Some sodic waters may be rich in nitrogen, potassium and sulphur. Such waters should be analysed and the fertilizer dose of nutrient should be adjusted accordingly.

Integrated nutrient management with heavy dressings of organic manures, regular incorporation of crop residues, application of organic materials such as rice husk, sawdust, sugar factory wastes, etc., have all been found useful in maintaining and improving soil physical properties and in counteracting the adverse effect of high levels of exchangeable sodium. Addition of organic materials mobilizes inherent Ca^{2+} from $CaCO_3$ and other minerals by organic acids and increases pCO_2 in soils. Although, some disagreement on short–term effects of organic matter on dispersion of sodic soil particles in soils undergoing sodication exists in literature (Gupta *et al.,* 1984), nevertheless, majority of the available reports still suggest overall beneficial positive role of organic manures. Recent observations have revealed that addition of organic amendments alone without gypsum is not capable to alleviate the harmful effects of sodic water. Organic manure (FYM) under sodic conditions enhances soil dispersion and decreases soil permeability and induces water stagnation. Therefore, its application should always be associated with some amendment. In that case, FYM not only supplies nutrients to the plants, but also plays an important role in improving soil physical properties. Besides, its application enhances leaching of salts accumulated in the root zone. The synergy of applying chemical amendment and organic manure in RSC water irrigated lands is depicted in Table 6.33 and Table 6.34. Its impact on soil physical properties is depicted in Table 6.35, which to some extent indicates that FYM alone is not much capable to improve soil properties although slight improvement in yields are noticed. The decrease in the infiltration rate at high FYM levels may be due to the dispersion of FYM under sodic condition which might have clogged the pores and resulted in reduced infiltration rate.

Table 6.33 Effect of gypsum with and without different organic manures on yield of rice and wheat and soil properties

Treatments	Mean Yield (t ha^{-1})		Soil pH$_2$	OC (%)
	Rice	Wheat		
N_{120} P_{26} K_{42} (100% recommended)	4.79	3.69	8.52	0.26
N_{120} P_{26} K_{42} + FYM	5.29	4.16	8.38	0.43
N_{120} P_{26} K_{42} + gypsum	5.23	4.10	8.18	0.37
N_{120} P_{26} K_{42} + Pressmud (PM)	5.31	4.46	8.29	0.42
N_{120} P_{26} K_{42} + FYM + gypsum	5.35	4.22	8.28	0.42
N_{120} P_{26} K_{42} + PM + gypsum	5.41	4.52	8.28	0.40
CD (P=0.05)	0.42	0.34	0.08	0.06
Initial soil properties			8.6	0.40

Source: Swarup and Yaduvanshi (2004); FYM = Farm yard manure; PM = Press mud

Table 6.34 Effect of gypsum and FYM treatments on mean yields (t ha^{-1}) of different crops

Treatments	Potato (3 years)	Tomato 3 years)	Brinjal (2 years)	Broccoli (4 years)	Cluster bean (3years)
FoGo	13.10	1.66	0.38	0.14	0.07
FoG1	21.10	23.60	16.73	4.32	9.57
FoG2	21.31	31.63	22.87	5.95	10.54
F1Go	19.09	2.29	1.35	1.45	0.19
F1G1	24.10	31.34	21.54	6.60	10.62
F1G2	24.86	40.07	26.43	9.29	11.44
F2Go	19.61	3.18	2.62	3.04	0.24
F2G1	25.28	30.42	23.65	8.13	11.10
F2G2	26.52	38.21	28.89	10.10	11.85
CD (5%)					
Gypsum	2.47	1.09	2.88	1.05	0.61
FYM	2.47	1.09	2.88	1.05	0.61
GXF	3.10	1.65	NS	NS	NS

Source: Phogat *et al.* (2010); Fo= no FYM; F1= 10 t ha^{-1}; F2= 20 t ha^{-1}; Go= no gypsum; G1= 50%; G2= 100% neutralization of RSC

Response of organic sources also varies with the nature of organic matter added. Sekhon and Bajwa (1993) observed that amongst various organic materials applied the response was in the order of paddy straw > green manure > FYM. Similarly, sodic water irrigation for 3 years increased the pH by 0.10 - 0.16 units where no or 100 % NP were used or only wheat residue was incorporated but the pH remained almost at the initial level when wheat residue was incorporated along with gypsum or sulphitation press mud (Table 6.36)

Table 6.35 Effect of gypsum and FYM treatments on infiltration and penetration rate of soil

Treatments	Infiltration (cm h^{-1})		Penetration rate
	2003-04	2005-06	2003-04
F0G0	0.21	0.18	0.27
F0G1	0.18	0.29	0.28
F0G2	0.32	0.36	0.35
F0G3	0.36	0.40	0.37
F0G4	0.41	0.48	0.38
F1G0	0.19	0.16	0.29
F1G1	0.25	0.24	0.32
F1G2	0.27	0.25	0.35
F1G3	0.31	0.32	0.37
F1G4	0.35	0.37	0.40
F2G0	0.18	0.13	0.33
F2G1	0.22	0.20	0.36
F2G2	0.24	0.22	0.38
F2G3	0.26	0.25	0.39
F2G4	0.29	0.27	0.41
CD (5 %)			
Gypsum	0.019	0.02	0.02
FYM	0.015	0.014	0.015
GXF	0.033	0.035	NS

Source: Phogat *et al.* (2010); Fo= no FYM; F1= 10 t ha^{-1}; F2= 20 t ha^{-1}; Go= no gypsum; G1= 25%; G2= 50%; G3=75% and G4=100% neutralization

Irrigation Water Management

Irrigation water management is another option that helps to manage sodic water. A method of irrigation must create and maintain favourable salt and water regime in the root zone so that water is readily available to plants for their growth without any adverse effect on the crop yield. The irrigation practices for sodic water must aim at preventing the build-up of soil sodicity in the root zone to levels that limit the land productivity as well as minimize the damaging effects of sodicity on crop growth. Any irrigation method that would reduce the irrigation requirement would be helpful.

Table 6.36 Effect of crop residue management on yield (mean of 3 years) and soil properties of sodic soil irrigated with poor quality water

Treatments	Mean Yield (t ha^{-1})		pH	OC
	Rice	Wheat		(%)
N_0 P_0	1.05	0.87	9.35	0.25
N_{90} $P_{19.5}$ (75% NP)	2.69	2.31	9.35	0.25
N_{120} P_{26} (100% NP)	3.49	2.94	9.31	0.26
100 % NP + wheat residue Burning	3.72	2.99	9.33	0.30
100 % NP + Incorporated wheat residue	4.34	3.19	9.20	0.35
100 % NP + Incorporated wheat residue + GM	4.45	3.35	9.18	0.35
100 % NP + Incorporated wheat residue + SPM	4.41	3.34	9.20	0.34
CD (P=0.05)	0.58	0.45	–	0.02

Source: Yaduvanshi and Sharma (2008); GM= green manure; SPM = Sulphitation press mud, OC= organic carbon

Irrigation practices

The surface irrigation including the border strip, check basin and furrows are the oldest and the most commonly practised methods. The properly designed and operated surface irrigation methods can maintain the salt balance and minimize the salinity/sodicity hazard. To meet these twin objectives, proper land levelling is necessary to apply shallow irrigation (less water per irrigation) and to ensure an even distribution of irrigation water. Since sodic water irrigated lands, development of ESP or quantity of gypsum applied to minimize development of ESP would be directly related with the amount of water applied, sprinkler and drip irrigation help to reduce the ESP build-up or gypsum needed to reclaim the sodic water irrigated lands. The yield data averaged over two years at Bawal (AICRP, Hisar) showed that the grain yield of wheat increased significantly with successive increase in the neutralization of RSC. Mean grain yield increased by 9.6, 18.3 and 21.9% in G_{50}, G_{75} and G_{100} neutralization of RSC over control. The yield also increased with increase in depth of water application, being significantly higher in 6 cm depth than 4 and 5 cm depth of irrigation (Table 6.37), the overall increase in 5 cm and 6 cm being 8.3 and 12.8%, respectively as compared to 4 cm depth. The increase is in spite of the fact that pH in 6 cm irrigation is higher than plots where 4 cm of irrigation water was applied (Fig. 6.9). Besides, Similar trend was observed in straw yield of wheat also. Benefits of sprinkler irrigation in case of green gram are also apparent at the Tiruchirapalli center of the AICRP (Table 6.38). Water saving in drip irrigation has also been reported over the furrow irrigation in sugarcane at Tiruchirapalli (Table 6.39), although yield differences remained insignificant.

Table 6.37 Wheat yield (t ha^{-1}) irrigated with sodic water through sprinkler irrigation

Gypsum levels	Depth of irrigation (cm)			Mean
	4	5	6	
G_0	2.92	3.16	3.26	3.11
G_{50}	3.19	3.46	3.59	3.41
G_{75}	3.43	3.71	3.90	3.68
G_{100}	3.54	3.83	4.00	3.79
Mean	3.27	3.54	3.69	-
CD (5%)	Gypsum (G): 0.11, Depth of irrigation (D): 0.10, G x D: NS			

Table 6.38 Effect of sodic water on yield of green gram

Treatments	Yield (kg ha^{-1})
Farmer's practice + gypsum @ 50% GR	576
Farmer's practice	456
Sprinkler irrigation with sodic water + gypsum @ 50% GR	684
Sprinkler irrigation with sodic water	602

Table 6.39 Amount of water used in sugarcane with drip and furrow irrigation for the second ratoon crop (2007-2008)

Treatment	Amount of water used (mm)	Water saving (%)
Drip irrigation with gypsum treated water (M_1)	1812	26
Drip irrigation with spent wash treated water (M_2)	1830	25
Drip irrigation with untreated sodic water (M_3)	1812	26
Furrow irrigation (M_4)	2280	-

Conjunctive Use of Fresh and Sodic Waters

Highly sodic water cannot be used for crop production directly. Such waters can be used in conjunction with fresh canal water as it will result in dilution effect. The successful strategies are through mixing or blending and cyclic use.

Blending of sodic waters with canal water

Blending is most promising conjunctive use mode in areas where fresh water is available in sufficient quantities on demand. The SAR or RSC of the mix water after mixing should be within the permissible limit, based on soil type, crop(s) to be grown and climate of the area. Since the technological and mixing principles are common to saline and sodic waters, no problems are foreseen in implementing such a technology for diluting sodic waters.

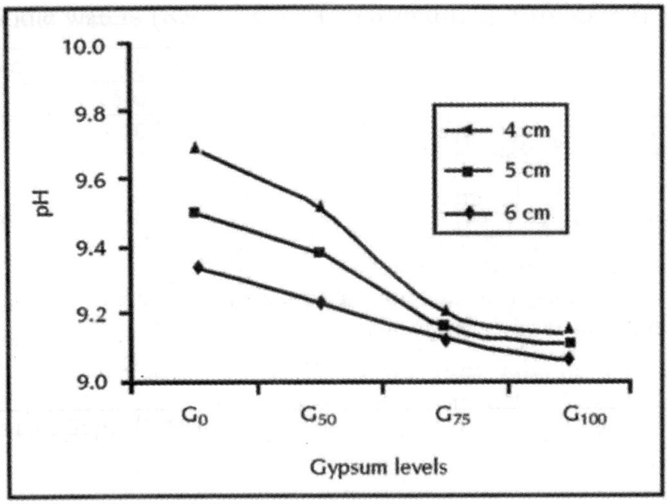

Fig. 6.9 Effect of gypsum levels on soil pH after wheat harvest in 2007-08

Calculation of mixing ratio: If two water supplies have RSC as RSC_a and RSC_b, then the RSC of the blended irrigation water is determined from the following relation:

$$RSC_m = (a\,RSC_a + b\,RSC_b)/(a+b) \tag{6.23}$$

Here RSC is the residual sodium carbonate of the water such that subscripts m, designate the RSC of blended water, a, b represent the relative fraction of each supply in the blended water. If a unit of mix water is prepared, then b can be replaced by (1-a) such that

$$RSC_m = RSC_b + a\,[RSC_a - RSC_b] \tag{6.24}$$

From eq. (6.24), fraction `a' can be estimated as follows:

$$a = \frac{RSC_m - RSC_b}{RSC_a - RSC_b} \tag{6.25}$$

From crop stand point of view, RSC_m can be replaced by RSC of the water that can be tolerated by the crops `RSC_c' such that:

$$a = \frac{RSC_c - RSC_b}{RSC_a - RSC_b} \tag{6.26}$$

The value of b then can be calculated by subtracting a from 1 that is b=(1-a). Although data on RSC_c values are unavailable in the literature, for practical standpoint, RSC_c for most crops can be taken as 2.5 although it can be increased to about 5.0 for light textured soils and crops that require less water for irrigation.

It may, however, be noted that eq. (6.24) is not valid for calculating SAR of the mixed water. For this purpose, we might use trial and error procedure. Calculate, concentration of Na, Ca and Mg ions in the mix water utilizing eq.

(6.24) to calculate SAR of the mix water. Else, SAR of the mix water is written as:

$$SAR_m = \frac{Na_b + a\,(Na_a - Na_b)}{\sqrt{(Ca_b + a(Ca_a - Ca_b) + Mg_b + a(Mg_a + Mg_b))}} \qquad (6.27)$$

Here Na, Ca and Mg designate the concentration of Na, Ca and Mg ions in meq L^{-1} and subscripts a and b denotes the waters of A and B type, `a' is the fraction of type A water in the mix and $SAR_m$ is the SAR of the blended water. When $SAR_m$ is known, then `a' can be calculated by trial and error.

Example: Two waters of RSC of 0.5 meq L^{-1} and 15 meq L^{-1} are available for mixing. Calculate in what ratio these waters should be mixed to obtain RSC of 5 meq L^{-1}. In case Na, Ca and Mg in the two waters are 1.3, 0.5, 0.5 and 156, 125, 48 meq L^{-1} respectively, calculate the mixing ratio such that SAR of the mix water is 10.0.

Solution

RSC of first water, RSC_a = 0.5

RSC of second water, RSC_b = 15.0

Desired RSC of mix water = 5.0

Incorporating these values in equation (6.26) we get:

$$a = \frac{5-15}{0.5-15} = \frac{-10}{-14.5} = 0.70$$

b = 1 - a = 1 - 0.70 = 0.30

Thus, the two qualities of waters should be mixed in a ratio, that the mix water contains 70% of water with an RSC of 0.5 meq L^{-1} and 30% of water with an RSC of 15 meq L^{-1}.

To achieve an SAR of 10.0, we make use of eq. (6.27) such that

$$10.0 = \frac{156 + a(-154.7)}{\sqrt{125 + a\,(-124.5) + 48 + a\,(-47.5)}}$$

Incorporating a = 0.3, we get

$$10.0 = \frac{109.6}{\sqrt{87.65 + 37.75}}$$

=109.6/11.02

=9.95

Thus, 30% of the low SAR water mixed with 70% of the high SAR water will yield SAR of the mixed water as 10.

On the basis of two calculations, it is recommended that 30% of the poor quality water be mixed with 70% of the good quality water to achieve an RSC of

5.0 meq L^{1}. Since the desired SAR is achieved by mixing 30% of the good quality water with 70% of the poor quality water, blending ratio as determined for RSC is recommended since in such a mix both SAR and RSC are within the desirable limits.

Rotational, cyclic or sequential mode

The rotational mode facilitates effective conjunctive use of fresh and sodic waters. In this strategy, canal water is replaced with sodic water in a pre-decided sequence/cycle. Bajwa and Josan (1989) reported that irrigation of sandy loam soil (18-26.8% clay) with sodic water (EC_{iw} 1.35 dS m^{-1}, RSC 10.1 meq L^{-1}, SAR 13.5 adj. SAR 26.7) increased the pH and ESP of the surface layers and reduced its infiltration rate to 14% (Table 6.40). The yields of rice and wheat decreased progressively with time and were 62 and 57%, respectively, of the potential yield, i.e., that obtained under canal irrigation, at the end of 6 years. However, when the sodic water was used in cyclic mode with canal water, yields of both the crops were maintained at par with canal water except in the 1 CW-2SW mode.

Table 6.40 Cyclic use strategies and their impact on rice and wheat yield

Water quality / Mode	Adj. SAR*	pH	ESP	RIR	Average yield (t ha^{-1})	
					Rice	Wheat
Canal water (CW)	0.3	8.2	4	100	6.78	5.43
Sodic water (SW)	22.0	9.7	46	14	4.17	3.08
2CW-1SW	8.9	8.8	13	72	6.67	5.22
1CW-1SW	12.8	9.2	18	59	6.30	5.72
1CW-2SW	18.5	9.3	22	34	5.72	4.85

*After accounting for 828 and 434 mm of irrigation and rainwater; RIR is relative infiltration rate; Source: Bajwa and Josan (1989)

Besides, conjunctive use strategies on seasonal basis for potato and sunflower crop and annual basis for the potato crop have also been tested and found useful. In the first case CW for potato and AW for sunflower proved much better than applying sodic water to both the crops but was lower in order when compared to cyclic use 1CW:1AW (Table 6.41). Similarly, one year canal and two year sodic water proved inferior to one year canal and one year sodic water (Table 6.42).

Some observations have also been made for the relative efficacy of conjunctive modes with sodic water irrigated lands. In an experiment where sodic water (EC 2.3 dS m^{-1}, RSC 11.3 meq L^{-1}) and good quality tube well water (EC 0.5 dS m^{-1}, RSC nil) were used for 6 years, cyclic modes (2TW: 1 AW, 1TW: 1AW, 1TW: 2AW) with decline in yield in the range of 16-20 and 6-12% in case of rice and wheat, respectively, performed slightly better than their counter mixing modes where the decline ranged between 19-23 and 9-14%, respectively. Thus, alternating sodic and canal waters can be considered to be a practical way to alleviate sodicity problems caused by the use of sodic water. Field observation in Kaithal area point out that farmers getting some canal water are able to sustain yields of rice-wheat crops. On the other hand, yield of these crops decline on farmer's fields having no access to canal water (Minhas *et al.*, 1996).

Table 6.41 Effect of various modes of irrigation with sodic and canal water on yield of potato crop

Mode	Yield (t ha^{-1}) during the year							
	1998-99	1999-00	2000-01	2001-02	2002-03	Mean	RY (%)*	SYI*
Potato tuber								
Canal water, CW	39.8	36.3	31.0	35.8	32.1	35.0	-	-
Sodic water, AW	25.1	18.7	3.7	9.0	2.9	11.9	34.0	0.063
Cyclic use								
1CW : 1AW	35.6	35.0	24.7	25.4	28.4	29.8	85.1	0.703
2CW : 2AW	35.0	34.4	22.6	23.6	26.5	28.4	81.1	0.642
2AW: 2 CW	30.6	30.2	18.7	19.4	14.9	22.7	64.9	0.442
4AW : 2CW	30.2	19.3	4.8	10.5	5.2	14.0	4.0	0.092
CW(P):AW(S)	38.7	33.7	22.6	23.2	21.6	28.0	80.0	0.579
Blending								
2CW :1AW	37.9	33.5	22.0	23.3	27.6	28.9	82.6	0.663
1CW :2AW	33.5	30.1	18.7	19.1	13.4	23.0	65.7	0.415
LSD(p=0.05)	4.8	2.7	1.1	2.0	1.6	2.4	-	-

*RY is the relative yield and SYI is the sustainability yield index; CW(P):AW(S) means canal water for potato and sodic water for sunflower

Table 6.42 Effect of modes of irrigation with sodic and canal waters on yield of crops

Treatments*	Yield (t ha^{-1})					
	2006-2007		2007-2008		Average of 5 years	
	Potato	Sunflower	Potato	Sunflower	Potato	Sunflower
T_1	31.20	1.50	25.70	1.10	27.10	1.30
T_2	23.60	1.20	16.00	0.60	19.40	080
T_3	13.70	0.70	8.40	0.60	14.00	0.70
T_4	24.90	1.10	23.40	1.00	22.60	1.00
T_5	13.70	0.60	14.00	0.70	14.00	0.80
T_6	12.70	0.80	09.40	0.70	12.70	0.80
T_7	16.60	0.70	12.90	0.60	15.20	0.70
T_8	24.00	1.10	21.30	0.90	22.10	1.00
T_9	2.60	0.20	2.30	0.10	3.90	0.20
CD (5%)	2.1	0.12	2.30	0.40	2.0	0.2

*The treatment combination were: T_1- All canal (control), T_2- One year canal : Two year sodic, T_3- Two year sodic : One year canal, T_4 – Two year canal : One year sodic, T_5- One year sodic : Two year canal, T_6- All sodic (potato) : All canal (sunflower), T_7- Mixing (One canal + Two sodic), T_8- Mixing (Two canal + One sodic) and T_9-All sodic (RSC 15 meq L^{-1})

According to a postulation made by Minhas *et al.* (1996) it was suggested that in the case of sodic water, a strategy that would either minimize the precipitation of calcium or maximize the dissolution of precipitated calcium is expected to do better. Usually both canal and ground water are in equilibrium with calcite, former at the pCO_2 of the atmosphere whereas the latter at a much higher pCO_2. The relation between concentration of Ca^{2+} and pCO_2 is not linear and is governed by the relation:

$$MCa^{2+} = \frac{Kh\,pCO_2\,K_1\,Kcal}{4K_2\lambda Ca^{2+}\,.\,\lambda HCO_3} \tag{6.28}$$

In the above equation, MCa^{2+} refers to concentration of Ca^{2+} (g L^{-1}), K_1 and K_2 represent first and second dissociation constant of carbonic acid and Kh is Henry's gas constant. λCa and λHCO_3 are the activity coefficients of ions while pCO_2 is partial pressure of CO_2. Therefore, it seems that mixing of surface waters with ground water of higher sodicity and low calcium will result in under–saturation with respect to calcite. Consequently, the blended water will have the tendency to pick up calcium through dissolution of native calcium (Minhas *et al.*, 2004). Kaledhonakar *et al.* (2006) used UNSATCHEM model to show that mixing of sodic and fresh water results in lesser build-up of ESP than the cyclic mode. It appears that benefits that accrue from cyclic and mixing modes with sodic water needs to be quantified and theoretically justified.

Application of Amendments

The sodicity hazard of the irrigation waters on the soil can be prevented or mitigated by neutralizing the RSC of irrigation water with gypsum including the ESP of the soil that has developed previously. When the RSC of the irrigation water exceeds 2.5, it may be necessary to add gypsum or any other amendment discussed in Chapter 3 such as gypsum, calcium chloride and calcium polysulfide directly supplying the calcium. Gypsum by virtue of its low cost is quite suitable for creating a favourable cationic ratio in the irrigation water or irrigated soil, improvement in Mg:Ca ratio and neutralization of RSC. The quantity of pure gypsum needed to neutralize one meq L^{-1} of RSC comes to 86 kg ha^{-1} per 10 cm of water or agriculture grade gypsum (70% purity) @ 12 kg ha^{-1} needs to be added for each cm of sodic water applied. The quantity of gypsum to be added is, thus determined by the RSC of the irrigation water and the quantity of water applied at each irrigation or during the crop growing season or on yearly basis. Thus, Gr to neutralize RSC of water is given as:

$$G_r = \frac{86 \times (\text{RSC of irrigation water} - \text{desirable RSC}) \times N_i \times D_i}{P_g \times 10000} \tag{6.29}$$

G_r is the amount of gypsum required (t ha^{-1}), D_i is the depth of each irrigation (cm); N_i is the number of irrigations and P_g is the purity of gypsum (%). If desirable RSC for a given crop is not known, it could be taken as 2.5 as per guidelines listed in Table 6.15.

Besides other amendments such as acid forming substances (notably pyrites), distillery spent wash (DSW), phospho-gypsum (PG) and sulphitation press mud (SPM) can be used either as soil application or through gypsum beds or mixing in water as in the case of DSW (Table 6.43; Anonymous, 2007, Anonymous, 2009) as per equivalence Table (Table 3.10). The beneficial effects of soil application of gypsum on rice and wheat yield are depicted in Table 6.44.

Table 6.43 Characteristics of amendments used to reclaim sodic water or sodic water irrigated soils

Amendment	Characteristics
DSW[1]	Byproduct of the molasses based distilleries is a nutrient rich wastewater with high biological oxygen and chemical oxygen demand (BOD and COD). It has 0.10 -0.15% N, 0.01 to 0.03% P_2O_5, 0.80-1.20% K_2O and appreciable quantities of Ca, Mg, S and micronutrients. DSW mixed in a ratio of 1:250 (DSW: Sodic water) proved beneficial to sugarcane crop with drip irrigation.
PG[2]	Phospho-gypsum contains 80-90% calcium sulphate and 1-2% fluorine. This fine, powder like material can be used to reclaim sodic lands and water. Some doubts on rocks containing fluoride and radio-active/heavy metals have been raised. Current researches revealed that one time application of phospho-gypsum for land reclamation (2) or small doses for water reclamation are unlikely to result in phytotoxic level of fluoride. Similarly fear of radioactive elements has been discounted although it may vary due to the source of rock phosphate used in the manufacturing process.
SPM[3]	Press-mud, a major byproduct of the sugar industry, is rich in easily decomposable organic matter and other primary plant nutrients. The PM produced in factories using sulphitation process is known as sulphitation press-mud, which has a good amount of sulphur varying from 1.7 to 2.1 ppm[3]. It has been used to reclaim sodic lands, to supplement plant nutrients and to add organic matter. Since the quantities required for barren land reclamation may be too large, it has a good potential to reclaim mildly sodic lands as well as to ameliorate sodicity build-up as a result of sodic water application.

Sources: [1] Srinivasamurthy *et al.* (2009); [2] Nayak *et al.* (2008); [3] Patel (1980) and Chand *et al.* (1976)

Table 6.44 Effect of gypsum application on grain yield of rice and wheat in farmer's fields

Wheat 1973-74			Rice 1974		
RSC of water	**Gypsum ($t\ ha^{-1}$)**	**Wheat yield ($t\ ha^{-1}$)**	**RSC of water**	**Gypsum ($t\ ha^{-1}$)**	**Rice yield ($t\ ha^{-1}$)**
12.5	Nil	0.95	10.0	Nil	0.61
7.5	2.0	2.10	7.5	2	1.00
5.0	3.0	2.63	3.0	4	1.62
2.5	4.0	2.63	2.5	6	2.02

Assuming application of 50 cm of water. Gypsum applied would neutralize the RSC to the level indicated. Source: Sharma *et al.* (1975a)

Gupta (1980) concluded that crop growth of wheat was not affected significantly under irrigation with water having RSC upto 10.0 meq L^{-1} provided that EC (<2 dSm^{-1}) and SAR (<10) were low for areas such as Karnal where the average total (700 mm) and the winter season rainfall (100 mm) were appreciable and effective and evaporation was less than 2000 mm.

However, rise in pH and SAR, on soil properties, if any, could be counteracted with the application of gypsum.

Some additional guidelines for application of gypsum are as follows:

- In the initial year, application of gypsum as per requirement of the irrigation water alone might be insufficient, in particular when the soil has been irrigated with sodic water previously or it is sodic in nature for some or the other reason. In that case, gypsum should be added both on the basis of soil requirement as well as to neutralize the RSC of the irrigation water. Subsequently, application of gypsum is sufficient on the basis of irrigation water only.
- Since, water is used year after year; application of gypsum needs to be repeated over the years.
- The best time for gypsum application is after the harvest of *rabi* crops, preferably in the month of May or June after some rains have occurred. After proper soil moisture conditions are attained, the soil should be ploughed to mix the applied gypsum in a shallow depth.
- From practical point of view, the irrigation water containing no residual sodium carbonate but having high SAR sometimes cause infiltration problems during rainy season which can be rectified by application of 1-2 tons of gypsum in the monsoon season.

Gypsum bed

Application of gypsum can either be made through soil application or added to the water to neutralize the adverse effect of RSC. Studies have shown that response of gypsum application is better when applied after each irrigation than the seasonal or annual application. Since soil application following each irrigation is likely to pose problems, therefore, technology to treat the irrigation water with gypsum has been developed so that there is no need to add gypsum to the soil. The easiest and the most practical way to treat irrigation water with amendment is to construct a gypsum bed (Fig. 6.10). Such a practice eliminates the costs involved in powdering, bagging and proper storage before its actual use. AICRP on Use of Saline Water, Karnal has also tested the performance of pyrites beds to compare the relative performance of gypsum and pyrites in treating high RSC water (Fig. 6.11). Initial results show that treatment with pyrite is also quite effective.

A number of alternatives to gypsum beds have been tried and perfected. Loose lining the irrigation channels with gypsum boulders is probably the cheapest and has been found quite effective. Pre-fabricated rings have also been used to construct a gypsum beds for small discharges usually required in drip irrigation. For the filling up of *diggis* in Rajasthan with reclaimed sodic water syntax tanks filled with gypsum boulders have been placed in the *diggy* and water allowed to pass from bottom upward (Fig. 6.12).

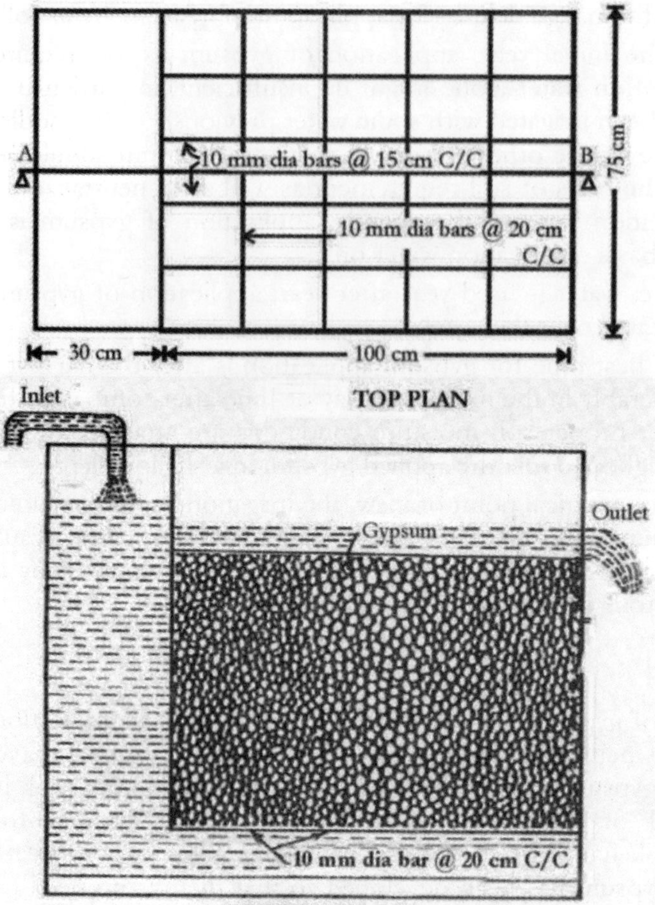

Fig. 6.10 A schematic view of the gypsum bed for treating sodic water

Fig. 6.11 A view of the gypsum and pyrites beds

Fig. 6.12 The gypsum bed to treat the sodic water for storage in a *diggy*

Design of Gypsum Beds

Dissolution of gypsum is governed by several factors such as distribution of gypsum fragments, flow velocity, salt content and chemical composition of water. In designing a gypsum bed, these factors should receive due consideration. The influence of various factors on gypsum dissolution is briefly discussed in the following section.

Electrolytic concentration

The effect of increasing electrolyte concentration on enhanced dissolution of gypsum can be accounted for by the use of an empirical relation given by Singh *et al.* (1986):

$$F = 1 / (1 + 0.0417\ C) \tag{6.30}$$

Where C is electrolyte concentration of irrigation water (meq L^{-1}). The empirical factor F is used to calculate the amount of gypsum dissolved at different electrolyte concentrations, relative to gypsum dissolved in salt-free water.

Ionic composition

Ions such as Ca^{2+} and SO_4^{-2} usually present in waters suppress the dissolution of gypsum through common ion effect. Presence of common ions decreases the saturation deficit and hence lowers the rates of gypsum dissolution.

Residual sodicity (RSC)

Singh *et al.* (1986) observed that at a given electrolyte and RSC level, the increase in dissolution of gypsum through variation in flow velocity from 0.564 to 347 cm s^{-1} was very small. However, an increase in RSC of water increased the dissolution of gypsum due to high association constant of $CaHCO_3^+$ ion pair as compared to $CaCl_2^+$. The effect of RSC on the amount of gypsum dissolution relative to RSC water can be evaluated from the empirical relation.

$$FRSC = 1 / (0.965 + 0.0013\ RSC) \quad (6.31)$$

The FRSC is used to calculate the amount of gypsum dissolved at different RSC, relative to gypsum dissolved in RSC free water.

Size distribution of gypsum fragments

Characteristic properties of assorted gypsum fragments in different particle size ranges are given in Table 6.45. An increase in fragment size or flow velocity or both decreases the dissolution of gypsum (Table 6.46).

Table 6.45 Characteristics of gypsum particles

Particle diameter (mm)	Av. wt. of fragments (g)	Estimated average surface area for fragment (cm^2)	Fragments/ 1000 cm^3 of bed volume	Estimated surface of gypsum/cm^3 of bed volume ($cm^2\ cm^{-3}$)	Pore space (%)
2–6	0.1	0.9	785	7.1	54
12–25	3.7	7.9	26	2.1	58
25–40	13.0	14.0	7	1.0	60
Singh *et al.* (1986) (Hanumangarh gypsum)					
2–6	0.2	1.2	514	6.3	54
12–25	6.3	11.8	15	1.8	60
25–40	28.0	31.7	3	0.9	65

Source : Kemper *et al.* (1975) : Singh *et al.* (1986)

Table 6.46 Effect of flow velocity and gypsum fragment size on gypsum dissolution and dissolution coefficient for canal water

	Gypsum fragment size					
	2-6 mm		12-25 mm		25-40 mm	
Flow velocity ($cm\ s^{-1}$)	Gypsum dissolution ($meq\ L^{-1}$)	Kx10^4 (s^{-1})	Gypsum dissolution ($meq\ L^{-1}$)	K x 10^4 (s^{-1})	Gypsum dissolution ($meq\ L^{-1}$)	Kx10^4 (s^{-1})
0.195	–	–	4.40	8.50	2.90	4.76
0.260	–	–	4.05	9.54	2.30	4.91
0.347	8.90	33.99	3.70	11.56	2.05	5.90
0.434	–	–	3.50	13.70	1.90	6.78
0.564	8.20	49.08	3.20	16.15	1.75	8.08

Source : Kemper *et al.* (1975)

Kemper *et al.* (1975) described the dissolution process by the following relationship:

$$K = -(1 / t_b) \ln [(C_s - C_b) / (C_s - C_o)] \quad (6.32)$$

Where K is dissolution coefficient (s^{-1}), C_s is the calcium concentration in saturated gypsum solution (30 $meq\ L^{-1}$), C_o and C_b refer to the concentrations of the gypsum (meq L^{-1}) in irrigation water entering (at t = 0) and water leaving the gypsum bed after time, t_b i.e. the contact time between flowing water and the gypsum particles. The value of t_b can be calculated as $t_b = (h \times \theta)/v$, such that h is the height of the gypsum bed with its porosity (θ) and v is the flow velocity through the bed.

The height of gypsum bed can be predicted by the following equation of Keiseling *et al.* (1978).

$$h = \ln \ [(C_s - C_b)/ (C_s - C_0)] \ (v - \mu/2D)^{-1} \quad (6.33)$$

In this expression D is the dispersion coefficient taken as 150 $cm^2 \ s^{-1}$ and the value of μ is calculated as:

$$\mu = v [1 + (4 D a / V^2)]^{0.5} \quad (6.34)$$

The term *a* refers to the mass transfer coefficient (s^{-1}) and signifies the reaction rate constant for the first order kinetics of gypsum dissolution. The value of *a* is invariant with flow velocity and is related to the surface area of the fragments (S) by the equation given by Singh *et al.* (1986).

$$a = 2.5 \times 10^{-4} (3.7 S-1) \quad (6.35)$$

The use of above equation for predicting the height of the gypsum bed matched well with the practical data from dissolution of gypsum from 70 cm gypsum beds using canal water (Singh *et al.*, 1986). The application of the equation is illustrated by the following example:

Example 3: Design a gypsum bed to reduce the RSC of a tube well water (discharge 8,000 L hr^{-1}) from 11.0 to 5.0 meq L^{-1} of following chemical composition : EC_{iw} 1.95 dS m^{-1}, Ca^{2+}+ Mg^{2+} 1.0 ; Na^+ 19.6; CO_3^{-2} 2.5; HCO_3^{-1} 9.5; Cl^{-1} 4.6 and SO_4^{--} 0.3 meq L^{-1} respectively.

Discharge (Q) = 8,000 L hr^{-1} or 2.22 x 10^3 mL s^{-1}

Table value of porosity (θ) for 2-6 mm gypsum fragments = 0.54 $cm^3 \ cm^{-3}$

If plinth area (A) of the proposed tank = 1.5 x 1.0 m^2 = 1.5 x 10^4 cm^2

Flow velocity, v = Q/A = 2.22 x 10^3 / 1.5 x 10^4 = 0.148 cm

Velocity through gypsum bed (v/θ) = 0.148/0.54 = 0.274 cm s^{-1}

For the Hanumangarh gypsum usually available in Haryana and Punjab in fragment sizes of 2-6 mm, the calculated value of a is :

$a = 2.5 \times 10^{-4} \ (3.7 \times 6.3-1) = 55.77 \times 10^{-4}$

At the observed velocity of 0.274 cm s^{-1} the value of (μ) will be

$\mu = 0.274 \ (1 + (4 \times 150 \times 55.7 \times 10^{-4} / (0.274)^2)^{0.5}$

$= 0.274 \ (1 + (3.342/0.075)^{0.5}$

$= 0.274 \ (6.7498) = 1.849$

Putting the value of μ = 1.849, the value of (h) can be calculated as:

$h = \ln [(30-6) / (30-1)] \ (0.274 - 1.849) / 2*150)^{-1}$

$= - 0.1892 \times (- 190.476)$

$= 36$ cm

However, it may be mentioned here that not more than 8 meq L^{-1} of Ca^{2+} can be dissolved in gypsum beds and hence may not suit to reclaim high RSC water.

Gypsum bed height and dissolution of gypsum

The dissolution of gypsum in water is quite limited as it has low solubility (1.9 g L^{-1}). Since the gypsum dissolution depends upon the gypsum bed height, size of the gypsum particles and the velocity of water etc., the effect of gypsum bed on the dissolution revealed that doubling the depth of the gypsum bed only reduced the RSC from 5.0-5.4 to about 3.3-3.6 in comparison to a 7 cm bed which reduced the RSC from 9.0 to 5.0-5.4 meq L^{-1} (Table 6.47). Similar results for a gypsum bed and addition of DSW are shown in Table 6.48 revealing that DSW is quite effective in reducing the RSC of the water although EC might be slightly more in DSW treated than with gypsum treated water.

Table 6.47 Changes in ionic composition of sodic water affected by gypsum bed treatment

Treatments	pH	EC (dS m^{-1})	Anions (meq L^{-1})			Cations (meq L^{-1})		RSC (meq L^{-1})
			HCO_3	Cl	SO_4^-	Ca+Mg	Na+K	
Kharif 2002								
Tube well water	8.0	1.1	10.4	1.1	0.3	1.4	8.8	9.0
Amended water after passing through gypsum bed								
7 cm	8.0	1.4	10.2	1.2	2.8	4.7	8.7	5.4
15 cm	7.9	1.4	10.2	1.2	4.0	6.6	8.6	3.6
Rabi 2002-03								
Tube well water	8.0	1.1	10.4	1.1	0.4	1.5	9.2	9.0
Amended water after passing through gypsum bed								
7 cm	7.8	1.4	10.0	1.2	3.0	5.0	8.9	5.0
15 cm	7.8	1.5	10.0	1.2	4.4	6.8	8.8	3.3

Source: Minhas *et al.* (1998)

Soil Application versus Gypsum Bed Treatment

Relative merit of soil application and gypsum bed treatment on crop yield revealed slight benefits of gypsum bed over soil application in all the test crops (Table 6.49). The beneficial effects on soil properties are apparent but again revealing slight advantage of gypsum bed treatment (Table 6.50).

Table 6.48 Effect of gypsum bed treatment and addition of DSW on RSC of sodic water

Treatments	pH	EC (dS m^{-1})	RSC (meq L^{-1})
Sodic water (untreated)	9.0	1.78	11.4
Gypsum bed treated water	8.4	1.85	5.0
Distillery spent wash treated water (1:250)	6.8	1.97	Nil

Source: Anonymous (2009)

Table 6.49 Yield (t ha^{-1}) of crops as affected by mode of gypsum application

Treatments	1993	1994	1995	1996	1997	1998	1999	2000	2001	2002	Mean
					Paddy						
T_1- Control	4.03	3.09	3.10	4.00	2.98	2.90	2.85	2.85	3.16	3.02	3.20
T_2- 7 cm gypsum bed	5.09	5.01	4.29	4.68	4.30	4.38	4.48	4.48	3.59	3.41	4.37
T_3-15 cm gypsum bed	5.19	5.29	4.56	4.80	4.65	4.65	4.70	4.70	3.74	3.89	4.62
T_4- Soil application equivalent to T_2	4.29	4.89	4.11	4.81	4.15	4.23	4.25	4.25	3.62	3.71	4.23
T_5- Soil application equivalent to T_3	4.23	5.01	4.39	5.03	4.45	4.48	4.5	4.5	3.84	3.85	4.43
CD at 5%	0.55	0.62	0.49	0.55	0.83	0.44	0.35	0.38	0.31	0.33	
					Sorghum*						
T_1- Control	2.44	2.56	2.28	2.11	*	*	2.10	2.08	2.25	2.19	2.25
T_2- 7 cm gypsum bed	3.36	3.20	2.96	3.10	*	*	3.15	3.15	2.99	3.11	3.13
T_3-15 cm gypsum bed	3.56	3.37	3.08	3.20	*	*	3.28	3.30	2.96	3.20	3.24
T_4- Soil application equivalent to T_2	2.33	3.09	2.79	2.9	*	*	2.91	2.97	2.83	2.89	2.84
T_5- Soil application equivalent to T_3	2.3	2.9	2.85	3.13	*	*	3.17	3.21	2.94	3.01	2.94
CD at 5%	0.53	0.26	0.52	0.71	*	*	0.3	0.29	0.31		
					Wheat						
Year	93-94	94-95	95-96	96-97	97-98	98-99	99-2k	2k-01	01-02	02-03	Mean
T_1- Control	3.27	2.98	2.91	3.06	2.50	2.43	2.09	2.00	2.90	2.83	2.70
T_2- 7 cm gypsum bed	3.44	3.54	2.86	4.46	3.71	3.78	3.80	3.81	3.45	3.57	3.64
T_3-15 cm gypsum bed	3.61	3.70	3.98	4.83	3.90	3.94	3.95	3.98	3.58	3.70	3.92
T_4- Soil application equivalent to T_2	3.23	3.40	2.96	4.60	3.60	3.70	3.75	3.76	3.48	3.49	3.60
T_5- Soil application equivalent to T_3	3.28	3.54	3.99	4.98	3.82	3.82	3.87	3.89	3.60	3.48	3.83
CD at 5%	NS	0.52	0.50	-	0.43	0.36	0.44	0.43	0.23	0.30	
					Mustard						
T_1- Control	1.23	1.38	1.28	1.26	1.15	1.13	1.08	1.05	1.30	1.14	1.20
T_2- 7 cm gypsum bed	1.63	1.97	2.10	1.94	1.90	2.10	2.27	2.30	1.69	1.71	1.96
T_3-15 cm gypsum bed	1.65	2.19	2.23	2.15	2.10	2.20	2.38	2.43	1.80	1.90	2.10
T_4- Soil application equivalent to T_2	1.51	2.05	2.09	2.01	1.81	1.96	2.01	2.15	1.71	1.71	1.90
T_5- Soil application equivalent to T_3	1.59	2.20	2.17	2.20	2.03	2.12	2.20	2.24	1.84	1.87	2.05
CD at 5%	0.25	0.44	0.27	0.27	0.26	0.26	0.25	0.25	0.17	-	

*Crop failed during 1997 &1998; Source: Minhas *et al.* (1998)

Table 6.50 Physico-chemical characteristics of soil as affected by successive irrigation with amended sodic water in unreclaimed soil

Treatments	Soil depth (cm)**	Rice harvest 2000			Rice harvest 2002		
		pH_s	EC_e ($dS\ m^{-1}$)	ESP	pH_s	EC_e ($dS\ m^{-1}$)	ESP
T1*	1	9.6	3.4	66.5	9.6	3.4	62.0
	2	9.5	2.5	74.1	9.5	2.6	61.3
T2	1	8.0	1.8	19.2	8.0	1.5	16.8
	2	7.9	1.9	19.2	8.0	1.5	17.8
T3	1	8.8	1.4	17.0	7.9	1.4	16.4
	2	8.8	1.7	17.7	7.8	1.6	18.0
T4	1	8.2	2.0	19.7	8.0	1.3	18.5
	2	8.2	2.0	21.0	8.0	1.6	20.8
T5	1	8.1	1.9	19.6	8.1	1.5	18.6
	2	8.0	2.0	21.0	8.0	1.8	18.6

*T1, T2 etc. have the same meaning as in Table 6.49;
**1 = 0-20 cm and 2 = 20-40 cm soil depth

PRODUCT QUALITY

Soil sodicity build-up as a result of irrigation with sodic waters besides affecting crops yield also results in adversely impacting the product quality in terms of grade, keeping quality, oil content and other quality parameters of vegetable crops (Anonymous, 2007). Lower grade potatoes (C grade) increased, fraction of such potatoes being the greatest (0.56) when irrigated with sodic water, compared to 0.13 for the canal water irrigation. Storage quality also deteriorated with potatoes getting shriveled with two-third-weight loss on storage for 90 days under sodic water whereas the weight loss was just about two-fifth under canal water. Similar differences in fruit size and weight for cabbage, tomato and brinjal are reported by Phogat *et al.* (2010). Similarly, the quality of sunflower produce was also poor with sodic water producing lower oil content and smaller seeds (lower 1,000-seed weight). Mean oil content was 3-5 % less in all sodic compared with all canal water irrigations (Table 6.51) while seed weight was less by about 8% in all sodic water irrigation (Table 6.52).

Varsha (2007) studied the effect of sodic water (RSC 11.6 meq L^{-1}) on nutritional and sensory characteristics of tomato, cabbage and brinjal. The parameters studied were total soluble solids, pH, firmness, moisture, titratable acidity, β-carotene, lycopene and ascorbic acid contents. Besides total soluble sugar, reducing sugar, non-reducing sugar, starch, total, soluble and insoluble dietary fibers, oxalic acid and polyphenols were studied (Table 6.53). An important observation, however, has been that upon neutralization of RSC, the size and weight increased approaching the size and weight of crops irrigated with canal water with 100% neutralization of RSC. In the study by Varsha (2007), most parameters were closer to canal irrigated treatment when irrigated with water with 50% neutralization of RSC and FYM @ 20 t ha^{-1} (G_1F_2) or with complete neutralization of sodicity and FYM @ 20 t ha^{-1} (G_2F_2). For some parameters, G_1F_2 treatment gave better values than G_2F_2 while in others G_2F_2 gave better values than G_1F_2.

Table 6.51 Effect of modes of irrigation with sodic and canal waters on oil content

Treatments	Oil content (%) in sunflower grains					
	2004	2005	2006	2007	2008	Mean
T_1	42.2	42.3	41.8	42.2	42.1	42.1
T_2	41.7	40.0	39.5	39.9	38.9	40.0
T_3	40.8	39.7	39.9	39.5	38.7	39.7
T_4	41.0	41.5	40.2	39.6	38.5	40.2
T_5	39.6	40.5	38.6	38.7	37.7	39.0
T_6	40.7	40.6	39.8	39.7	38.9	39.9
T_7	41.0	40.2	39.7	39.8	38.7	39.9
T_8	41.7	41.5	40.5	40.2	39.9	40.8
T_9	38.6	38.3	37.8	36.9	35.9	37.5
CD (5%)	1.4	1.5	1.4	1.4	1.5	-

The treatment combinations were: T_1- All canal (control), T_2- One year canal : Two year sodic, T_3- Two year sodic : One year canal, T_4 – Two year canal : One year sodic, T_5- One year sodic : Two year canal, T_6- All sodic (potato) : All canal (sunflower), T_7- Mixing (One canal + Two sodic), T_8- Mixing (Two canal + One sodic) and T_9-All sodic (RSC 15 meq L^{-1})

Table 6.52 Effect of various treatments on 1000 seed weight of sunflower

Mode of water use	1000-seed weight (g)
Canal water, CW	31.2
Sodic water	23.0
Cyclic use	
1CW:1AW	27.7
2CW:2AW	27.7
2AW:2AW	25.4
4AW:2CW	23.8
Blending	
2CW:1AW	28.1
1CW:2AW	26.7
LSD(p=0.05)	2.6

For example, tomato, cabbage and brinjal in G_1F_2 treatment had significantly ($P<0.05$) lower protein content than those with G_2F_2 and canal water irrigated plots. The decrease in protein content might be due to the competition between ions present in sodic water (chloride, carbonates and bicarbonates) and nitrates present in soil at the same absorption site, which ultimately caused reduced uptake of NO_3 and nitrogen. Reports that the acidity of tomato increased with increase in the salinity of water have also appeared (Nangare and Singh, 2010). On the whole it appears that for vegetables, crop quality can be maintained with neutralization of RSC by 50-100% and application of organic manure (Phogat *et al.*, 2010).

Table 6.53 Effect of gypsum and FYM on proximate composition of vegetables grown under sodic water irrigation (%, on dry matter basis)

Treatments	Protein	Fat	Crude fibre	Ash	Carbohydrate
Tomato					
Control (Canal water)	15.84±0.21	3.17±0.23	10.31±0.23	8.73±0.26	61.90±0.18
G_1F_2	11.42±0.43	2.28±0.15	8.85±0.19	10.71±0.50	66.70±0.57
G_2F_2	16.90±0.25	3.00±0.24	10.00±0.14	8.46±0.26	61.60±0.20
CD (P<0.05)	1.07	0.73	0.66	1.24	1.28
Cabbage					
Control (Canal water)	21.57±0.79	3.25±0.08	12.29±0.17	9.05±0.09	53.80±0.38
G_1F_2	16.66±0.27	2.55±0.17	10.50±0.32	11.10±0.20	59.15±0.40
G_2F_2	22.45±0.29	3.15±0.18	12.16±0.30	8.88±0.19	53.33±0.23
CD (P<0.05)	1.77	0.53	0.93	0.60	1.21
Brinjal					
Control (Canal water)	18.66±0.21	3.33±0.19	17.33±0.25	7.33±0.25	53.33±0.33
G_1F_2	13.92±0.13	2.91±0.22	15.10±0.16	9.74±0.15	58.30±0.31
G_2F_2	19.37±0.27	3.31±0.18	17.21±0.17	7.28±0.25	52.83±0.14
CD (P<0.05)	0.73	NS	0.68	0.76	0.94

Values are Mean ± SE of three independent determinations; G_1 : 50% neutralization of RSC with gypsum; G_2 : Complete neutralization of RSC with gypsum, F_2 : FYM @20 t ha^{-1}.

GUIDELINES FOR USING SODIC WATER

Based on the field experience and experimental evidences emerging from all over India, guidelines for using sodic ground water have been prepared by a team of experts drawn from AICRP Saline Water; CSSRI, CCSHAU and PAU (Table 6.54).

WATR QUALITY AND MANAGEMENT OPTIONS FOR USE

In spite of our expanded knowledge on soil water chemistry, use of poor quality irrigation water has remained an art rather than a science. A major area of concern is that while art is developing, science behind the use has taken a back seat causing hardships in generalizations and upscaling of field results. As discussed in this chapter, the adverse impacts of the water depend upon the type of crop and its salt tolerance, the characteristics of the soil, soil and water management practices being adopted and climate especially the rainfall and its management (Fig. 6.14). Besides, the effects of salinity and sodicity of irrigation waters is very site specific, making it difficult to set some rigid numbers for irrigation or some other use. Therefore, a stepwise procedure to determine the suitability of poor quality water for crop production is proposed in Fig. 6.13.

Table 6.54 Sodic waters (RSC > 2.5 meq L^{-1} and EC_{iw} < 4 dS m^{-1})

Soil texture (% clay)	Upper limits of		Remarks
	SAR	RSC	
Fine (> 30)	10	2.5 – 3.5	• Limits pertain to *kharif* fallow/*rabi* crop rotation when annual rainfall is 350-550 mm.
Moderately fine (20-30)	10	3.5 – 5.0	• When the water have Na < 75% (Ca + Mg > 25%) or rainfall is > 550 mm, the upper limit of the RSC range become safe.
Moderately coarse (10-20)	15	5.0 – 7.5	• For double cropping, neutralization of RSC with gypsum is essential
Coarse (< 10)	20	7.5 –10.0	

Source: Minhas and Gupta (1992)

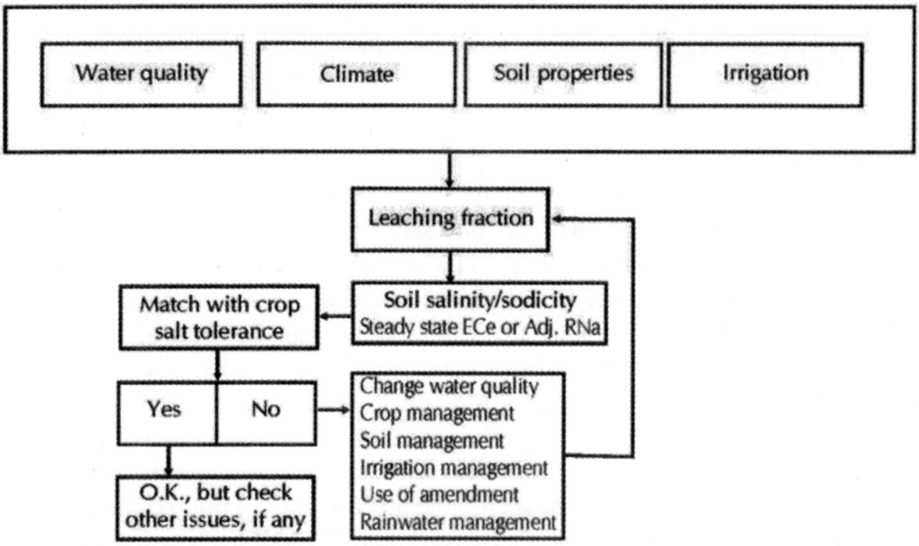

Fig. 6.13 Flow diagram for evaluating salinity and sodicity impacts of irrigation water

The steps include collection of data on water quality, climate (rainfall), identification of soil properties and irrigation application rates/methods. The information on these parameters is used to estimate the leaching fraction and the average root zone salinity and sodicity (Adj. R_{Na}). In the next step, the information is used to estimate relative crop yield and compared with the target yield. If inappropriate, one or a combination of the management options are identified that may help change the salt and water balance in the root zone. The process of assessment continues until target yield is achieved. If adverse impacts are expected on broader catchment issues such as regional water table, ground water pollution and surface water quality etc., these can also be included in the assessment.

7

SOCIO-ECONOMIC ISSUES

India is confronted with a challenging task of increasing agricultural production to feed its ever-increasing population that is likely to touch 1.4 billion mark by 2025. By then the land to man ratio is likely to decline to 0.1 ha from 0.34 ha in 1950. With shrinking land under plough and keen competition between various demands on land and water, it appears that agriculture has to exploit marginal and poor quality soil and water resources. The trend over the last few decades indicate that horizontal expansion of the agriculture on good soil resources is likely to be negligible as our net cultivated area has remained within 140± mark and is likely to remain around this level. On the other hand large agriculturally productive lands would be taken out of agriculture for domestic and industrial sectors. Therefore, the only possibility to increase agricultural production is through the vertical expansion i.e., by increasing food production per hectare of land under plough including increase in the cropping intensity. A redeeming feature, however, is that nearly 6.73 M ha of salt affected land could prove to be an opportunity if brought under cultivation. More than 3 M ha of this land is sodic in nature, which upon reclamation have the potential to contribute as much as 10 t ha^{-1} annually. Even at a modest scale of 8 t ha^{-1}, it can contribute 24 M tones per annum to the food basket of India. Such a strategy requires huge investments in the private and public sectors. Primary investments in this case will be mainly done by the private sector (farmer) whereas large-scale secondary investments have to be made by the private and the public sectors (industrial houses and the government). As such, economic and financial analysis becomes an integral part of formulating a land reclamation project as well as for post reclamation assessment and management. This chapter focuses on this aspect of land reclamation of sodic lands.

NEED FOR RECLAMATION

Problems of water logging and soil salinity are an integral part of irrigated agriculture world over and India is no exception. But the problems in India assume significance because

- India has only 2.4% of the global share of land but is required to feed 17% of the world population

- The land holdings are small and any loss of land directly impacts the household food security
- The problem areas are increasing rapidly especially the irrigated lands in the green revolution belt of the nation.
- Enormous demographic and economic pressures are building-up requiring an increase in crop production by about 50% in a sustainable manner representing one of the most important challenges in the 21st century

The land loss causes immense loss to the national wealth and as such will not allow us to leave an inch of land and a drop of water without its economic use. The productivity loss in India in general is estimated at Rs. 10,000 ha^{-1}. The data from the Chambal command, however, put the figure at Rs. 21,000 ha^{-1}. According to a conservative estimate made recently, loss of crop production due to water logging and salinity in the Tungabhadra project in Karnataka is assessed at about Rs. 250-300 million per year. Besides losses of Rs. 300-500 million per year are anticipated as a result of inequitable distribution and loss of water. Sharma *et al.* (2015) assessed that monitory losses due to soil sodicity in India are Rs. 150175 million on account of losses in production alone. For salinity the loss is assessed at Rs. 80021 million making a total of Rs. 230196 million from the salt affected soils of India. Even a simple calculation on the anticipated losses in irrigation command could be quite mind-boggling. For example if one ha of irrigated land goes out of cultivation, the investment of more than Rs. 200,000 is blocked. The interest on this investment alone is Rs. 20,000 $annum^{-1}$. Assuming a production of about 6 t ha^{-1}, a loss of about Rs. 30,000 ha^{-1} would be entailed if the irrigated area is taken out of plough. Assuming this area to be about 5.76 M ha, annual loss is thus, estimated at Rs. 290,000 million. Besides, huge damage to infrastructure and flora and fauna has been experienced due to increased extent and degree of water logging and soil salinity. Rural housing is particularly vulnerable as *kutcha* structures are easily damaged. Both human and animal population is prone to increased incidences of water borne diseases. Since, it is well established that investments on health related programs can decidedly be brought down through reclamation; it is sufficient justification for investments of about Rs. 290,000 million on prevention and reclamation of water logged salt affected lands. Besides the economic losses, other social and health related issues being realized now calls for an action plan to reclaim water logged salt affected lands.

CONSTRAINTS IN LAND RECLAMATION

Technical feasibility of sodic lands reclamation has been proved beyond doubt at various locations in India. It will be shown later in this chapter that the proposition is also economically sound and financially lucrative. Positive features that have helped this technology to gain the support of the farmers are listed in Table 7.1. In spite of these positive features, the pace of sodic land reclamation is not as per expectation as one would believe with such a wonderful technology.

Such a scenario can be attributed to financial, social, administrative and political constraints. It is believed that if some of the constraints being discussed in this section are addressed, the technology might gain further support and even the lands under *Panchayats* can be turned from white deserts to green fields.

Table 7.1 Adoption potential of sodic land reclamation technology

Technologies likely to be adopted easily	Sodic land reclamation program
Simple technology	√
Direct effects visible	√
Technically flawless	√
Individual farmer oriented	√
Economically viable under farmers practices and inputs	√
Low initial investments	√
Policy incentives are available	√
Institutional support	√
In line with the existing social norms	√

Financial Constraints

Any reclamation program by the farmers can be initiated only if he has substantial funds for initial investment, this technology requiring about Rs. 50,000 to meet the cost of gypsum, land preparation, tube well installation, rice production and other subsidiary works. The farmers owning only sodic lands or those with small holdings besides such barren lands do not have much savings to invest on reclamation. In spite of the much publicized loan facilities available through banks, farmers fail to take advantage because of illiteracy and bank loans failing to offer long-term, low-cost financial solutions. The loan amount is also meagre as some of the components of the technology are not financed by the banks. Since, this technology is quite effective only if it is implemented as a package, small farmers fear that they may not be able to implement the whole package with the meagre amount of loan. Moreover, tedious and time-consuming nature of processing of loan applications is also one reason for the lack of interest in taking loans. To overcome this hurdle, it may be useful to associate agriculture development officers of the lead banks in the training programs so that they are sensitized about the resources and farmers plight.

Social Constraints

Since the reclamation technology is within the practised social norms of the society, it has wide acceptance amongst the farmers. Since, there are no competitive traditional technologies of land reclamation, it seems that the spread of new technology is only limited due to the lack of knowledge of this technology (Parsad and Singh, 1980). The movement of resource poor farmers being limited to their communities and villages, they are unable to learn from the experiences of other farmers. They also do not respond to the extension workers because of their inhibitions. Probably an extensive mobile training network with some agricultural institute as the focal point may help in spreading the message quickly and effectively.

Administrative and Political Constraints

Governments in the states of Punjab, Haryana and Uttar Pradesh have well defined programs on sodic land reclamation. Land Reclamation Corporations have been set up which undertake the reclamation work on behalf of the farmers. They also have their own pilot projects to demonstrate the feasibility of the technology and to identify technological gaps. Subsidy on gypsum is provided and priority is given for electrical connection. As such these three states are leading the others in reclamation of sodic soils. Even in these states, some of the constraints need to be addressed with regard to *Panchayat* lands. Since these lands are given on short-term lease to an individual farmer, he does not take interest in the reclamation of these lands. It seems that for sodic lands, at least 10 years lease should be granted so that a farmer not only recovers the cost but also enjoys the benefits of labour that has been put in land reclamation.

Another constraint especially for big farmers is that surplus land beyond a certain threshold needs to be declared. If a big farmer reclaims the land, it may result in crossing this threshold. A suitable strategy to deal with such problems should be evolved so as to encourage land reclamation.

Based on the above discussions as well as from discussions with the farming community at various points of time, the major handicaps in extending the technology are:

- Small and scattered land holdings
- Low investment capacity of farmers
- Widespread illiteracy amongst the farming communities
- General lack of awareness
- Cost intensive reclamation technology
- Inadequate extension network especially to extend knowledge based technologies
- Need for a adequate and reliable source of water
- Inadequate political will

ECONOMIC AND FINANCIAL ANALYSIS

No distinction is often made between the economic analysis and the financial analysis of land reclamation although these two are quite distinct forms of analysis (Table 7.2). While the economic analysis of a project aims at assessing the viability of a project from a societal perspective and includes broader benefits and costs of a project, financial analysis, on the other hand is concerned with the ability of the project developer to raise funds and repay the project costs. While the economic analysis considers all benefits and costs of a project, and determines the project's overall value to the society, a financial analysis determines the ability of a project to generate enough revenue to pay back all the incurred financial costs. The real problem arises when an economic analysis is confused with financial analysis or vice-versa. It may be noted that an economically viable project may fail in financial analysis and a financially viable project may fail in economic analysis, calling for both kinds of analysis with clear distinctions between them. A financial analysis must follow the economic analysis before any recommendation is made on the project implementation (Fig. 7.1, Adapted from De Souza *et al.*, 2011).

Table 7.2 Key Aspects of economic and financial analysis

Aspect	Economic analysis	Financial analysis
Analysis perspective	Accounts for all benefits and costs "to whomsoever they shall accrue"	Accounts only for benefits and costs realized by a project developer
Inflation adjustment	Uses real prices which exclude general inflationary effects but accounts for individual resource price changes different from the general inflation rate (labor, energy)	Uses nominal prices and revenues which include inflationary effects
Discount rate	Calculates net present value using two real (no inflation) discount rates: 1) a low consumption discount rate and 2) a high return to private capital discount rate	Not Applicable
Interest rate	Not Applicable	Nominal (includes inflation) cost of borrowing money

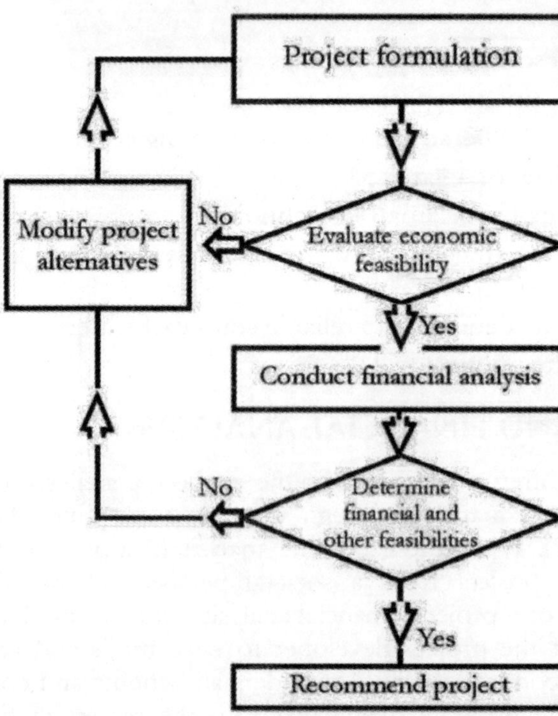

Fig. 7.1 Economic and financial analysis process for a project

Economics of Land Reclamation

The economics of land reclamation can be calculated by working out the cost and benefit arising out of the investments. The various actions required to perform the economic analysis are given in Fig. 7.2 (Adapted from De Souza *et al.*, 2011).

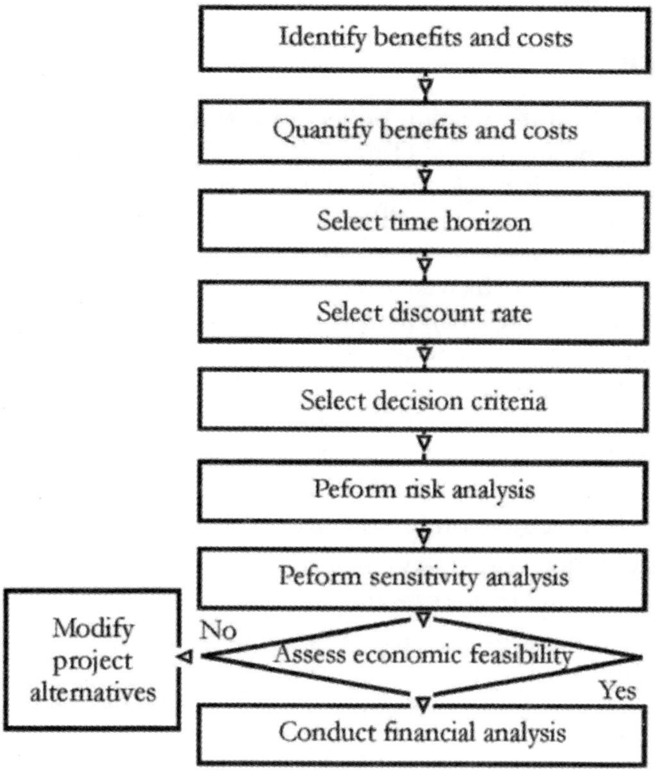

Fig. 7.2 Economic analysis processes

Decision Criteria

A project is successful and worthy of investment, only if it is economically and financially viable. The judgment of economic viability is made on the basis of one of the following criteria:

- Net benefits or financial feasibility
- Simple rate of interest (SRR)
- Benefit-cost ratio
- Rate of internal return
- Payback period
- Expected net present value (ENPV)

Though there are no hard and fast rules to select a particular criterion, benefit-cost analysis is arguably the most commonly used technique for economic assessment. It is expected that benefits should exceed cost or benefit-cost ratio should be more than one.

In the financial analysis, the investor must be able to get enough financial returns in the short-term so that he is able to pay the equated monthly/seasonal installment assuming that he has borrowed the money from the bank at the prevailing interest rates.

ITEMS OF COST

The cost of a sodic land reclamation project can be divided into two groups i.e. the capital cost and the annual cost (Table 7.2 and 7.3). The capital cost, which is of non-recurrent nature, is made once in the first year. In subsequent years only interest on this investment is included towards expenditure. Though investment on tube well is of non-recurrent nature, sometimes it does not form the part of investment on reclamation. Instead irrigation charges are worked out and are included suitably in the reclamation and crop production program. In some cases, it is included as investment and the operational and maintenance cost is included in the crop production programs. The seasonal/annual cost in reclamation programs is also due to cultivation cost of rice and wheat. In some cases, green manuring crop is raised to improve the physical condition of the soil.

Items of cost in a reclamation project are:

- Land acquisition cost, if any
- Depreciation and interest on the investments
- Initial investment in the project including chemical amendments, field collector and main ditches, control works, pipes and pumps, cross drainage works, land development, roads and re-allocation of existing structure
- Investment on replacements in future
- Cost of operation, management and maintenance of the project
- All other associated costs with the project such as outlays for fertilizers, pesticides and implements. Even in the on-going cultivated lands, additional costs may be incurred on account of better varieties of seeds, additional cost of fertilizers, micro-nutrients and even on labor.
- Cost of damage to the environment of the area, if anticipated. For example, marine life in creeks can be severely affected with large saline effluent coming out of drainage projects.

In the cost calculations, each item of cost is assigned a value and time. It is an accepted practice to base the calculations on a constant price level. Future inflation is neglected mainly because it is assumed that the value of money declines by that amount. If relative deviations in price or costs are expected, this should be reflected in the calculations. An accounting procedure is used to convert all prices to a common base by discounting over the economic life of the project.

The relative costs on various components of land reclamation reveal that maximum investments are needed for amendment (Table 7.3). Besides, all indirect costs must be evaluated and included in cost calculations.

Table 7.2 Work components as envisaged in reclamation of sodic soil

Sr. No.	Kind of work	Quantity/length/number
Recurrent cost		
1	*Bunding* (m ha^{-1})	470
2	Irrigation channels (m ha^{-1})	100
3	Drainage channels (m ha^{-1})	100
4	Outlet (small)	1
5	Land levelling (Tractor hrs)	10
6	Tube well	1 per 4 ha
7	Pump set (no. per tube well)	1
8	Soil amendment and application(t ha^{-1})	8-12
Annual cost		
9	Crop production (Rice)	Per ha
10	Crop production (Wheat)	Per ha
11	Crop production (green manuring)	Per ha

Source: Tripathi and Sharma (2011)

Table 7.3 Initial reclamation and operational cost of sodic land reclamation

Particulars	Percentage of total cost
On-farm land development	3.64
Tube well	15.76
Gypsum application	32.93
Rice cultivation	24.72
Wheat cultivation	22.95

BENEFITS

Possible benefits of sodic land reclamation projects as observed in various countries are:

- Increase in crop yield
- Improved product quality
- Reduced cost of farm operations and inputs
- Reduced fallow land and increased cropping intensity
- A change in cropping pattern with introduction of more remunerative crops
- Reduced inputs especially seeds
- Reduced plant diseases and pest control
- Reduced floods, soil erosion and reduced damage due to siltation in and outside the area
- Benefits to public and private properties

- Indirect benefits such as employment generation and improvement in health etc.
- Overall socio-economic improvement and environmental benefits
- Carbon sequestration.

All direct and indirect benefits must be evaluated and included while calculating the benefits of the project (Table 7.4). Usually, it is difficult to assign numbers (benefits in terms of economic value) to the indirect benefits. Both indirect costs and benefits can be left out from initial economic analysis, if it is difficult to assign the numbers.

Table 7.4 Direct and indirect benefits of land reclamation

Agriculture	Socio-economic	Natural	Regional
Increased agricultural productivity	Reduced socio-economic disparity	Increase in land value	Reduced expenditure on health services
No restriction on crops	Reduced migration from rural to urban areas	Improved soil health	Reduced damage to infrastructure
Increased cropping intensity	Reduced gender disparity	Better eco-system health	Minimal landscape degradation
Better product quality	Improved law and order situation		
Sustainability of production	Employment generation		

The principal benefit is reflected in yield increase. The total income resulting from direct increase in yields could be calculated by the following relation:

$$\sum_{i=1}^{n} (\text{Aio Yio Pio} - \text{Aio Cio}) / \sum_{i=1}^{n} \text{Ao} - \sum_{i=1}^{n} (\text{Ai Yi Pi} - \text{Ai Ci}) / \sum_{i=1}^{n} \text{Ai} \quad (7.1)$$

Where Ai is the area under i_{th} crop, Yi is the yield of the i_{th} crop, Pi is the market price of the i_{th} crop and Ci is the cost of cultivation of the i_{th} crop per ha before reclamation. Aio, Yio, Pio and Cio have the same meaning but after reclamation. The above equation needs to be used for all the years of the project life.

Besides this direct increase, incremental benefits are also obtained in the yield through increased cropping intensity (Table 7.5; Thimmappa *et al.*, 2015), showing an average increase of about 62.2% on four kinds of lands. It reached 200% even on severely affected sodic lands and in some cases it can reach as high as 300%. The benefits also accrue on account of change in crop mix. These can be suitably reflected in framing eq. (7.1). It may be noted that the full anticipated benefits cannot be obtained in the first year of the project. Usually, it takes 2-4 years before the full benefits of the project emerge. Under this situation and to be practical, the anticipated benefits should be spread over 2-4 years as per the expected pace of reclamation. After this period, it will be safe to assume that the benefits achieved are sustained for the remaining period.

Table 7.5 Cropping intensity (%) by sodicity class in Santaraha village, Hardoi, UP

Land quality (ESP)	Cropping intensity	
	Before reclamation (Average of 2009-2012)	After reclamation (Average of 2012-2014)
Normal (<15)	198.5	198.5
Slight (<15)	193.3	199.7
Moderate (15-40)	100.0	199.9
Severe (>40)	0.0	200.0
Annual average	123.0	199.5

ECONOMIC ANALYSIS

Time Period

While considering time period, one must distinguish between the economic life and the financial life of a project. The actual life of a well managed sodic land reclamation project may be more than 100 years. In fact, there is no depreciation but appreciation of the land resource that has been reclaimed. However, in economic terms future costs and benefits occurring say after 30 years are insignificant. For example; the discounted benefit for the 12% discount rate in the 30th year would be 0.033 times the actual benefit. It would further go down thereafter suggesting that the economic life of projects can be taken as 30 years.

Another important issue in an economic evaluation is the financial life of the project often imposed by the financing organizations. A farmer having obtained the loan from a financial institution must pay back the loan in the pre-decided duration say in 10 years. In that event, cash flow must exceed the costs during the financial life of the project. Although the project has a terminal value in this case, yet it has little meaning to the farmer who is concerned with cash flow during the financial life of the project. In projects where farmers' contribution is limited or they are not supposed to contribute to the project, one may forego the financial evaluation part and restrict to the economic analysis of the project. For the sodic land reclamation program both the economic and financial analysis is being done considering a life of 10 years.

Discounting

The costs and benefits of a reclamation project occur at different times during the project period. On the other hand, present worth of costs and benefits received say 10 years after would be less than received today. In other words, an amount available today is worth more than the same sum of money available 10 years after. Such variation is taken into account by following the discounted cash flow method of analysis. Usually discount rates vary from a low of 6% to about 12%. Current bank rates on fixed deposits are good indicators of the discount rates to be adopted in economic evaluation. Since bank deposits rates fluctuate, it may be worthwhile to make the analysis using few discount rates (Less and more) to derive most appropriate information. For example; for the current rate of interest of 10.5% on fixed deposits, one can carry out the economic analysis for discount rates of 8.0%, 10.5% and 12.0%. In financial analysis prevailing interest rates are usually adopted. The discount factors (DF) are calculated for a given discount rate as per the following relation:

$$DF = 1/\ (1+ Dr/100)^{t} \qquad (7.2)$$

Where Dr is the discount rate (%) and t is the year say 1, 2, 3 .....30. For t=1 and Dr = 10%, DF would be:

$$DF = \frac{1}{1+.1} = 0.909$$

Discount factors for a discount rate of 10.5% over a period of 30 years are given in Table 7.6.

Table 7.6 Discount factors at a discount rate of 10.5%

Year	Discount factor	Year	Discount factor	Year	Discount factor	Year	Discount factor
1	0.905	9	0.407	17	0.183	25	0.082
2	0.818	10	0.368	18	0.166	26	0.075
3	0.741	11	0.333	19	0.150	27	0.067
4	0.671	12	0.302	20	0.136	28	0.061
5	0.607	13	0.273	21	0.123	29	0.055
6	0.549	14	0.247	22	0.111	30	0.050
7	0.497	15	0.224	23	0.101		
8	0.450	16	0.202	24	0.091		

Inflation

Since the future costs and benefits are usually calculated in terms of present prices, it may be useful to take care of inflation. Although inflation is often avoided, it can be accounted for by increasing the costs and benefits by an assumed inflation rate.

Simple Rate of Return (SRR)

SRR of a project is used to compare with the rate that has been fixed as cut off criterion or the minimum rate of return. The SRR is calculated by the following formula:

$$r = \frac{E-D}{I} * \frac{100}{n} \qquad (7.3)$$

Where, r is the simple rate of return, E is the annual net benefit, D is the annual depreciation including O & M costs, I is the initial capital investment and n are the number of years.

Net Present Worth (NPW)

It is the difference between the present values of benefits over the present values of costs. A positive value indicates the worthiness of the project.

$$NPW = \sum_{t=1}^{T} (B_t - C_t)\ (1+Dr)^{-t} \qquad (7.4)$$

Here B is the value of the benefit, C is the value of the cost, Dr is the discount rate and t is time (as years since the reference year).

Benefit-Cost Ratio (B-C ratio)

It is the ratio of the present value of benefits divided by the present value of costs. For an economically viable project, the ratio should be more than one.

$$\text{BC ratio} = \sum_{t=1}^{T} B_t\ (1+Dr)^{-t} \ / \sum_{t=1}^{T} C_t\ (1+Dr)^{-t} \qquad (7.5)$$

Payback Period (PBP)

It is the time period for an investment to generate sufficient incremental cash to recover its initial capital outlay in full. Thus, the PBP measures number of years it will take for the net undiscounted benefits to repay the investment. If the cash flow is uniform over the years, the payback period, P is given by

$$P = (I/E) \qquad (7.6)$$

Where I is the initial investments and E is the annual net benefits.

Internal Rate of Return (IRR)

IRR is used to find out the rate of return which a project is likely to earn over its useful life. IRR is that rate of discount at which NPW is zero and B-C ratio one.

$$IRR = \sum_{t=1}^{T} \frac{B_t - C_t}{(1+i)^t} = 0 \qquad (7.7)$$

Thus, if the NPW of a project is positive at a given discount rate, its IRR would be more than the assumed discount rate. On the other hand, if the NPW is negative, the discount rate would be less than the assumed discount rate. As such, when the two discount rates at which NPW is positive and negative are known, then IRR could be approximated by the following relation:

$$IRR = LDR + DTDR \times \frac{NPWLDR}{DNPW\ TDR} \qquad (7.8)$$

Where, LDR is the lower discount rate, DTDR is the difference between the two discount rates, NPWLDR is the NPW at the lower discount rate and NPWTDR is the difference between NPW's at the two discount rates. A check on NPW should be made with the assessed IRR using eq. (7.8). Minor adjustments might be needed in the assessed value.

IRR is compared with the minimum acceptable rate of return (say prevailing market rate). If the IRR is higher or equal to the minimum acceptable rate of return, the project is worthy of investment. In most projects, all the three indices IRR, NPW and B-C ratio are determined to assess the economic viability of the project.

In few situations, it may be difficult to assess the benefits. In that case it may be useful to approach the problem from the other end as to what should be the minimum benefits that must be obtained to cover the investments. In other words, for the project to be viable, benefits must exceed this minimum. For this purpose following two parameters are commonly used.

Expected Net Present Value (ENPV)

By specifying risk in the form of estimated probabilities, NPV becomes Expected Net Present Value (ENPV). The benefit of using ENPV is that it accounts for risk by using probabilities for sets of parameter values (scenarios). Estimating these probabilities may require specialized statistical skills. Using the probabilities estimated for each scenario, a single value for ENPV can be calculated

$$\mathrm{NPW} = \sum_{i=1}^{N} P_i \sum_{t=1}^{T} (B_{it} - C_{it})(1+Dr)^{-t} \tag{7.9}$$

Here, P is the probability of an event.

Break Even Benefits (BEB)

Break even benefit is the minimal benefit required to meet the costs. It is calculated as follows (Smedema and Rycroft, 1983):

$$\mathrm{BEB} = (\text{NPW of costs} / \sum_{i=1}^{n} \text{discount factors}) \tag{7.10}$$

For a project to be viable, the real undiscounted average benefits per year should exceed the BEB.

Equated Seasonal Instalment

Cost of the borrowed capital can be readily estimated from the known interest rate and the period with in which the loan has to be repaid. The **e**quated seasonal/annual instalment, (ESI) is calculated by the following formula.

$$\mathrm{ESI} = (L * I) * [(1+I)^M / \{(1+I)^M - 1\}] \tag{7.11}$$

Here, M is the loan period in seasons (Years*2), I is the interest rate (Interest rate per annum / 2) /100, L is the borrowed capital in Rs.

If the capital is invested from self savings, the cost might be higher depending upon the alternative investment opportunities. It should be assessed from the likely returns from the alternative use of the capital. Sometimes, an average interest rate of the borrowed and estimated interest rate of self funds is taken as the interest rate on the total capital to calculate the virtual ESI. For a financially viable project, the seasonal/annual benefits must exceed the seasonal/annual instalment. For example; for a soft loan of Rs. 70,000 @ 6% per annum, ESI for 10 years period would be Rs. 8206/-. Thus, the farmers must earn this much amount per season over and above the current income to repay the loan instalment.

Sensitivity and Risk Analysis

Uncertainty and risk are present throughout the economic analysis as there are no assurances that predictions are accurate. Therefore, to address uncertainty and risks, sensitivity and risk analyses are carried out. When doing the analysis, areas which are likely to be sensitive are clearly identified and addressed (Water Resources Council, 1983). A sensitivity analysis varies values of key variables, such as the discount rate, to determine the effect on the final outcome. For some projects a risk analysis may be required. Risk analysis uses probability distribution functions to show a range of outcomes. As demonstrated with ENPV, probabilities can be incorporated into the calculation. Methods to perform sensitivity and risk analysis in connection with an economic appraisal of reclamation projects are not included in this book, being out of the scope.

Calculation of Indices

Example: Calculate NPC, NPB, NPW, BCR, IRR and PBR for the costs incurred and benefits received in a sodic land reclamation program as given in Table 7.7. While the initial investment on reclamation, cost of operation and maintenance and total costs are given in Columns 2, 3 and 4 of the Table 7.7, the benefits accrued including in the first year are given in column 8.

With the information given in Table 7.7, following step wise procedure is adopted to calculate the various economic parameters.

- Select and fix the discount rate. For this example, a discount rate of 10% is chosen (Column 5).
- With the help of eq. (7.2), discount factors are calculated as reported in column 6. The discount factors decrease with increasing number of years. It may be noted that sometime first year is counted as year zero and the calculations end at required time say 10 years for this case. However it would make only minor variations. The purpose is to convert all values to the same base level.
- Discounted costs are calculated for each year by multiplying the respective cost with discount factor of the respective year (Column 7). Similarly, discounted benefits are calculated by multiplying the benefits with respective discount factors of that year (column 9).
- Cash flow is calculated by subtracting the costs from the net benefits (Column 10). Note the negative value for the first year when major investments are made to reclaim the land.
- Summation of all the values in column 7 will give the net present value of the costs (NPC) while the summation of values in column 9 will give the present value of net benefits (NPB).

Once these calculations are made, NPW is calculated as follows:

$$\text{NPW} = \text{NPB} - \text{NPC} \tag{7.12}$$

$$\text{BCR} = \text{NPB/NPC} \tag{7.13}$$

For this example, PBR is 4 years as the negative cash flow for the first year is fully paid by the fourth year.

To calculate IRR for the same example as illustrated in Table 7.7, we can either use the trial and error procedure or the financial function IRR of the Microsoft Excel sheet. Following stepwise procedure is followed if trial and error procedure is used.

- Select and fix a discount rate. Since the NPW in the above example is positive, the discount rate is more than 10%.
- Calculate NPW with the new discount rate.
- If it is positive again, further increase the discount rate and continue till the NPW is zero or negative.
- If zero, the discount rate at which NPW is zero is the IRR. If negative, then either make another trial for a discount rate between this value and the previous highest value of the discount rate at which NPW was positive. Alternatively, make the interpolation between these two discount rates using eq. (7.8). Check the NPW with this discount rate and this should be nearly zero.

For the given example, the discount rate turns out to be 36%.

Table 7.7 Calculations of B:C ratio, NPV, IRR and payback period for a drainage project

Year	Cost	O& M cost	Total cost	DR	DF	DC	Net benefits	DB	Cash flow
1	2	3	4=2+3	5	6 (Eq. 7.2)	7=4*6	8	9=8*6	10=8-4
1	40000	0	40000	10	0.909	36300	11000	9090	-29000
2	0	500	500	10	0.826	413	11000	9086	10500
3	0	500	500	10	0.751	375.5	12000	9012	11500
4	0	500	500	10	0.683	341.5	12000	8196	11500
5	0	500	500	10	0.621	310.5	12000	7452	11500
6	0	500	500	10	0.565	282.5	12000	6780	11500
7	0	500	500	10	0.513	256.5	12000	6156	11500
8	0	500	500	10	0.467	233.5	12000	5604	11500
9	0	500	500	10	0.424	212.0	12000	5088	11500
10	2500		2500	10	0.386	193.0	12000	4632	9500
			57500			38917		71096	
									IRR=36%

O& M: Operation and maintenance, DR, DF, DC and DB are the discount rate, discount factors discounted costs and discounted benefits
Benefit –cost ratio = discounted benefits/discounted costs = 71096/38917 = 1.82
NPW=discounted benefits -discounted costs = 32179
Payback period= 4 years
IRR as per calculation with financial function IRR is 36%

Calculation of IRR using Microsoft Excel Sheet

If the data of Table (7.7) is entered into an Excel sheet, then the cash flow (column 10) is used to calculate the IRR (Tripathi, 2007). Following syntax is used to calculate the IRR:

IRR (values, guess)

Values are an array or a reference to cells that contain numbers for which one wants to calculate the internal rate of return. Values must contain at least one positive value and one negative value to calculate the internal rate of return. IRR uses the order of values to interpret the order of cash flows. All payments and income values are entered in the sequence these are paid or received. If an array or reference argument contains text, logical values, or is empty then all those values are ignored.

Guess is a number that is close to the expected IRR and is given as the first guess. In most cases one need not give the guess. In that case, it is assumed to be 0.1 (10%). Microsoft Excel uses an iterative technique to calculate IRR. Starting with guess, IRR cycles through the calculation until the result is accurate within 0.00001%. If IRR can't find a result even after 20 tries, the #NUM! error value is returned. If IRR gives the # NUM ! error or if the result is not close to the one expected, try again with a different value of guess than the previous value.

Using the cash flow data of Table (7.7), and a guess of 12, the IRR for the example is calculated as 36%.

The average annual break even benefit is calculated as

Present value of costs for 10 years	38917
$\sum_{I=1}^{10}$ Discount factors	6.15
Annual break even benefit = 38917/6.15 =	6334

Since the real average benefits during the 10 years are 11,800 (column 8), more than the break even benefits, the project is financially viable.

The equated seasonal instalment can be calculated with the help of eq. (7.11) which says that on a loan of Rs. 40000 taken at the rate of interest of 10% for a period of 10 years to be paid in two seasonal instalments per year, the ESI will be Rs. 3210 per season. Since the net profit in the first year is more than Rs. 6420, the proposition is again financially viable from this point of view as well.

CASE STUDIES OF FINANCIAL AND ECONOMIC ANALYSIS

Cost benefit of sodic land reclamation has been worked out at different places and under different agro-climatic conditions. Typical examples from the three states of Punjab, Haryana and Uttar Pradesh are included.

Punjab

Chhiba and Singh (1978) worked out the economics of sodic land reclamation in the Operational Research Project, Kapurthala. The initial pH_2 of the soil was 10.5 indicating that land under reclamation was highly deteriorated. Gypsum @ 10 t ha^{-1} was applied. A statement of expenditure and returns show that cost of reclamation was recovered within 2 years (Table 7.8). Although, a marginal loss of Rs. 170/- occurred in the first year when the land reclamation cost is also included in the total cost, but in the second year, not only the total cost is recovered but a net gain of Rs. 3530/- is also realized. Apparently, land reclamation in Punjab is a viable proposition.

Table 7.8 Expenditure and returns (Rs. ha^{-1}) statement for ORP Kapurthala

Particulars	Period	
	1975-76	1976-77
Cost of operations		
Bunding and levelling	540	–
Gypsum application and leaching	1230	–
Green manuring (Growing and burying)	1070	560
Production cost of rice	1990	1340
Production cost of wheat	1970	1260
Gross cost	6700	3160
Sale of rice and its by-products	3730	3950
Sale of wheat and its by-products	2800	2910
Total	6530	6860
Net returns	(-)170	3700
Net gain in 2 years	3530	

Figures are nearest to ten rupees

Haryana

Economics of land reclamation in Haryana calculated on the basis of data collected from 60 farmer's fields shows that the net profit from cultivation alone is much less as compared to Punjab. The net profit estimated at Rs. 2420/- is insufficient to recover the cost of reclamation which is estimated at Rs. 3260/- (Table 7.9). However, in these calculations, subsidy on gypsum @ 50% of the total cost of gypsum has not been considered. Besides, it is assumed that the land has been taken on rent @ Rs. 500/- per annum, and is included in the cost. If the farmer owns the land, a sum of Rs. 1000/- can be deducted from the total cost making the net profits for 2 years as Rs. 3420/-. An additional feature of land reclamation brought out in this analysis is increase in the land value. If it is taken into account, the total cost of reclamation is more than offset by this increase alone (Table 7.9). In another study, it has emerged that the whole cost of reclamation and crop cultivation is fully recovered within three years of land reclamation (Table 7.10).

Table 7.9 Cost and returns per ha on sodic soils under reclamation in the first two years of crop production

Cost	Amount (Rs.)	Returns	Amount (Rs.)
1975-76			
Cost of land preparation and amendment application	3260	Increase in land value after one year of reclamation	4620
Cost of rice cultivation	1690	Returns from rice crop	2410
Cost of wheat cultivation	1640	Returns from wheat crop	1720
Total cost of cultivation	3330	Total returns	4130
Net Returns			800
1976-77			
Cost of rice cultivation	1950	Returns from rice crop	3310
Cost of wheat cultivation	1650	Returns from wheat crop	1890
Total cost of cultivation	3600	Total returns	5220
Net returns			1620

Table 7.10 Initial establishment and operational costs for reclamation of sodic lands

Particulars	Cost (Rs. ha^{-1})	Percentage of total	Returns (Rs. ha^{-1})
Ist year			31,600
Bunding and levelling	2,050	1.85	
Tube well installation	8,925	8.07	
Gypsum application	18,650	16.85	
Rice cultivation	14,000	12.66	
Wheat cultivation	13,000	11.75	
2nd year			39,500
Rice cultivation	14,000	12.66	
Wheat cultivation	13,000	11.75	
3rd year			42,850
Rice cultivation	14,000	12.66	
Wheat cultivation	13,000	11.75	
Total (3 years)	**110,626**	**100.00**	**113,750**

The cost-benefit of land reclamation on lands adopted by the ICAR-Central Soil Salinity Research Institute, Karnal and the land reclaimed by the farmers reveal that the benefit- cost ratio even at the farmer's level of adoption is 3 or more than 3 depending upon the component of the subsidy. With a subsidy of 70%, the B-C ratio is 4.6 (Table 7.11). The B-C ratio is still higher if the complete package is implemented compared to when it is partially implemented as it happens under real field situations. Financial studies carried out by Haryana Land Reclamation and Development Corporation brings out that a fair return on investments including repayment is possible even in the very first year.

Table 7.11 Benefit-cost (B:C) ratio, net present worth (NPW), net benefit investment ratio and payback period of sodic soil reclamation technology

Indicators	Subsidy level		
	Without subsidy	50% subsidy	70% subsidy
As per CSSRI recommendation			
Benefit-cost ratio	4.6	6.0	7.0
Net present worth (10^3 Rs. ha^{-1})	66.0	70.4	72.5
Net benefit investment ratio	7.0	11.4	16.7
Payback period (years)	2.0	1.8	1.6
As per farmer's level of adoption			
Benefit-cost ratio	3.0	3.9	4.6
Net present worth (10^3 Rs. ha^{-1})	36.7	41.0	43.0
Net benefit investment ratio	3.0	4.6	5.6
Payback period (years)	3.4	2.0	2.6

Uttar Pradesh

Pyrites is commonly used as an amendment in the state of Uttar Pradesh. Singh (1980) worked out the land reclamation cost at Kanpur on the basis of pyrites and gypsum applied at the rate of 40% gypsum requirement (Table 7.12). It was observed that increasing the rate of gypsum application from 40% GR to 60% GR, involved an additional expenditure of Rs. 1432. Differences in the rice yield as a result of higher dose of the amendment were insignificant but the yield of wheat increased by 0.3-0.4 t ha^{-1} in gypsum treated plots. Pyrites was observed to be economically viable proposition.

Table 7.12 Reclamation cost of sodic lands

Item	Total cost (Rs. ha^{-1})	
	Pyrites (22% sulfur)	Gypsum
Land preparation and layout	403	403
Amendment	2254	3150
Amendment application and leaching	198	265
Total	2855	3818

Long-term Effect on Farm Income

Land reclaimed during 1974-75 continued to display a dynamic behavior in cropping pattern, farm income and employment (Agnihotri, 1985). Substantial changes in profit, costs of production and yield variation were observed over the years. Relative data for the two years provide enough evidence of these changes (Table 7.13). It may be seen that after reclamation, barley was the unique crop, which yielded more returns while cost of production reduced over the years. The yield of *basmati* rice and *berseem* crops stabilized during this period.

Table 7.13 Effect of land reclamation on farm-income, cost and yield variability

Particulars	IR8		Basmati		Wheat		Barley		*Berseem*	
Returns (Rs. ha^{-1})	N	S	N	S	N	S	N	S	N	S
1975-76	1581	796	1714	633	1398	387	479	380	1665	739
1980-81	2154	1228	2330	1682	1869	1523	565	541	2431	2178
Cost (Rs. ha^{-1})										
1975-76	925	836	687	937	1030	792	625	528	816	293
1980-81	2614	2746	2219	1914	2091	2075	364	393	2987	2516
Instability (CV)										
1975-76	0.28	0.28	0.27	0.62	0.17	0.36	0.14	0.32	0.20	0.31
1980-81	0.28	0.28	0.27	0.42	0.17	0.36	0.14	0.32	0.20	0.29
Change in returns (%)	36	54	36	166	34	293	18	42	46	195
Change in cost (%)	183	228	223	104	103	162	-42	-26	266	270

N= Normal, S= Sodic. The change in returns and costs is from 1975-76 to 1980-81

SOCIAL CRITERIA

A question often arises whether an economically unviable project should be taken up by the government. The answer is Yes. If social benefits of a project outweigh the financial benefits, such projects can be approved even if these do not stringently satisfy the economic and financial criteria. Recent guidelines from various funding organizations indicate that even those projects which have a benefit-cost ratio less than one can be sanctioned provided the project helps in poverty alleviation or meets one or more of social objectives of the state. With emphasis on employment generation, a project even with low BC ratio can be approved if it results in generating substantial employment. National and international funding organizations finances many schemes with low BC ratio provided such schemes generate employment, rehabilitate the poor and bridge gender disparity. Some projects are approved if these improve the hygienic or the environmental conditions of an area. It also emphasizes the role of indirect benefits in the economic analysis.

Poverty Alleviation and Reduced Disparity

Most agricultural lands affected by water logging and soil salinity are in the possession of small and marginal farmers. Besides, many holders belong to schedule caste or schedule tribes including economically weaker sections. If these strata of the people get an opportunity to reclaim their lands, it would not only alleviate poverty but also save them from neglect by the society. Datta *et al.* (2004) calculated the 'Gini Concentration Ratio' of a drainage project and observed that it decreased from 23% before drainage to 3% after drainage. The reduced value shows increased income parity amongst the farmers. Similar results are expected even from sodic land reclamation projects.

In rural India, poor men are socially neglected lot. Particularly hurt are those, who have land but cannot bring them under plough because of excessive salts. Sometime they lose their savings in a bid to get something out of the lifeless barren land, and forces them to take loans at exorbitant interest rates. Land reclamation helps to enhance the status of the poor in rural setting. Recent studies have revealed a substantial increase in land value upon land reclamation in Uttar Pradesh (Fig. 7.3). Such a change in the land value would raise the status of the poor farmers in the villages.

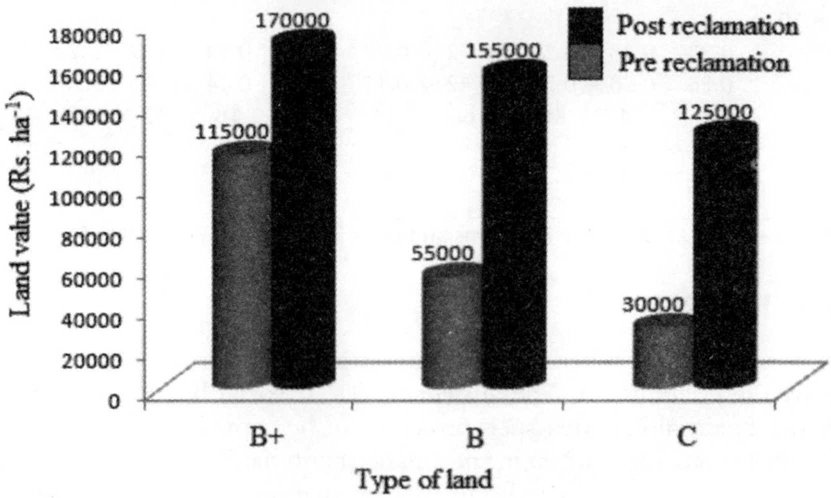

Fig. 7.3 Pre and post reclamation land value of 3 categories of sodic lands in UP

Employment

A social benefit that must be included in any evaluation of a project is its capacity to generate employment opportunities (Singh and Gupta, 1979). Sodic land reclamation programs generate huge employment opportunities. This alone could be a good ground for the approval of land reclamation projects. Skilled and unskilled human power is required to accomplish various activities in these projects. The employment opportunities in a reclamation project can be classified into two groups:

- On-farm employment generation such as for land preparation, amendment application and crop production.
- Indirect employment opportunities such as at mining sites, tube wells installation, in new industries that may come up in the region to meet additional requirements of the project and for handling additional produce in the market.

Capacity to generate employment opportunity at the field level due to reclamation of sodic lands is given in Table 7.14. In the first year, the employment generation is due to land reclamation as well as due to agricultural operations on the reclaimed lands. In subsequent years, it is mainly due to agricultural operations alone. It is apparent that employment generation in terms of human labour, bullock days, farm machinery hours were higher in case of plots under the Institute's demonstration than in areas reclaimed by the farmers. It is because of the fact that farmers while adopting the technologies miss or leave many finer points of the technology. As a result the work generated is less and so are the yields of crops, compared to the demonstration plots. If the one time employment generation opportunities are deducted from the total, 148 and 110 person days of employment per annum are generated in the demonstration plots and the plots reclaimed by the farmers respectively. It may be understood that with time the employment generation in the plots reclaimed by the farmers would also approach that of the demonstration plots.

Table 7.14 Employment generation potential of sodic lands

Particulars	Human labour (days)	Bullocks (days)	Farm machinery (hr)		
			Tractor	Tube well	Thresher
Institute demonstration plots					
Land reclamation	54.3	11.7	8.0	23.0	–
Rice production	94.2	1.8	0.4	138.7	–
Wheat production	53.9	12.2	3.7	78.2	10.0
Total	202.4	25.7	12.1	239.9	10.0
Area reclaimed by the farmer					
Land reclamation	34.0	4.3	11.3	20.3	–
Rice production	74.2	1.0	0.1	124.4	–
Wheat production	35.5	7.5	5.6	61.3	8.7
Total	143.7	12.8	17.0	206.0	8.7

With reclamation, poor farmers become proud owners of lush green fields and are able to engage their family members in some useful tasks besides giving employment to some of the landless labourers. It helps in halting the migration of people from rural to urban areas, a major cause of ill-treatment to women folk in the absence of male members of the family. They are also not required to work on somebody else's field, another cause of ill-treatment. Thus, with reclamation, they can play an independent role in socio-politico-economic functions of the village.

Human Health

Canal irrigation has often been related with increased human health problems due to increased incidences of water borne diseases. While running water could be responsible for this situation in some cases, in others stagnating water is responsible. Increased incidences of malaria are reported from waterlogged areas as Anolpheles mosquitoes breed in the standing water. Documented evidence of such a situation is a report on migration of people living in and around Amritsar in the first decade of the 20th century. The vertical drainage scheme designed to halt the migration of people from Amritsar (India) in 1917 can be quoted on this issue. The main aim of the scheme was to improve the general hygienic conditions of the area by lowering the water table. As per IWMI reports, government expenditure on health services could be curtailed in the same order of magnitude as the investments made in the land reclamation programs.

Inter-sectoral Terms of Trade

Resource mobilization from one sector to another is an indication of strengthening inter-sectoral terms of trade and economic development. Amendment is a major input in any sodic land reclamation program. While land reclamation programs provide opportunities to use low quality gypsum, mining of additional gypsum provides employment opportunities. As the current thinking is to use phospho-gypsum, distillery spent wash and press mud, all byproducts of the industrial sector, it should help the industries to dispose of their byproduct in an environmental friendly manner. Besides, it should minimize the pollution of soil and water resources at the disposal sites. Moreover, if the employment generated under these indirect employment opportunities is also added to employment generation figures quoted earlier, it turns out to be an enormously high figure. A simple calculation is made to show that enormous work is generated as a result of reclamation program on 2.0 million ha land. It will require 400,000 tube wells and tube well rooms, 400,000 pumping sets, 30,000 tractors and matching implements, 4.5 million tons of additional nitrogen, fertilizers and 20 million tons of amendments. As the reclamation may take around 15 years or so, it is understood that employment generation will be spread over a sufficiently long period of time.

Effective Utilization of Rainfall and Reduced Flood Hazard

Most of the rainfall occurring over barren sodic lands is converted into run-off causing long-term water stagnation and imposes heavy load on under designed drainage systems of the area. In some cases, even if water were removed from the area, congestion occurs at the tail end of the watershed. Short-term water stagnation is a perpetual problem in these soils (Fig. 7.4). Experimental evidences now available reveal that with reclamation, a major part of this run-off is conserved in the catchment itself. Not only the total run-off is reduced, peak is also reduced and has a delaying effect on its appearance at the outlet. The water retention capacity of the watershed is also increased, thereby reducing flood hazard. Land reclamation results in increased ground water recharge which is badly needed in areas where water table is depleting. A study at the Central Soil Salinity Research Institute, Karnal has indicated that with good management more than 90% of the rainfall can be conserved within the catchment. Naturally it contributes to reduction in drainage requirement and thereby reduced investments on drainage. Flood hazard is reduced substantially as a result of rainwater conservation.

Fig. 7.4 Short-term water stagnation in a sodic soil following rainfall

Improved Environment and Quality of Life

Barren sodic lands where water stagnates for few weeks to months serve as breeding places for mosquitoes. With reclamation, barren fields being the breeding places for the mosquitoes are no more there. Although there has not been any direct evidence on the reduction of malaria hazard in the surroundings where land has been largely reclaimed, people certainly are happier and agree that they have a much healthier life than before. If *Panchayat* lands are put under trees, it should go a long way in improving the environment in the village as a whole.

OTHER ISSUES

Subsidy on Land Reclamation

There is ample justification for granting subsidy to the farmers on land reclamation. Reclamation of lands leads to an overall community and regional development the fruits of which will be shared by all whether he is an owner of the land or not. As others will be sharing the benefits and not the cost, compensation in the form of subsidy is justified. Besides this, reclamation of sodic lands will enhance the prospects of achieving national goals and thus, it is a priority area for investments. Appropriate amount of subsidy should encourage the farmers to become partner in this task of national importance. Many times it is argued that subsidy is misused by the farming community. This alone cannot be a sufficient ground to deny the subsidy. To avoid misuse, subsidy could be in the form of adjustment in land revenue or in terms of cash against purchases made.

Progress of Reclamation

With the advent of green revolution and a technology of land reclamation made available at about the same time by CSSRI, many state governments realized the potential of land reclamation for increasing food grain production and established Land Reclamation Corporations in their states. A notable mention of Punjab, Haryana and Uttar Pradesh can be made in this regard (Parsad, 1989). Government of India came out with many schemes providing incentives and subsidies on several components of land reclamation programs. These attempts lead to reclamation of large areas. Gupta (2001) assessed the reclaimed areas in these states by deducting the area reported as salt-affected in 1971 and during the last five years (Table 7.15). It is assessed that more than 1 million ha of sodic lands in these three states have been reclaimed. Current reconciliation of data on area reclaimed by different organizations, gypsum distributed for land reclamation and subsurface drainage installed in different states, it has emerged that 1.73 M ha of barren salt affected soils have been reclaimed by 2006-07, major reclamation activity being in the states of Punjab, Haryana and Uttar Pradesh (Table 7.16). Current estimate is that nearly 1.95 M ha area has been reclaimed by March 2015. Rajasthan leads in reclamation of saline lands through subsurface drainage followed by Haryana and Maharashtra. Currently, on an average 50,000 ha of salt affected soils are being reclaimed per annum.

The massive efforts that have gone in land reclamation has ensured household food security and changed the socio-economic profile of more than 9 million people directly. The land reclaimed so far and direct and indirect benefits in terms of production, income and employment generation are given in Fig. 7.5. This shows that it has contributed around 15 million tons of food grains valued at 13,500 million Rs. It has generated 8.33 million person months of employment (Tripathi and Sharma, 2011). In spite of the achievements made so far, there is a need to enhance the pace of land reclamation as total salt affected soils currently are at about the same level as reported by Abrol and Bhumbla in 1971. Indirectly, it means that the pace of reclamation and degradation were of the same order of magnitude during the last 40 years. Attempts are now being made in Bihar to reclaim sodic lands. It is believed that with site specific technologies being made available and the experiences so far, land reclamation and management program in the country would get the desired fillip. By providing institutional support and policy backstopping, Government can provide the necessary incentive to this activity. It is believed that the time is not far when India would be free from the problem of sodic wastelands.

Table 7.15. Extent of the salt-affected area (million ha) and area reclaimed (million ha) in different states

State	Area affected (1971)	Area affected (1995-2001)	Area reclaimed
Punjab	0.530	0.164	0.366
Haryana	0.700	0.153	0.547
Uttar Pradesh	1.200	1.060	0.140
Total		1.053	

Source: Gupta (2001)

Table 7.16 Sodic land reclaimed during 1969-70 to 2006-07 in India

Decade	Area reclaimed (ha)
1971-72 to 1980-81	190,348
1981-82 to 1990-91	563,926
1991-92 to 2000-01	571,509
2001-02 to 2006-07	410,161
Total	1,735,944 or 1.73 M ha

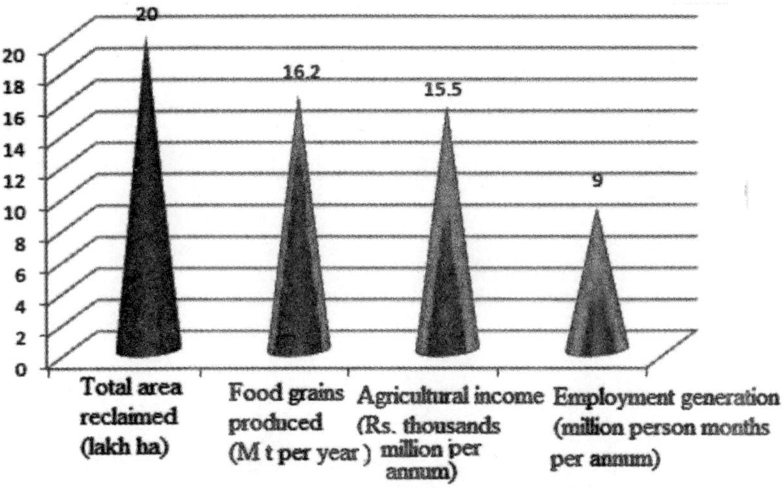

Fig. 7.5 Area reclaimed and some direct and indirect benefits at the national level

SETTING AN APEX ORGANIZTION

To meet the ever-increasing food requirement of the nation due to burgeoning population as well as due to improvement in the living standard, higher and higher targets of food grain production are being fixed from one plan to another. On the other hand, increasing extent and degree of problems of water logging and soil salinity are posing big constraints in achieving the optimum production and productivity in the irrigation commands. Reclamation and management of these lands is presently being looked after by a number of organizations at state and central levels. Integration of activities related to reclamation of salt-affected lands is absolutely essential to avoid duplication or even triplication of efforts. Since the resources are scarce, a close review in a coordinated manner by Ministries of Water Resource and Agriculture is a must for the success of such programs. There is a need to have a unified apex organization that could integrate the activities such as monitoring of the problems, plan measures for prevention, take up surface or subsurface drainage works and undertake reclamation of saline/sodic lands. Such an apex organization should be set-up at the central level and have equally supportive organizations at the state and the project levels. Depending upon the availability of funds, a reasonably good time frame can be fixed to achieve the objective of land reclamation. This challenging task, spread over large number of irrigation commands through the length and breadth of the nation, will need trained human manpower. To meet this requirement, a nodal agency for Human Resource Development should also be identified. This agency should be provided enough funds to take up this activity on war footing. Central Soil Salinity Research Institute with its manpower and infrastructure can act as the nodal agency and train the manpower in collaboration with existing Water and Land Management Institutes that have been established in many states.

Annexure 1

WORLD DISTRIBUTION OF SALT AFFECTED AREAS

Continent	Country	Area, 000 ha		Total
		Saline/Solonchaks	Sodic/Solonetz	
North America	Canada	264	6 974	7 238
Mexico and Central America	USA	5 927	2 590	8 517
	Cuba	316	-	316
	Mexico	1 649	-	1 649
South America	Argentina	32 473	53 139	85 612
	Bolivia	5 233	716	5 949
	Brazil	4 141	362	4 503
	Chile	5 000	3 642	8 642
	Colombia	907	-	907
	Ecuador	387	-	387
	Paraguay	20 008	1 894	21 902
	Peru	21	-	21
	Venezuela	1 240	-	1 240
Africa	Afars and Issas	1 741	-	1 741
	Algeria	3 021	129	3 150
	Angola	440	86	526
	Botswana	5 009	670	5 679
	Chad	2 417	5 850	8 267
	Egypt	7 360	-	7 360
	Ethiopia	10 608	425	11 033
	Gambia	150	-	150
	Ghana	200	118	318
	Guinea	525	-	525
	Guinea-Bissau	194	-	194
	Kenya	4 410	448	4 858
	Liberia	362	44	406
	Libyan Arab Jamahiriya	2 457	-	2 457
	Madagascar	37	1 287	1 324
	Mali	2 770	-	2 770
	Mauritania	640	-	640
	Morocco	1 148	-	1 148
	Namibia	562	1 751	2 313
	Niger	-	1 389	1 389

Continent	Country	Area, 000 ha		Total
	Nigeria	665	5 837	6 502
	Rhodesia	-	26	26
	Senegal	765	-	765
	Sierra Leone	307	-	307
	Somalia	1 569	4 033	5 602
	Sudan	2 138	2 736	4 874
	Tunisia	990	-	990
	United Rep. of Cameroon	-	671	671
	United Rep. of Tanzania	2 954	583	3 537
	Zaire	53	-	53
	Zambia	-	863	863
South Asia	Afghanistan	3 103	-	3 101
	Bangladesh	2 479	538	3 017
	Burma	634	-	634
	India	23 222	574	23 796
	Iran	26 399	686	27 085
	Iraq	6 726	-	6 726
	Israel	28	-	28
	Jordan	180	-	180
	Kuwait	209	-	209
	Muscat and Oman	290	-	290
	Pakistan	10 456	-	10 456
	Qatar	225	-	225
	Sarawak	1 538	-	1 538
	Saudi Arabia	6 002	-	6 002
	Sri Lanka	200	-	200
	Syrian Arab Rep.	532	-	532
	United Arab Emirates	1 089	-	1 089
North and Central Asia	China	36 221	437	36 658
	Mongolia	4 070	-	4 070
	USSR	51 092	119 628	170 720
South-East Asia	Democratic Kampuchea	1 291	-	1 291
	Indonesia	13 213	-	13 213
	Malaysia	3 040	-	3 040
	Socialist Rep. of Vietnam	983	-	983
	Thailand	1 456	-	1 456
Australasia	Australia	17 269	339 971	357 240
	Fiji	90	-	90
	Solomon Islands	238	-	238

Source: Massoud, 1977

Salt affected soils of Europe (000, ha)

Continent	Country	Saline/ Solonchaks	Sodic/ Solonetz	Potential salt affected soils	Total
Europe	Europe Czechoslovakia	6.2	14.5	85.0	105.7
	France	175.0	75.0	-	250.0
	Hungary	1.6	384.5	885.5	1 271.6
	Italy	50.0	-	400.0	450.0
	Rumania	40.0	210.0	-	250.0
	Spain	/	/	/	840.0
	USSR	7 546.0	21 998.0	17 781.0	47 325.0
	Yugoslavia	20.0	235.0	-	255.0

Source: Szabolcs, 1974

ANNEXURE II

CONVERSION TABLES

Length (Linear)

To convert from	To	Multiply by
micrometers	millimeters	1.0×10^{-3}
	meters	1.0×10^{-6}
	angstrom units (A)	1.0×10^{4}
	mils	0.03937
	inches	$3.937\ 01 \times 10^{-5}$
millimeters	micrometers	1.0 x 103
	centimeters (cm)	0.1
	meters	1.0×10^{-3}
	mils	39.370 08
	inches	0.039 37
	feet (ft)	$3.280\ 84 \times 10^{-3}$
centimeters	millimeters	10.0
	meters	0.01
	mils	0.3937×10^{3}
	inches	0.3937
	feet	0.032 81
inches	millimeters	25.40
	meters	0.0254
	mils	1.0×10^{3}
	feet	0.083 33
feet	millimeters	304.8
	meters	0.3048
	inches	12.0
	yards(yd)	0.333 33
yards	meters	0.9144
	inches	36.0
	feet	3.0
meters	millimeters	1.0×10^{3}
	Kilometers (km)	1.0×10^{-3}
	inches	39.370 08
	yards	1.093 61
	miles (mi)	$6.213\ 71 \times 10^{-4}$
kilometers	meters	1.0×10^{3}
	feet	$3.280\ 84 \times 10^{3}$

To convert from	To	Multiply by
	miles	0.621 37
miles	meters	1.609 34 x 10^3
	kilometers	1.609 34
	feet	5280.0
	yards	1760.0

Area

To convert from	To	Multiply by
square millimeters	square centimeters (cm^2)	0.01
	square inches (in^2)	1.550 x 10^{-3}
square centimeters	square millimeters (mm^2)	100.0
	square meters (m^2)	1.0 x 10^{-4}
	square inches	0.1550
	square feet (ft^2)	1.076 39 x 10-3
square inches	square millimeters	645.16
	square centimeters	6.4516
	square meters	6.4516 x 10^{-4}
	square feet	69.444 x 10^{-4}
square feet	square meters	0.0929
	hectares (ha)	9.2903 x 10^{-6}
	square inches	144.0
	acres	2.295 68 x 10^{-5}
square yards	square meters	0.836 13
	hectares	8.3613 x 10^{-5}
	square feet	9.0
	acres	2.066 12 x 10^{-4}
square meters	hectares	1.0 x 10^{-4}
	square feet	10.763 91
	acres	2.471 x 1^{-4}
	square yards (yd^2)	1.195 99
acres	square meters	4046.8564
	hectares	0.404 69
	square feet	4.356 x 10^4
hectares	square meters	1.0 x 10^6
	acres	2.471
square kilometers	square meters	1.0 x 10^6
	hectares	100.0
	square feet	107.6391 x 10^5
	acres	247.105 38
	square miles (mi^2)	0.3861
square miles	square meters	258.998 81 x 10^4
	hectares	258.998 81
	square kilometers (km^2)	2.589 99
	square feet	2.787 84 x 10^7
	acres	640.0

Volume-capacity

To convert from	To	Multiply by
cubic millimeters	cubic centimeters (cm^3)	1.0×10^{-3}
	liters (L)	1.0×10^{-6}
	cubic inches (in^3)	$61.023\ 74 \times 10^{-6}$
cubic centimeters	liters	1.0×10^{-3}
	milliliters (ml)	1.0
	cubic inches	$61.023\ 74 \times 10^{-3}$
	fluid ounces (fl oz)	33.814×10^{-3}
milliliters	liters	1.0×10^{-3}
	cubic centimeters	1.0
cubic inches	milliliters	16.387 06
	cubic feet (ft^3)	$57.870\ 37 \times 10^{-5}$
liters	cubic meters (m^3)	1.0×10^{-3}
	cubic feet	0.035 31
	gallons	0.264 17
	fluid ounces	33.814
gallons	liters	3.785 41
	cubic meters	$3.785\ 41 \times 10^{-3}$
	fluid ounces	128.0
	cubic feet	0.133 68
cubic feet	liter	28.316 85
	cubic meters	$28.316\ 85 \times 10^{-3}$
	cubic decameters (dam^3)	$28.316\ 85 \times 10^{-3}$
	cubic inches	1728.0
	cubic yards (yd^3)	$37.037\ 04 \times 10^{-3}$
	gallons (gal)	7.480 52
	acre-feet (acre-ft)	$22.956\ 84 \times 10^{-6}$
cubic miles	cubic decameters	$4.168\ 18 \times 10^{6}$
	cubic kilometers (km^2)	4.168 18
	acre-feet	3.3792×10^{6}
cubic yards	cubic meters	0.764 55
	cubic feet	27.0
cubic meters	liters	1.0×10^{3}
	cubic decameters	1.0×10^{-3}
	gallons	264.1721
	cubic yards	1.307 95
	acre-feet	8.107×10^{-4}
acre-feet	cubic meters	1233.482
	cubic decameters	1.233 48
	cubic feet	43.560×10^{3}
	gallons	325.8514×10^{3}
cubic dekameters	cubic meters	1.0×10^{3}
	cubic feet	$35.314\ 67 \times 10^{3}$
	acre-feet	0.810 71
	gallons	$26.417\ 21 \times 10^{4}$
cubic kilometers	cubic decameters	1.0×10^{6}
	acre-feet	$0.810\ 71 \times 10^{6}$
	cubic miles (mi^3)	0.239 91

Velocity

To convert from	To	Multiply by
feet per second	meters per second (m s^{-1})	0.3048
	kilometers per hour (km h^{-1})	1.097 28
	miles per hour (mi h^{-1})	0.681 82
meters per second	kilometers per hour	3.60
	feet per second (ft s^{-1})	3.280 81
	miles per hour	2.236 94
kilometers per hour	meters per second	0.277 78
	feet per second	0.911 34
	miles per hour	0.621 47
miles per hour	kilometers per hour	1.609 34
	meters per second	0.447 04
	feet per second	1.466 67
feet per year (ft/yr)	millimeters per second	9.665 14 x 10^{-6}

Volume per unit Time Flow

To convert from	To	Multiply by
cubic feet per second	liters per second (L s^{-1})	28.316.85
	cubic meters per second (m^3 s^{-1})	0.028 32
	cubic decameters/day (dam^3 d^{-1})	2.446 57
	gallons per minute (gal min^{-1})	448.831 17
	acre-feet per day (acre-ft d^{-1})	1.983 47
	cubic feet per minute (ft^3 min^{-1})	60.0
gallons per minute	cubic meters per second	0.631 x 10^{-4}
	liters per second	0.0631
	cubic decameters per day	5.451 x 10^{-3}
	cubic feet per second (ft^3 s^{-1})	2.228 x 10^{-3}
	acre-feet per day	4.4192 x 10^{-3}
acre-feet per day	cubic meters per second	0.014 28
	cubic decameters per day	1.233 48
	cubic feet per second	0.504 17
cubic decameters per day	cubic meters per second	0.011 57
	cubic feet per second	0.408 74
	acre-feet per day	0.810 71

Volume per Area per Unit Time

To convert from	To	Multiply by
cubic feet per square foot per day	cubic meters per square meter per day ($m^3\ m^{-2}.d^{-1}$)	0.3048
	cubic feet per square foot per minute ($ft^3\ ft^{-2}.\ min^{-1}$)	0.6944 x 10^{-3}
	liters per square meter per day ($L\ m^{-2}.d^{-1}$)	304.8
	gallons per square foot per day ($gal\ ft^{-2}.d^{-1}$)	7.480 52
	cubic millimeters per square millimeter per day ($mm^3\ mm^{-2}.d^{-1}$)	304.8
	cubic millimeter per square millimeter per hour ($mm^3\ mm^{-2}.h^{-1}$)	25.4
	cubic inches per square inch per hour ($in^3\ in^{-2}.h^{-1}$)	0.5
gallons per square foot per day	cubic meter per square meter per day ($m^3\ m^{-2}.d^{-1}$)	40.7458 x 10^{-3}
	liters per square meter per day ($L\ m^{-2}.d^{-1}$)	40.7458
	cubic per square foot per day ($ft^3\ ft^{-2}.d^{-1}$)	0.133 68

Volume per length per Unit Time

To convert from	To	Multiply by
Cubic feet per foot per day	cubic meters per meter per day ($m^3\ m^{-1}.d^{-1}$)	0.0929
	gallons per foot per day ($gal\ ft^{-1}.d^{-1}$)	7.480 52
	liters per meter per day ($L\ m^{-1}.d^{-1}$)	92.903

Salt Concentration

To convert from	To	Multiply by
TDS ($mg\ L^{-1}$)	TDS (ppm)	1.0
Salinity ($dS\ m^{-1}$)	salinity(mmho cm^{-1})	1.0
Salinity (mmho cm^{-1})	salinity (μmho cm^{-1})	1000
Salinity ($dS\ m^{-1}$) (3-30)	osmotic pressure	0.36
Salinity ($dS\ m^{-1}$) (0.1-5.0)	ppm	0.64
Parts of salt per million parts of irrigation water	tonnes $ha^{-1}.cm^{-1}$	0.001323
Salinity ($dS\ m^{-1}$) ($< 5\ dS\ m^{-1}$)	TDS ($mg\ L^{-1}$)	640
Salinity ($dS\ m^{-1}$) ($> 5\ dS\ m^{-1}$)	TDS ($mg\ L^{-1}$)	800
Water, NO_3-N, SO_4-S, B (ppm)	kg $ha^{-1}.cm^{-1}$ of water applied	0.1015

TDS = Total dissolved salts

Elemental conversions

Percent or kg	Multiply by	Percent or kg
P_2O_5	0.437	P
K_2O	0.826	K
CaO	0.710	Ca
MgO	0.600	Mg
$CaCO_3$	0.400	Ca
NO_3-N	0.226	N
NH_4-N	0.777	N

ANNEXURE III

CHEMICAL COMPOSITION OF PYRITES (PERCENT)

Constituent	Pyrites (powder)	Pyrites (shales)
Sulphur	22-24	11.7
Iron	20-22	11.7
Al_2O_3	6-8	11.3
SiO_2	35-40	46.9
CaO	0.1	4.3
MgO	0.5-0.6	0.9
Cu	0.05	–
Mn	0.01	–
C	–	3.2
Moisture	–	14.7

GLOSSARY OF TERMS

1:1 or 1:2 soil solution: 1 part water is added to 1 or 1/2 part soil (weight:weight) to measure pH or electrical conductivity. The measured EC/pH is designated as EC_1/pH_1 or EC_2/pH_2.

Acid: A substance that has a pH value between 0 and 7. A compound resulting in a pH less than 7 when a molecule can give up a proton to a base, accept an unshared pair of electrons from a base or react with a base to form a salt, a substance that has more free hydrogen ions, H^+, than hydroxyl ions, OH^-.

Acidity: The quantitative capacity of a water or water solution to neutralize an alkali or base. It is usually measured by titration with a standard solution of sodium hydroxide, and expressed in ppm or mg L^{-1} of its calcium carbonate equivalent.

Acid soils: All soils with a pH below 7 are acidic in nature. However, the growth and yield of most plants is unaffected until the pH drops below about 5. As pH level drops, degree of soil acidity increases.

Adjusted R_{Na}: A parameter similar to SAR describing a sodic water to take care of precipitation of Ca due to presence of HCO_3.

Adjusted SAR: A water quality parameter similar to SAR. In this case, Ca concentration of the water is adjusted for the HCO_3 present in the water and the EC of the water. Also see SAR.

Adj. SAR = SAR (1 + 8.4 – pH_c)

The values of pH_c above 8.4 indicate a tendency to dissolve lime from the soil and values below 8.4 indicate a tendency to precipitate lime from the applied water.

Aggregate: A unit of soil structure, usually formed by natural processes and generally <20 mm in diameter. Primary soil particles (predominantly silt and clay) combine to form secondary units (aggregates or clods).

Alkali soil: A soil that contains sufficient exchangeable sodium (ESP> 15) and pH_s >8.2 to interfere with the growth of most crop plants, without appreciable quantities of soluble salts. Also called sodic soils.

Alkaline: Water or soils containing an amount of alkali (strongly basic) substances sufficient to raise the pH value above 7.0.

Alkaline soil: A soil with a pH above 7.

Alkali water: Water having SAR more than 15 and RSC more than 2.5. EC may or may not be high. Also called sodic water.

Alkalization: A process whereby the exchangeable-sodium content of the soil is increased. Also called sodification.

Amendment (soil): An alteration in the properties of a soil and thereby of the soil by the addition of substances such as lime, gypsum and sawdust to make the soil more suitable for the growth of plants; Any substance used for this purpose. Fertilizers constitute a special group of soil amendments.

Anion: Anions are negatively charged particles (ions) such as chloride (Cl^-), sulphate ($SO4^{2-}$), carbonate ($CO3^{2-}$), or bicarbonate (HCO_3^-).

Anion exchange capacity: The total amount of exchangeable anions that a soil can absorb. It is expressed as milli-equivalents per 100 g of soil.

Anthropogenic: Occurring because of, or influenced by, human activity such as pollution of soil/water.

Arable soil: A soil suitable for ploughing and cultivation.

Aromatic: A general term that includes the **organic compounds** containing at least one **benzene** ring.

Available nutrients: The portion of any element or compound in the soil that can be readily absorbed and assimilated by growing plants.

Available water: The portion of water in a soil that can be readily absorbed by plant roots. Defined as the quantity of water retained in the soil that is smaller than field capacity (FC) and larger than the permanent wilting point (PWP). Available water is the difference between the water content at FC and PWP.

Bar: A unit of pressure. One bar equals 14.5 pounds per square inch (psi) or about 0.987 standard atmospheres.

Base: A compound resulting in a pH greater than 7 when in aqueous solution, a molecule that can accept a proton from an acid, donate an unshared pair of electrons to an acid or react with an acid to form a salt, a substance that has fewer free hydrogen ions, $H+$, than hydroxyl ions, OH^-.

Benefit:Cost ratio (B:C ratio): The discounted values of benefits divided by the discounted values of costs adjusted to a common time basis.

Bio-drainage: A drainage method that uses a biological material usually plants as the drainage material to control the water table.

Black alkali: An alkali soil often referred as such because of black patches that appear due to accumulation of organic matter at the soil surface.

Blending: Also referred as mixing; a term used in conjunctive use of multi-quality waters when two or more qualities of waters are mixed for irrigation.

Break Even Benefits (BEB): It is the minimal benefit required to meet the costs.

Calcareous soil: Soil containing sufficient free calcium carbonate or calcium-magnesium carbonate to effervesce visibly when treated with cold 0.1 N hydrochloric acid.

Calcite: Calcium carbonate ($CaCO_3$). A trade name for finely ground limestone, very high in calcium carbonate, which is used to raise the pH of acidic water. The insoluble form of calcium precipitated by waters high in Ca, HCO_3^- and CO_3^{2-}. Sometimes naturally occurring in calcareous soils in arid regions. Insoluble until reacted with an acid

Calcium: An important cation in soils and an important plant nutrient. Calcium has two positive charges, and is attracted strongly to clay. This attraction means that calcium holds clays together. It causes flocculation (clumping together) of wet clays resulting in good soil structure. Gypsum and lime are both calcium salts often used to obtain this effect. Calcium works against the dispersion that occurs when a soil is high in sodium. Calcium must be present in excess of 20 ppm for good soil flocculation to take place. This assumes low levels of carbonates and bicarbonates. High levels of carbonates and bicarbonates require higher levels of calcium for retention of good physical properties.

Capillary fringe: The zone above the free water table where some or all of the capillary interstices are filled with water; this water is continuous with the water in the saturated zone, but is held above that zone by capillarity against gravity.

Capillary rise: The upward movement of water from a free water table due to adhesion of water to the tubular soil pores (capillaries) and the cohesion of water molecules.

Cation: Cations are positively charged particles of atoms or molecules. (Ca^{2+}), magnesium (Mg^{2+}), sodium (Na^{+}), potassium (K^{+}), or ammonium (NH^{4+}). The ratios of these cations as well as the amounts of the various cations present are important for plant nutrition and the structure of the soil.

Cation exchange: The interchange between a cation in solution and another cation on the surface of any surface-active material such as clay colloid or organic colloid.

Cation exchange capacity (CEC): A measure of the net negative charge of a soil. Measured as the total quantity of cations that can be exchanged on a unit of soil material, expressed as milliequivalents per 100 grams of soil (meq 100 g^{-1}), millimoles of charge per kilogram of soil (mmol [+ charge] kg^{-1}) or centimoles of charge per kilogram of soil (cmole [+ charge] kg^{-1}). These cations are held on the exchange sites of clay minerals, or of organic matter.

Compaction: A process that changes the pore sizes in the root zone of the growth media. This in turn reduces the degree of aeration, restricts root growth and reduces absorption of phosphorus and other nutrients.

Concentration: Quantitative amount of a solute, chemical or pollutant in a specified volume or weight of solvent, air, water, soil or other medium, accumulating a level of some material over and above the level found in the ambient environment, generally applied to an organism.

Conjunctive use: Use of multi-quality, multi-sources water in conjunction to get better returns in comparison to when these waters are used separately.

Crop tolerance: Tolerance of crops to adverse abiotic stresses such as water logging, soil salinity, sodicity, heat and cold waves etc.; it may be to the biotic stresses as well

Crust: A surface layer of soil, from a few millimeters to 2.5 cm thick, caused by rain drops impact, clay dispersion and clay translocation. When dry, it is much more compact, hard, and brittle than the material just under it.

Cyclic Use: A term used in conjunctive use of multi quality waters when the two or more qualities of waters are used to irrigate the crop in some cycle or sequence, to avoid stress at salt sensitive stages.

Degradation: The process of soil change from one type to another resulting in increased production constraints. Processes of soils being salinized or sodified or even the replacement of sodium by hydrogen during leaching.

Denitrification: The gaseous loss of nitrogen by either biological or chemical mechanisms, but exclusive of ammonia volatilization.

Desalinization: Removal of excessive soluble salts from the saline soil, usually by leaching.

Discounted benefits: Actual benefits occurring in future adjusted through the use of discount rates to a common time base.

Discounted costs: Actual costs occurring in future adjusted through the use of discount rates to a common time base.

Discounting: The use of interest factors to adjust cash flows to a common time bases.

Dispersed soil: Soil in which clay readily forms a colloidal solution, usually have low permeability, tend to shrink, crack and become hard on drying and become plastic on wetting.

Dispersion: Dispersion is the breakdown of soil clay minerals into single particles in solution. The opposite is flocculation. Dispersion occurs when the bonds in a soil are not strong enough to keep the particles together. It is usually because a soil has too great a proportion of exchangeable sodium in it, or is too low in organic matter. However, working a soil too wet or too dry can also increase its tendency to disperse on wetting. Soil aggregates form larger, more continuous soil pores than do individual soil particles. The larger pores provide better water and air movement.

Dissolution: The process by which a salt dissociates in a liquid (water) becoming its component ions with uniform distribution throughout the solution.

Dissolved solids: Amount of minerals, such as salt, that are dissolved in water; amount of dissolved solids is an indicator of salinity or hardness.

Distillery spent wash: A byproduct of the molasses based distilleries, nutrient rich wastewater with high biological oxygen and chemical oxygen demand (BOD and COD). Used as an amendment to reclaim sodic soils and water.

Dolomite ($CaMg(CO_3)_2$) or calcium magnesium carbonate: insoluble calcium/magnesium combination precipitated from waters high in Ca, Mg, HCO_3^-, and CO_3^{2-}. Sometimes occurring naturally in calcareous soils in arid regions. Insoluble until reacted with an acid.

Drainage: Unimpeded downward movement of water beyond the root zone. To improve soil drainage, it needs to be artificially strengthened by human actions.

Drainage system: Collection of open and/or closed drains, used to collect and dispose of excess surface or subsurface water.

Duplex soil: A duplex soil is one that has a sand, silt or loam topsoil over a clay subsoil. These soils may have restricted drainage or lateral drainage at the top of the subsoil because of this texture change.

Economic analysis: Assessment of an investment's profitability to a whole society or economy regardless of who in the society contributes to the costs and receives the benefits, conventionally measured in 'opportunity' rather than 'market' prices.

Edaphic: Of or pertaining to the soils, is the condition resulting from or influenced by factors inherent in the soil or other substrata, rather than by climatic factors.

Electrical conductivity: It is the property of a medium expressing its capacity to carry electrical charge from one point to another point. It is the reciprocal of the electrical resistivity. The resistivity is the resistance in ohms of a conductor which is 1cm long and has a cross-sectional area of 1cm^2. This property, usually expressed in milli or micro mhos/cm, is used to express the salinity of the soil extracts, drainage water, irrigation water, etc. International units of EC are deci-Simens per meter or dS m^{-1}. One dS m^{-1} is equal to 1 mmhos cm^{-1}.

Elemental sulphur (S): A yellow, inert crystalline mineral that is finely ground. In soil, elemental S is oxidized to sulfuric acid via microbial activity. The rate of elemental S oxidation in soil is most rapid in warm, moist soils. Complete oxidation of elemental S to sulfate often takes one to several years. Elemental sulphur as an amendment is applied whenever sodium is in excess and there is an excess of lime. Gypsum can be substituted for elemental sulphur. It is also recommended if available iron is low, elemental sulphur is recommended to help acidify the soil and make more iron available.

Environmental impact: The effect on the environment of a certain human interference; both positive and negative impacts are evaluated e.g. irrigation, artificial drainage, soil reclamation etc.

Evapotranspiration: The quantity of water used for transpiration by vegetation and lost by evaporation from the soil.

Excess lime (Calcium Carbonate): An excess of lime (1.0 or more) will reduce the availability of phosphorus and micronutrients and will cause the soil to crust. An excess of lime can also decrease the decomposition of some herbicides. Nearly all soils with an excess of lime will have pH values between 7.0 and 8.4. The application of elemental sulphur or acid forming fertilizers can improve the overall detrimental effects of excess lime.

Exchangeable-Cation Percentage (ECP): The extent to which the adsorption complex of a soil is occupied by a particular cation. See Exchangeable Sodium Percentage.

Exchange complex: Active surface constituents of soils which are capable of ion exchange.

Exchangeable Sodium Percentage (ESP): The exchangeable sodium percentage is simply the amount of exchangeable sodium in a soil expressed as a percentage of the total exchangeable ions in the soil. An ESP of more than 15 defines a sodic soil.

ESP = Exchangeable sodium (meq 100 g^{-1} soil)/Cation-exchange capacity (meq 100 g^{-1} soil) * 100

Financial analysis: Assessment of an investment's profitability to those people or entities contributing capital and directly benefitting from the returns, conventionally measured at market prices.

Flocculation: The joining together of smaller individual particles of soil, especially clay, into larger units or flocs. In soils, the presence of calcium tends to flocculate clays, maintaining structure, while sodium tends to disperse clays in solution. Gypsum and lime, both calcium salts, are used to achieve flocculation in some sodic soils.

Fluoride: Fluoride is a compound which occurs naturally in both groundwater and surface water. At low levels, fluoride can provide protection against tooth decay by increasing the durability of tooth enamel. The USEPA has set a maximum drinking water standard for fluoride of 4 parts per million (ppm). EPA has also set a secondary fluoride standard of 2 ppm to protect against dental fluorosis (tooth staining or pitting in developing teeth of children under 9 years old). The optimum fluoride level for drinking water is 0.8 ppm. Where the fluoride level of drinking water is less than 0.6 ppm, supplements may be worth-while depending on the individual situation.

Flushing: A process to remove salts from the top few cm by impounding and draining the water through surface drainage. Used for heavy textured or sodic soils where leaching may pose problems. See leaching.

Ground water: Water that occurs beneath the water table in soils and in geologic formations that are fully saturated.

Ground water quality: Quality of ground water in relation to various purposes. For the discussions in this book purpose is irrigation.

Gypsum: Gypsum (calcium sulfate) occurs naturally as crystals and is mined. This gypsum is of variable quality and contains impurities. Gypsum is also available as a by-product of phosphate fertilizer production (phosphogypsum), or of hydrofluoric acid production (calcium sulfate anhydrite). Calcium sulfate anhydrite has the highest purity of commercially available gypsum. It is used as an amendment to reclaim sodic soils.

Gypsum bed: A structure used to dissolve gypsum in irrigation water before it is applied for irrigating the fields.

Gypsum Requirement (GR): The approximate amount of gypsum needed per hectare to lower the ESP of the soil to a desired level. It is calculated as:

GR = (present ESP - desired ESP) x CEC x 0.052.

The factor of 0.052 assumes CEC is in meq100g^{-1} or cmol (+charge) kg^{-1} units. If CEC is in mmols (+ charge) kg^{-1}, the factor is 0.0052. These factors assume 90% reclamation efficiency and a low SAR of the irrigation water.

Hardpan: A hardened layer of soil (often clay-rich) below the zone of cultivation, caused by cementation (chemical process) and/or extreme compaction (mechanical process) of soil particles. Hardpans may inhibit root growth and trap water, resulting in a waterlogged soil layer.

Hard water: Water containing a high level of calcium, magnesium, and other minerals; hard water reduces the cleansing power of soap and produces scale in hot water lines, boilers and appliances.

Humus: Organic matter, also called 'humus', forms from the decay of leaves, plants and other life.

Hydraulic conductivity: The rate of flow of water through a porous material of a unit cross-sectional area under a unit hydraulic gradient, at a specified temperature. It is also called unit of permeability, transmission constant or coefficient of transmission. It is the coefficient, K in the Darcy's equation v = -Ki, such that v is the flow velocity of a specified liquid under saturated conditions, and i is the hydraulic gradient. Thus, unit of K is $L\,T^{-1}$.

Illite: Illites are clay minerals also known as clay micas. They are the dominant clay minerals in most of the red brown earths, solonised brown soils, calcareous red earths, and terra rossa soils. In general, they are very stable clay minerals. Illites vary in their ability to swell, and are generally more prone to dispersion than kaolins and smectites.

Indicator plants: Plants characteristic of specific soil or site condition. Used to visually assess the problems.

Infiltration: The process of water entry into the soil, generally (but not necessarily) through the soil surface and vertically downward; The flux density of water passing through the (or a) soil surface and flowing into the soil is known as infiltration rate.

Intangible benefits or intangible effects: All the benefits that cannot be expressed in monetary terms.

Integrated nutrient management: Application of organic and inorganic nutrients in a judicious manner to sustain productivity and maintain soil physical condition.

Internal rate of return: The rate of discounting at which the present value of benefits becomes equal to the present value of costs.

Ion: Ions are electrically charged particles made up of atoms or molecules. Cations are positively charged ions. Anions are negatively charged ions.

Ion exchange: The interchange between an ion in solution and another ion on the surface of any surface-active material such as clay or humus.

Ionic toxicity: The ions, which can be absorbed by the crops in excess quantity to affect the crop adversely. The examples of such ions are: copper, cadmium, boron, arsenic, iron, lead, manganese, etc.

Isograms, or Isohaline lines: The locus of points of equal salt concentration.

Kaolin: Kaolin is the most weathered of the clay minerals, virtually having no swelling potential. This type of clay is commonly called china clay and is used in ceramics and pottery because of its inert nature. Soils dominated by kaolin clay occur mainly in areas of high temperature and rainfall. These soils are generally acidic due to the weathering process. Kaolin clay soils are frequently strongly structured, with good infiltration and drainage.

Land forming: Land forming is an earthmoving process whereby the natural topography is altered to suit a particular purpose, such as surface irrigation, surface drainage etc.

Land Reclamation: Changing general physico-chemical characteristics of the land to make it capable of more intensive use for agriculture.

Leaching: The process by which soluble materials in the soil, such as salts, nutrients, pesticide chemicals or contaminants, are washed into a lower layer of soil or are dissolved and carried away by water.

Leaching fraction: The fraction of water applied on the soil that actually passes through the root zone as against the leaching requirement which must pass through the root zone to prevent soil salinity from exceeding a specified value.

Leaching requirement: The leaching fraction necessary to keep soil salinity, chloride, or sodium (whichever is the most limiting factor) from exceeding a tolerance level for the crop rotation. Leaching requirement refers to long-term average conditions.

Lime: Strictly, calcium oxide (CaO), but as commonly used in agricultural terminology, calcium carbonate ($CaCO_3$) and calcium hydroxide $Ca(OH)_2$ are included. Agricultural lime refers to any of these compounds, with or without magnesia, used as an amendment for acid soils to make them more alkaline. The carbonate in lime neutralizes the acidity in soils and raises the pH. However, lime may also be useful on some acid sodic soils, because of its calcium content.

Magnesium: Magnesium is an important part of chlorophyll, the green colouring material of plants. It is also vital for photosynthesis, the process that converts the sun's energy into food for the plant. An excess of exchangeable magnesium in a sodic soil can worsen the dispersion caused by exchangeable sodium.

Matric Potential: The energy by which soil water is held by soil particles. This represents the energy required to be expended by plants to take up water from the soil. Matric potential is always negative indicating that energy has to be applied to remove water from the soil.

meq L^{-1}: Milliequivalents per liter.

Mhos: Reciprocal of Ohms. A unit of Electrical Conductivity to measure the quality of water, See Electrical Conductivity.

Micrograms per liter ($\mu g\ L^{-1}$): A unit expressing the concentration of constituents in solution as weight (micrograms) of solute per unit volume (liter) of water; equivalent to one part per billion in most stream water and ground water. One thousand micrograms per liter equals 1 mg L^{-1}.

Micronutrient: A chemical element necessary in only small amounts for the growth of the plants, usually less than 1 ppm in plant. Examples of these elements are zinc, manganese, copper, iron, boron, molybdenum etc.

Milliequivalents per litre (meq L^{-1}): Equal to ppm / molecular weight.

Milligram per liter (mg L^{-1}): see micrograms per liter. Equalt to parts per million (ppm).

Mixing: See blending

Montmorillonite: An alumine silicate clay mineral with a 2:1 expanding crystal structure that is with two silicon tetrahedral layers enclosing an aluminium octahedral layers. Considerable expansion may be caused along the C axis by water moving between silica layers of continuous units.

Natric horizon: A mineral soil horizon that is developed in a subsurface position in the profile, that satisfies the definition of an argillic horizon, and that also has a columnar structure and more than 15% saturation of the exchangeable cation sites by sodium.

Net present value: The present value of benefits minus the present value of costs.

Non-saline alkali soil: A soil that contains sufficient exchangeable sodium to interfere with the growth of most crop plants and contains variable quantities of soluble salts. The exchangeable sodium percentage is greater than 15, the conductivity of the saturation extract is usually less than 4 dS/m (at 25°C) as against sodic soils where it is variable and the pH of the saturated soil usually ranges between 8.2 to 10.0 or even more.

Organic matter: Any living or dead plant and animal material in the soil is soil organic matter. Organic matter is important for both the chemical and the physical properties of the soil. Organic matter makes bonds between soil particles, creating and maintaining soil structure. It also contains nutrients that are essential for plant growth.

Osmotic head: The head equivalent, in metres, of the osmotic potential or osmotic pressure.

Osmotic Effect: Reduction in the ability of plants to extract water from the soil as the osmotic pressure of the soil water solution increases with increasing salinity.

Osmotic potential (ϕo): The osmotic potential (ϕo) of the constituent water in soil at a specified temperature is the amount of useful work per unit mass of pure water, in J kg^{-1}, that must be done to transfer reversibly and isothermally an infinitesimal quantity of water from a pool of pure free water to a pool of free soil solution (identical in composition with the soil liquid phase at the point under consideration, and at the same height). The potential may be expressed in terms of the experimentally accessible osmotic pressure of the solution (π). Osmotic stress/tension/suction is identical to osmotic pressure as defined above. It can be interpreted as the negative of the negative osmotic pressure, in pascal.

Parent material: Soil parent material refers to the rocks which were weathered to form the soil in the first place. Usually the parent material is below the soil, but it can be distant if glaciers translocated the soils during the ice ages. Parent material is the focus of the study of geology.

Parts per million (ppm): The concentration of a substance expressed as the number of unit weights of that substance per million units of weight of solution. Also expressed as mg kg^{-1} in a solid matrix or mg L^{-1} in solution.

Payback period: It is the time period for an investment to generate sufficient incremental cash to recover its initial capital outlay in full.

Pedology: The science of studying soils and their interaction with the wider environment.

Permeability: the capacity or ability of a porous rock, sediment, or soil to allow the movement of water through its pores.

Permeability Index: The permeability Index (PI) given by Doneen is calculated by the following equation:

$PI = 100 * (Na + \sqrt{HCO_3}) / (Na + Ca + Mg + K)$.

The plot of PI and total salt concentration describes the problems of permeability if any with the continuous use of waters.

pH: See soil pH.

Phospho-gypsum: A kind of gypsum available as a by-product of phosphate fertilizer industry.

Press mud: A by-product of sugar industry, used in agriculture as a source of nutrient as well as an amendment to reclaim sodic soils if press mud is from industries that use sulphur to clarify sugarcane juice, called sulphitation press mud.

Primary benefits or primary effects: Benefits from project including the value of any reduction in costs due to the project and of any increase of farm products marketed or consumed by the farm family.

Recycle: Using same water several times in the same sector i.e. using irrigation return flow or drainage water for irrigation

Residual Sodium Bicarbonate (RSBC): It is defined as the excess of HCO_3 ions over the Ca ions present in water. It is calculated as, HCO_3 – Ca, ions expressed as meq L^{-1}.

Residual Sodium Carbonate (RSC): It is defined as the excess of CO_3+HCO_3 ions over the Ca+Mg ions present in the water. The RSC is calculated with the following relation having units of meq l^{-1}.

$RSC = (CO_3 + HCO_3) - (Ca + Mg)$

Resodification: Sodification of reclaimed lands; the process commonly encountered in areas where water table is shallow, irrigation water is sodic and where lands are poorly managed by the farmers.

Reuse: Use of same water several times often in other sectors such as use of domestic wastewater in agriculture. Often recycle and reuse are used synonymously.

Saline-sodic soil: A soil containing both sufficient exchangeable sodium and soluble salts to adversely affect the growth of most crop plants. The ESP is >15, the conductivity of the saturation extract > 4 dS m^{-1} (at 25°C) and the pH is usually 8.5 or more in the water saturated soil.

Saline soil: A non alkali (sodic) soil containing soluble salts in such quantities that they interfere with the growth of most crop plants. The conductivity of the saturation extract is greater than 4 dS m^{-1} (at 25°C), the exchangeable sodium percentage is less than 15, and the pH of the saturated soil is usually less than 8.5.

Saline water: Water that contains significant amounts of dissolved solids and its EC is more than 2.0 dS m^{-1}.

Salinity: Salinity refers to the concentration of all salts in the soil solution. Any readily soluble salt (including fertilizers) may contribute to soil salinity. If the soil solution is saline, plants may have difficulty taking up both water and nutrients. Some plants may be sensitive to specific ions, such as sodium and chloride and may suffer even in the absence of high salinity.

Salinization: The process of accumulation of salts in the soil particularly in the root zone.

Salt-affected soil: A generic term used to describe soils having an excess of soluble salts, or an excess of exchangeable sodium, or both, such that the growth of most crop plants is adversely affected.

Salt balance: The salt balance of an irrigated area is commonly defined as the difference between the total dissolved solids brought to the land in a pre-decided duration by the irrigation water and the total solids carried away by the drainage water during the same period.

Salt tolerance: The average soil salinity required to produce a specified decrease in plant yield; the ability, expressed qualitatively or quantitatively, of a plant species to withstand salt concentrations in soil/water.

Saturated soil paste: A mixture of soil and pure water such that all the voids between the soil particles are filled with water, while at the same time there is no accumulation of free water on the surface. At saturation the soil paste glistens as it reflects light, flows slightly when the container is tipped and the paste slides freely and cleanly off a spatula for all soils, except those with a high clay content.

Saturation extract: The solution obtained by pressure or vacuum filtration of a soil paste that has been made up to a saturated condition by adding water while stirring.

Scraping: The process of removal of the salt incrustation accumulated at the surface due to excessive evaporation.

Sedentary soil: A soil formed *in situ* from underlying parent rock, in contrast to one derived from transported parent material. Now obsolete.

Sediment (sediment): Any material carried in suspension by water, which settles to the bottom if the water velocity is reduced; fine water-borne matter deposited or accumulated in beds. Sediment is ordinarily transported as suspended sediment, by saltation or as bed load.

Sedimentary rock: A rock formed from materials deposited from suspension or precipitated from solution and usually, but not necessarily, consolidated (i.e. cemented). The principal sedimentary rocks are sandstones, shales, limestones and conglomerates.

Sedimentation: The process of subsidence and deposition or settling of suspended matter carried by water or other liquids, by gravity.

Seepage: The loss of water by leakage from a canal, reservoir or other body of water or from a field. It is generally expressed as flow volume per unit time. Seepage into a body is referred to as influent seepage and away from a body, as effluent seepage.

Selectivity coefficient: A value describing equilibrium conditions for an exchange reaction. It may be compared with the thermodynamic equilibrium constant, but differs from it in that assumptions are made regarding the activities of the ions participating in the exchange reaction.

Sensitive crops or plants: A qualitative term used to designate the plants which are sensitive to water, salts or some toxic element. Also see tolerant.

Sesquioxide: A binary compound of a metal and oxygen in the proportion of 3 to 2, as Al_2O_3 and Fe_2O_3. The term sesquioxides is also used generally to describe free iron, aluminium and manganese oxides in the soil.

Shale: A fine-grained, indurated, detrital sedimentary rock formed by the consolidation (as by compression or cementation) of clay, silt or mud and characterized by finely stratified structure and/or fissility that is approximately parallel to the bedding (along which the rock breaks readily into thin layers) which is commonly most conspicuous on weathered surfaces. It has an appreciable content of clay minerals or derivatives from clay minerals and of detrital quartz as mudstone; slate.

Shear strength: The maximum resistance to shearing stresses which a specimen or element of soil can withstand before failure occurs. The shearing strength is generally considered to be made up of (i) internal friction, or resistance due to interlocking of the particles and (ii) cohesion, or resistance due to the forces tending to hold the particles together in a solid mass. The law governing the shear failure of soils is generally known as Coulomb's law.

Siemens: The SI unit for electrical conductance; the reciprocal of Ohm.

Silica: The chemically resistant dioxide of silicon: SiO_2. It occurs naturally in five crystalline polymorphs (the minerals quartz, tridymite, cristobalite, coesite and stishovite); in cryptocrystalline form (as chalcedony); in amorphous and hydrated forms (as opal); in less pure forms (such as sand, diatomite, tripoli, chert and flint) and combined in silicates as an essential constituent of many minerals.

Siliceous: Relating to or containing silica, e.g. a siliceous clay; siliceous deposits.

Slaked lime: Calcium hydroxide, $Ca(OH)_2$; used in agriculture as a liming material for the amendment of acid soils.

Slickenside: Refers to a polished or grooved ped surface within the soil. It results from part of the soil mass sliding or moving against adjacent material along a plane, which defines the extent of the slickenside. They occur only in clayey materials with relatively high smectite content.

Slick spots: Barren areas having puddled or crusted, very smooth, nearly impervious surfaces, usually because of high salinity or sodicity.

Smectite: Smectites are the clay minerals that shrink and swell on drying and wetting. They are an important component of black earths and other cracking clays. A soil that is dominated by smectites is a relatively young soil - as it weathers with time, the smectites will break down into the other, less reactive clay minerals. Members of the group include dioctahedral montmorillonite (Mg-rich), beidellite (Al-rich) and nontronite (Fe-rich) and trioctahedral hectorite (Mg, Li-rich), saponite (Mg-rich) and sauconite (Zn-rich). CEC ranges from 80-100 cmolc kg^{-1} and surface area from 6×10^5 to 8×10^5 l^{-1} (kg^{-1}). Cf. bentonite.

Sodication or Sodification: See Alkalization

Sodic soils: A term commonly used synonymously to describe alkali soils. These soils generally have excess of sodium resulting in a dispersed or deflocculated condition. The exchangeable sodium percentage (ESP) is greater than 15. In this condition the water cannot enter or is impaired from entering the soil.

Sodic water: A water with residual sodium carbonate more than 2.5 and SAR more or less than 15. EC may or may not be high. Sodic water often has more carbonate and bicarbonate ions.

Sodium: Sodium is a chemical element with symbol Na and atomic number 11. It is a soft, silver-white, highly reactive metal. an element from which many substances are formed, including common salt. An excess of sodium causes the soil to exhibit sodic conditions such as soil crusting, slow water intake and poor aeration. These conditions reduce the productivity of the soil. Sodium ions are essential to numerous biological processes in animals. Also see Exchangeable Sodium Percentage.

Sodium Adsorption Ratio (SAR): A ratio for soil extracts and irrigation waters used to express the relative activity of sodium to calcium and magnesium. SAR is calculated as follows:

$SAR= Na/((Ca+Mg)/2)^{1/2}$ (all ions expressed in meq L^{-1})

$SAR= Na/(Ca+Mg)^{1/2}$ (All ions expressed in mmol L^{-1})

Expressed in units of (mmol $L^{-1})^{1/2}$.

Sodium to Calcium Activity Ratio (SCAR): A ratio for soil extracts and irrigation waters used to express the relative activity of sodium to calcium SCAR = $Na/\sqrt{Ca}$, all ions expressed as meq L^{-1}.

Soil aeration: The process by which air in the soil is replaced by air from the atmosphere. In a well-aerated soil, the soil air is very similar in composition to the atmosphere above the soil. Poorly aerated soils usually contain a much higher percentage of carbon dioxide and a correspondingly lower percentage of oxygen than the atmosphere above the soil. The rate of aeration depends largely on the volume and continuity of airfilled pores within the soil.

Soil aggregation: The cementing or building together of several soil particles into a secondary unit or ped, aggregate or granule. Water-stable aggregates, which will not disintegrate easily, are of special importance to soil structure.

Soil amendment: A substance used to alter the properties of soil for the purpose of making it more suitable for a particular purpose. Examples are lime and gypsum. The term fertilizer is preferably used for those amendments which provide elements essential for plant growth. Other terms are: ameliorating agent; ameliorant; soil conditioner.

Soil conditioner: A soil amendment, particularly one that promotes aggregation. Usually refers to the synthetic polymers used for this purpose, e.g. Krilium, PVA, VAMA, HPAN, etc.; Any material added to a soil for the purpose of improving its physical condition.

Soil crust: A thin surface layer on soil, ranging in thickness from less than one mm up to about 25 mm, which is hard and brittle when dry and much more compact than the material immediately beneath it.

Soil degradation: The physical, chemical or biological deterioration of soil such as soil erosion, soil salinization and loss of biological life, respectively.

Soil extract: The solution separated from a soil suspension or from a soil by filtration, centrifugation, suction or pressure.

Soil genesis: The mode of origin of the soil with special reference to the processes and soil-forming factors responsible for the development of the solum or true soil from the unconsolidated parent material; That branch of pedology concerned with the origin of soils.

Soil horizon: A layer of soil or soil material approximately parallel to the land surface and differing from adjacent genetically related layers in physical, chemical and biological properties or characteristics such as colour, structure, texture, consistence, kinds and numbers of organisms present, degree of acidity or alkalinity/sodicity.

Soil management: The sum total of all tillage operations, cropping practices, fertilizer, lime and other treatments conducted on, or applied to, a soil for the production of plants.

Soil map: A map showing the distribution of soil types or other soil mapping units in relation to the prominent physical and cultural features of the earth's surface.

Soil pH: Soil pH refers to the acidity or alkalinity of the soil with a pH of 7.0 being neutral. A pH below 7.0 being acid, a pH above 7.0 being alkaline and a pH above 8.4 being alkali because of excess sodium. Most soils with a pH above 7.0 will usually have an excess of undissolved free lime. The optimum pH for most plants is between 6.3 and 7.5, and pH's below 6.0 require lime for best growth. pH_s refers to the pH of soil saturation paste while pH (1:1 or 1:2 soil solution) refers 1 part water added to 1 or 1/2 parts soil (weight : weight) in order to measure pH.

Descriptive terms commonly associated with certain ranges in soil pH measured in distilled water are:

Extremely acid :	< 4.5
very strongly acid :	4.5 – 5.0
strongly acid :	5.1 – 5.5
medium acid :	5.6 – 6.0
slightly acid :	6.1 – 6.5
neutral :	6.6 – 7.3
mildly alkaline :	7.4 – 7.8
moderately alkaline :	7.9 – 8.4
strongly alkaline :	8.5 – 9.0
very strongly alkaline:	> 9.0

Soil profile: A vertical section of the soil through all its horizons and extending to the underlying material. Cf. soil horizon.

Soil separate: Mineral particles < 2.0 mm in diameter, ranging between specified limits. Synonymous with soil particle size class. Refer soil texture.

Soil series: The basic unit of soil classification, being a subdivision of a soil family and consisting of soils which are essentially alike in all major profile characteristics. It is the lowest category in a formal classification system.

Soil structure: The combination or arrangement of primary soil particles into secondary units or peds. These secondary units may be, but usually are not, arranged in the profile in such a manner as to give a distinctive characteristic pattern. The secondary units are characterized and classified on the basis of size, shape and degree of distinctness into classes, types and grades, respectively.

Soil survey: The systematic examination, description, classification and mapping of soils in an area for a specific purpose. Soil surveys are classified according to the purpose of, kind and intensity of field examination.

Soil texture: The texture is the term indicating the coarseness or fineness of the soil. The amount and quantity of each of the grain size group of particle that constitute the soil. Given by the relative proportion (percent) of the three soil separates (sand, silt, and clay) in a soil.

Solonchaks: An intra-zonal group of soils, having a high concentration of soluble salts, usually light coloured, without characteristic structural form; developed under salt-loving grass or shrub vegetation mostly in arid, semiarid and sub humid regions.

Solonetz: An intra-zonal group of soils, having a variable surface, horizon of friable soil underlain by hard soil, ordinarily with columnar structure; usually highly alkali, developed under grass or shrub vegetation mostly in semiarid and sub humid climates.

Soluble Sodium Percentage (SSP): The proportion of sodium ions in solution in relation to the total cation concentration, defined as follows :

$$SSP = \frac{\text{Soluble sodium concentration (meq L}^{-1})}{\text{Total cation concentration (meq L}^{-1})}$$

Sorption: General term for the interaction (binding or association) of a solute ion or molecule with a solid

Specific gravity: The density of a material related to that of water whose specific gravity is defined as one

Specific ion effect: Any effect of a salt constituent in the substrate on plant growth which is not due to the osmatic properties of the substrate.

Subsoil: That portion of the normal soil profile underlying the A horizon. In humid climates it is lower in organic matter content, is lighter in colour, usually of finer texture, of higher bulk density, and of lower fertility than the surface soil.

Subsoiling: Tillage operation that is primarily concerned with altering the soil below plough depth. This includes chiselling or other means of breaking into subsoil.

Subsurface drainage: Drainage methods commonly employed to control water table and facilitate leaching of salts e.g. pipe drainage, mole drainage and vertical drainage etc.

Sulphitation press mud: See press mud.

Surface drainage: The diversion or removal of excess water from the surface of the land by means of improved natural or constructed channels, supplemented when necessary by shaping and grading of land surfaces to such channels.

S-value: The sum of exchangeable (as opposed to soluble) Ca, Mg, Na and K ions expressed in cmolc kg^{-1} soil.

Swelling: The increase in bulk volume of a substance, usually upon addition of water. A swelling soil is one that contains expanding-lattice clay minerals which cause the soil to swell or heave upon wetting and shrink (with the formation of cracks) upon drying.

Synergism: The simultaneous actions of two or more agencies that together have a greater total effect than the sum of their individual effects.

Tangible benefits or tangible effects: All the benefits that can be expressed in monetary terms.

TDS, total dissolved solids: A quantitative measure of the total dissolved organic and inorganic solids concentration in water, an indicator test used for water analysis and also a measure of the mineral content of bottled water and ground water, one can monitor electrical conductivity quickly in the field and estimate total dissolved solids or TDS. also sum of all dissolved materials such as salts, which are non-filterable and remain following evaporation of the water.

Tilth: The physical condition of soil as related to its ease of tillage, fitness as a seedbed and its impedance to seedling emergence and root penetration.

Tolerant: Capable of withstanding a condition usually un-favourable. Some plants are tolerant to soil acidity and others are tolerant of soil salinity or alkalinity. See crop tolerance

Toxic: Poisonous; tending to reduce the yield of crops below the normal.

Toxic water: A water containing one or more of toxic elements. Yield reduction is attributed to toxic effects of the ions rather than the osmotic potential of the soil solution.

Trace element: An element found in only minor amounts (concentrations less than 1.0 mg L^{-1}) in water or soil or sediment; includes arsenic, cadmium, chromium, copper, lead, mercury, nickel, and zinc

Vertical drainage: Disposal of drainage water through wells into a porous layer of earth or an open-rock formation; the term vertical drainage is also used for a drainage system in which tube wells are used to lower the water table.

Wasteland: Land not suitable to produce material or services of value. A miscellaneous land type.

Water logging: It refers to the condition of land in which it cannot be put to its normal use as a result of either or both of shallow water table and surface ponding. In arable lands depending upon the extent and duration of water logging, the crop yield may reduce substantially or even become zero.

Water management: The planning, monitoring and administration of water resources for various purposes.

Water quality: A judgment of the chemical, physical and biological characteristics of water for its suitability for a particular purpose.

Water-stable aggregates: A soil aggregate which is stable to the action of water such as falling drops or agitation as in wet sieving analysis.

Water table: The elevation at which the pressure of the water is zero with respect to the atmospheric pressure. The water table is the upper boundary of groundwater.

REFERENCES

Abrol, I.P. and Bhumbla, D.R. 1971. Saline and alkali soils in India: their occurrence and management. In: FAO World Soil Resource Report. 41: 42-51.

Abrol, I.P. and Bhumbla, D.R. 1979. Crop response to differential gypsum application in a highly sodic soil and tolerance of several crops to exchangeable sodium under field conditions. Soil Science. 127: 79-85.

Abrol, I.P., Chhabra, R. and Gupta, R.K. 1980. A fresh look at the diagnostic criteria for sodic soils. In: Int. Symp. on Salt Affected Soils. Central Soil Salinity Research Institute, Karnal. February 18-21, 1980. 142-147.

Abrol, I. P., Chhabra, R. and Gupta, R. K. 1981. Diagnosing alkali soils. Indian Farming. 31:22-24, 37.

Abrol; I.P., Dahiya, I.S. and Bhumbla, D.R. 1975. On the method of determining gypsum requirement of soils. Soil Science. 120: 30-36.

Abrol, I.P., Dargan, K.S. and Bhumbla, D.R. 1973. Reclaiming Alkali Soils. Bull. 2. Central Soil Salinity Research Institute, Karnal.

Abrol, I. P., Saha, A. K. and Acharya, C. L. 1978. Effect of exchangeable sodium on some physical properties. J. Indian Society Soil Sci. 26: 98-105.

Abrol, I.P. and Verma, K.S. 1978. A comparative study of the effect of gypsum and pyrites on yield of rice and wheat grown in a highly sodic soil. Proc. FAI-PPCL Sem. Use of Sedimentary Pyrites in Reclamation of Sodic Soils, Lucknow, India. 136a-136c.

Acharya C.L. and Abrol, I.P. 1975. Comparative study of soil water behaviour in a sodic and adjacent normal soil. J. Indian Society Soil Sci. 23: 391-400.

Acharya, C.L. and Abrol, I.P. 1991. Soil Water Reclamation and Root Water Extraction in Alkali Soils. Bull. 16. Central Soil Salinity Research Institute, Karnal. 96 p.

Agarwal, R. R. 1957. Alkali soils can be reclaimed. Indian Farming. 7: 6-8.

Agarwal, R. R. 1961. Soil classification for the alluvium derived soils of the Indo-Gangetic plains in Uttar Pradesh. J. Indian Society Soil Sci. 9: 219-232.

Agarwal, R.R. and Gupta, R.N. 1968. Saline Alkali Soils in India. Tech. Bull. (Agric. Series), Indian Council of Agricultural Research, New Delhi. 221p.

Agarwal, R. R., Mehrotra, C. L. and Gupta, C. P. 1957. Spread and intensity of soil alkalinity with canal irrigation in Gangetic alluvium of Uttar Pradesh. Indian J. Agric. Science. 27: 363-373.

Agarwal, R. R. and Yadav, J. S. P. 1956. Diagnostic techniques for the saline and alkali soils of the Indian Gangetic alluvium in Uttar Pradesh. J. Indian Society Soil Sci. 7: 109-121.

Agarwal, R.R. and Yadav, J.S.P., Gupta, R.N. 1979. Saline and Alkali Soils of India, ICAR, New Delhi. 286 p.

Agnihotri, A. K. 1985. Effect of sodic soil reclamation on income, employment and cropping pattern. Annual Report, CSSRI, Karnal. 110-111.

Alperovitch, N., Shainberg, I. and Rhoades, J.D. 1986. Effect of mineral weathering on the response of sodic soils to exchangeable magnesium. Soil Science Society America J. 50: 901-904.

Alva, A.K., and M.E. Sumner. 1989. Alleviation of aluminum toxicity to soybeans by phosphogypsum or calcium sulfate in dilute nutrient solutions. Soil Science. 147: 278–285.

Alva, A.K., Sumner, M.E. and Miller, W.P. 1990. Reactions of gypsum or phosphor-gypsum in highly weathered acid sub-soils. Soil Science Society of America Journal. 54: 993–998.

Ambast, S.K., Gupta, S.K. and Sharma, D.K. 2015. Laser Land Levelling. Technical Bull. CSSRI/Karnal/2015/01. CSSRI, Karnal. 53 p.

Anonymous. 2000. Drainage Master Plan. A Report Prepared by High Powered Committee of the Govt. of Haryana.

Anonymous, 2002. Indo- Dutch Network Project (IDNP). Recommendations on Water Logging and Salinity Control Based on Pilot Area Drainage Research. CSSRI, Karnal and Alterra-ILRI, Wageningen, 100 p.

Anonymous, 2007. Biennial Report (2004-06). AICRP on Management of Salt Affected Soils and Use of Saline Water in Agriculture. CSSRI, Karnal. 171p.

Anonymous, 2009. Biennial Report (2007-09). AICRP on Management of Salt Affected Soils and Use of Saline Water in Agriculture. CSSRI, Karnal. 212p.

ANZECC (Australian and New Zealand Environment and Conservation Council). 2000. Australian and New Zealand Guidelines for Fresh and Marine Water Quality: The Guidelines. Paper No. 4. Agriculture and Resource Management Council of Australia and New Zealand, Canberra. 314 p.

Arora, S., Singh, A.K., Mishra, V.K., Singh, Y.P. and Sharma, D.K. 2015. Kinetics of gypsum and native $CaCO_3$ dissolution and nutrient transformations mediated through organic amendments and microbial inoculants for crop production in sodic soils. Annual Report 2014-15. Central Soil Salinity Research Institute, Karnal. 95-100.

Ashgar, A.G. and Dhawan, C.L. 1947. Effect of irrigation and growing rice on saline soils. I. Manganese, nitrogen and phosphate status of soil. Indian J. Agricultural Science. 17: 199-202.

Ashgar, A.G. and Dhawan, C.L. 1948. Quality of irrigation water. Effect on pH and exchangeable bases of soils. J. Indian Chemical Society. 25: 179-184.

Ashok Kumar and Abrol, I.P. 1982. Note on the effect of gypsum levels on the boron content of soil and its uptake by five forage grasses in a highly sodic soil. Indian J. Agricultural Science. 52: 615-617.

Arubnandam, K. and Heinzer R.T. 1977. Factors influencing erosion in dispersive clays and methods of identification. Proc. Paris Sem. Erosion and solid Matter Transport in Inland Waters. IAHS – Alsh. Publication 122. 75-88.

Ayers, R.S. and Westcot, D.W. 1976. Water Quality for Agriculture. Irrigation and Drainage Paper, 29. FAO, Rome.

Ayers, R.S. and D.W. Westcot, 1985. Water Quality for Agriculture FAO Irrigation and Drainage Paper No 29 (Rev. 1): FAO Rome. 109 p.

Ayyappan, D., Balakrishnan, V. and Ravindran, K.C. 2013. Potentiality of *Suaeda monoica* Forsk. a salt marsh halophyte on restoration of saline agricultural soil. World Applied Sciences Journal. 28: 2026-2032.

Bajwa, M.S., Choudhary, O.P. and Josan, A.S. 1992. Effect of continuous irrigation with sodic and saline sodic water on soil properties and crop yield under cotton-wheat rotation in Northern India. Agricultural Water Management. 22: 345-350.

Bajwa, M.S., Choudhary, O.P. and Josan, A.S. 1998. Sodic waters and their management. In: Agricultural Salinity Management in India (Tyagi and Minhas, Eds.). Central Soil Salinity Research Institute, Karnal. 431-450.

Bajwa, M.S., Hira, G.S. and Singh, N. T. 1983. Effect of sodium and bicarbonate irrigation waters on sodium accumulation and maize and wheat yields in northern India. Irrigation Science. 4:91-99.

Bajwa, M.S. and Josan, A.S. 1989. Effect of gypsum and sodic water irrigation on soil and crop yields in a rice-wheat rotation. Agricultural Water Management. 16 : 53-61.

Bajwa, M. S., Josan, A.S. and Choudhary, O.P. 1993. Effect of frequency of sodic and saline-sodic irrigation and gypsum on the build-up of sodium in soil and crop yields. Irrigation Science. 13: 21-26.

Bal, A.R. 1976. Salinity tolerance through seed treatment with proline. Biology Plantarum. 18: 227-229.

Balpande, S.S., Deshpande, S.B., and Pal, D.K. 1996. Factors and processes of soil degradation in Vertisols of the Purna valley, Maharashtra, India. Land Degradation & Development. 7: 313-324.

Balpande, S.S., Kadu, P.R., Pal, D.K., Bhattacharyya, T. and Durge, S.L. 1997. Role of swelling mineral in fixing lower limit of exchangeable sodium percentage for sodic soils. Agropedology. 7: 54-61

Banerjee, S. 1959. Some studies of salt-affected soils of Canning, West Bengal. Plant Soil. 10: 207-219.

Barbiéro, L. and Van Vliet-Lanoe, B. 1998. The alkali soils of the middle Niger valley: origin, formation and present evolution. Geoderma. 84: 323-343

Barnes, R.F. and Baylor, J.E. 1995. Forages in a changing world. In: Forages: An Introduction to Grassland Agriculture. State University Press, New York, NY, USA. 3-13.

Basu, J. K. 1950. Problems of saline soils in the Bombay state, India. Trans. 4th Int. Congress Soil Science. 3: 169-172.

Basu, J. K. and Tagare, V. D. 1943. Soils of the Deccan canals. IV. The alkali soils, their nature and management. Indian J. Agricultural Science. 13: 157-181.

Batra, L., Kumar, A. and Manna, M.C. 1997. Microbiological and chemical amelioration of alkaline soils by growing Karnal grass and gypsum application. Experimental Agriculture. 33: 389-394.

Beek, C.G.E.M. van, and Breemen, N. van 1973. The alkalinity of alkali soils. Journal Soil Science. 24: 129-136.

Bhargava, G. P. 2003. Training Manual for Undertaking Studies on Genesis of Sodic/Alkali Soils. Central Soil Salinity Research Institute, Karnal. 111 p.

Bhargava, G.P. and Abrol, I.P. 1978. Characteristics of Some Typical Salt Affected Soils of Uttar Pradesh. Report No. 6. CSSRI, Karnal. 31p.

Bhargava, G.P. and Bhattacharjee, J.C. 1982. Morphology, genesis and classification of salt-affected soils. In: Review Soil Research in India Vol. II, Indian Soc. Soil Sci. 508-529.

Bhargava, G.P., Singhla, S.K. and Gupta I.C. 1974. Distribution of boron in a salt affected and contiguous normal soils. Annals Arid Zone. 13: 32-38.

Bhargava, G.P., Sharma, R.C., Pal, D.K. and Abrol, I.P. 1980. A case study of the distribution of salt affected soils in Haryana state. Int. Symp. Salt Affected Soils. CSSRI, Karnal. 83-91.

Bhatt, M.N., Goel, K.N. and Uprety, M.C. 1978. Use of pyrites in reclamation of sodic soils in Uttar Pradesh. Proc. FAI-PPCL Sem. Use of Sedimentary Pyrites in Reclamation of Sodic Soils, Lucknow, India. 107-114.

Bhattacharyya, T., Srivastava, R., Sharma, J.P. and Sehgal, J.L. 1994. Classification of saline sodic Vertisols in the coastal plain of Gujarat. J. Indian Society Soil Science. 12: 306-309.

Bhumbla D.R. and Dhingra D.R. 1964. Micro-nutrient status of saline and alkali soils of the Punjab. J. Indian Society Soil Science. 12 : 255-260.

Bhumbla D.R. and Abrol I.P. 1975. Effect of application of different levels of gypsum on the yield of rice, wheat and barley grown in a saline soil. Proc. Int. Symp. New Development in the Field of Salt Affected Soils, Cairo, 1972. 537-544.

Borchardt, G. 1989. Smectites. In: Minerals in Soil Environments. (Eds. Dixon and Weed). Soil Science Society of America: Madison WI. 675-727.

Bower, C.A. 1961. Prediction of the effect of irrigation water on soil. Proc. Tehran Symp. Salinity Problems in Arid Zones. 1-8.

Bower C.A. and Massland, M. 1963. Sodium hazard of Punjab groundwater. West Pakistan Engg. Congress Proc. 49-61.

Camberato. 2001. http://www.scnla.com/Irrigation_Water_Quality.pdf

Center, W. 1880. Note on alkali or *reh* soils and saline water. Records Geological Survey, India. 43: 253-273.

CGWB. 1997. Inland Groundwater Salinity in India. GOI, Ministry of Water Resources, Central Ground Water Board, Faridabad. 62p.

Chakravarti, P. and Chakravarti, S. 1957. Soils of West Bengal. Annals Indian National Science Academy. 23: 93-101.

Chand, M., Abrol, I.P. and Bhumbla, D.R. 1980. A comparison of the effect of eight amendments on soil properties and crop growth in a highly sodic soil. Indian J. Agricultural Science. 47: 348-354.

Chandra, S. 1979. Genetics and plant breeding. In: A Decade of Research. CSSRI, Karnal. 80- 98.

Chandraju, S., Siddappa, C. S., Chidan Kumar and Shree Harsha Kumar, S.S. 2013. Studies on the germination and growth of neem (*Azadirachta indica*) and honge (*Pongamia pinnata)* seeds irrigated by distillery spent wash. International Journal of Pharmaceutical, Chemical and Biological Sciences. 3: 102-106.

Chang, C.W. and Dregne, H.E. 1955. The effect of exchangeable sodium on soil properties and on growth and cation content of alfalfa and cotton. Soil Science Society America Proceedings 19: 29-35.

Chatterji R.K. and Sachdev C.B. 1979. Mineralogical characteristics of salt affected and adjacent normal soils. Current Agriculture. 3: 43-52.

Chauhan, R.K.S. 1978. Note on comparative study of pyrites and gypsum as soil amendments in sodic soils. Proc. FAI-PPCL Sem. Use of Sedimentary Pyrites in Reclamation of Sodic Soils, Lucknow, India. 127-129.

Chawla, K. L. and Chhabra, R. 1991. Physical properties of a gypsum amended sodic soil as affected by long-term use of fertilizers. J. Indian Society Soil Science. 39: 40-45.

Chhabra, R. 1996. Soil Salinity and Water Quality. Oxford & IBH Publication Pvt. Ltd., New Delhi.

Chhabra, R. 2003. Chemistry of salt-affected soils. Lecture Notes for UPLDC Training Programme, ICAR Research Complex for Eastern Region. Patna, India.

Chhabra, R and Abrol, I.P. 1977. Reclaiming effect of rice grown in sodic soil. Soil Science. 124: 49-55.

Chhabra, R., Singh, A. and Abrol, I.P. 1980. Fluorine in sodic soils. Soil Science Society America Journal. 44: 33-36.

Chhabra, R. Singh, N.T. and Hansen, A. 1989. Interaction between type of amendment and P availability. Annual Report. Central Soil Salinity Research Institute, Karnal. 38.

Chhiba, I. M. and Singh, N.T. 1978. An ex-army man reclaims his alkali land. Indian Farming. 28: 15-17.

Choudhary, O.P., Ghuman, B.S., Brar, J.S. and Singh, R.P. 2009. Response of three wheat varieties to irrigation with high RSC waters in semi-arid regions. Platinum Jubilee symposium of ISSS Soil Science in Meeting the Challenges to Food Security and Environmental Quality (Abstracts), 22-25 Dec. 2009, New Delhi. 276.

CSSRI. 2013. Vision-2050. Indian Council of Agricultural Research. Central Soil Salinity Research Institute. Karnal, India. 29 p.

Dahiya, I.S. 2009. Origin, classification, distribution and characteristics of salt-affected soils. http://isdahiya-soil-water-reclamation.blogspot.in/2009/05/origin-occurrence-and-characteristics. html (Opened on 02.05.2015).

Dagar, J.C., Yadav, R.K., Dar, S.R. and Ahamad, S. 2015. Liquorice (*Glycyrrhiza glabra*): a potential salt-tolerant, highly remunerative medicinal crop for remediation of alkali soils. Current Science. 108: 1683-1688.

Damodaran, T., Sharma, D.K., Mishra, V.K. and Jha, S.K. 2014. Rhizosphere engineering: an innovative approach for sustainable crop production in sodic soils. In: 4th National Seminar on Innovative Saline Agriculture in Changing Environment. 12-14 December at RVSKV, Gwalior.

Dargan, K. S. 1979. Agronomic and cultural practices for crop production in salt-affected soils. Current Agriculture. 3: 1-20.

Dargan, K.S., Chandra, H. and Bhumbla, D.R. 1973. In saline alkali soils-urea spraying for more paddy. Indian Farming. 22: 9-10.

Dargan, K.S. and Gaul, B. L. 1974. Optimum nitrogen levels for dwarf varieties of paddy in semi-reclaimed alkali soils. Fertilizer News. 19: 24-25.

Dargan, K. S., Gaul, B. L. and Abrol, I. P. 1976a. Effect of gypsum, farmyard manure and method of sowing on the yield of sugarcane in a sodic soil. Indian J. Agronomy. 21: 349-353.

Dargan, K. S., Gaul, B. L., Abrol, I. P. and Bhumbla, D. R. 1976b. Effect of gypsum, farmyard manure and zinc on the yield of berseem, paddy and maize in a highly sodic soil. Indian J. Agricultural Science. 46: 335-341.

Dargan, K.S., Nitant, H.C. and Bhumbla, D.R. 1975. In alkali soils-urea spraying for saving nitrogen and more wheat. Indian Farming. 27: 9-11.

Dargan, K.S., Singh, O.P. and Gupta, I.C. 1982. Crop Production in Salt-affected Soils. Oxford and IBH Publishing Co. New Delhi. 276 p.

Datta, K.K., Tewari, L. and Joshi, P.K. 2004. Impact of subsurface drainage on improvement of crop production and farm income in northwest India. Irrigation and Drainage Systems. 18: 43-55.

Datta, S., Thibault, Y. Fyfe, W.S., Powell, M.A., Hart, B.R., Martin, R.R, and Tripthy, S. 2002. Occurrence of trona in alkaline soils of the Indo-Gangetic Plains of Uttar Pradesh (UP), India. Episodes. 25: 236-239.

De"Sigmond, A. A. J. 1924. The alkali soils in Hungary and their Reclamation. Soil Science. 18: 379-381

De'Sigmond A.A.J. 1925. (Translated Waksman, S.A.). Contribution to the Theory of the Origin of Alkali Soils. U. S . Department of Agriculture, Washington, D. C.

De'Sigmond A.A.J. 1938. The Principles of Soil Science. Thomas Murby and Co., London. 84-116.

De Souza, S., Azuara, J.M., Burley, N., Lund, J.R. and Howitt, R.E. 2011. Guidelines for Preparing Economic Analysis for Water Recycling Projects. Center of Watershed Sciences University of California, Davis.

Dhar, N. R. 1939. Molasses and press mud in *usar* (alkali) land reclamation. Journal Indian Chemical Society. 2:105-111.

Dhar, N. R. and Mukherji, S. K. 1936. Reclamation of alkali soils by molasses and press mud. Proc. Sugarcane Technologist Association, India. 5: 15-24.

Dhawan, C.L., 1964. Land Reclamation (Reclamation of Saline and Alkali Soils). Central Board of Irrigation and Power, New Delhi. 140 p.

Dhawan, C. L. and Dhand, A. D. 1950. The occurrence and significance of trace elements in relation to soil deterioration. II boron. Indian J. Agricultural Science. 20: 479-485.

Dhawan, C. L., Malhotra, M. M. L. and Singh, J. 1953. Reclamation of alkali soils- *Jantar* (*Sesbania cannabina*) root system in black alkali soils treated with different reclaiming agents. Bulletin International Commission Irrigation Drainage. 76-82.

Dhawan, C. L., Malhotra, M. M. L. and Singh, J. 1955. Reclamation of alkali soils. Proceedings National Academy Science, India. 24A: 625-630.

Dhawan, C. L. and Singh, J. 1958. Reclamation of alkali soils by different crop rotations. Proceedings National Academy Science, India. 27A: 238-252.

Dhillon, K.S. and Dhillon, S.K. 2000. Selenium accumulation by sequentially grown wheat and rice as influenced by gypsum application in a seleniferous soil. Plant and Soil. 227: 243-248.

Dhir, R.P. 1998. Origin and accumulation of salts in soils and ground waters. In: Agricultural Salinity Management in India (Tyagi and Minhas Eds.). Central Soil Salinity Research Institute, Karnal. 41-60.

Dixit, S.P. 1971. Quality of irrigation waters from different sources of western Uttar Pradesh. All India Symposium Soil Salinity. UP Inst. Agric. Sci, Kanpur. 107-111.

DNR 1997a. DNR Water Facts: Irrigation Water Quality, Salinity and Soil Structure Stability, No. W55, Department of Natural Resources, Brisbane.

DNR 1997b. Salinity Management Handbook. Department of Natural Resources, Brisbane.

Doneen, L.D., 1964. Notes on Water Quality in Agriculture. Published as a water science and engineering paper 4001, Department of Water Science and Engineering, University of California.

Dregne, H., Kassas, M. and Razanov, B. 1991. A new assessment of the world status of desertification. Desertification Control Bulletin (United Nations Environment Programe). 20: 6-18.

Dregne, H.E., Xiong, Z. and Xiong, S. 1996. Soil salinity in China. Desertification Control Bulletin. 28: 28-33.

Dudal, R. and Purnell, M. F. 1986. Land Resources: salt affected soils, reclamation and Revegetation Research. 5: 1-10.

Duthi, J. F. 1896. Reclamation of *reh* or *usar* land. Agricture Ledger. 3:1-8.

Dwivedi, R.S. 1979. Plant Physiology: A Decade of Research. CSSRI, Karnal. 99-116.

Eaton, F.M., 1950. Significance of carbonate in Irrigation waters. Soil Science. 67: 128-133.

El-Gabaly, M.M. 1971. Reclamation and management of salt affected soils. In: Salinity Seminar, Baghdad. Irrigation and Drainage Paper, 7, FAO, Rome.

El Mahi, Y.E. and Mustafa, M.A. 1980. The effect of electrolyte concentration and sodium adsorption ratio on phosphate retention by soils. Soil Science. 130:321-325

Emerson, W.W. and Bakker, A.C. 1973. The comparative effects of exchangeable calcium, magnesium, and sodium on some physical properties of red-brown earth sub-soils. II. The spontaneous dispersion of aggregates in water. Australian Journal of Soil Research. 11: 151-157.

FAO. 1997. Production Yearbook, Food and Agricultural Organization, Rome, v. 48.

FAO-AGL. 2000. Extent and causes of salt-affected soils in participating countries. FAO/AGL Global Network on Integrated Soil Management for Sustainable Use of Salt-affected Lands. Available on http://www.fao.org/waicent/faoinfo/agricult/agl/agll/

Galvani, A. 2007. The challenge of the food sufficiency through salt tolerant crops. Reviews in Environmental Science and Biotechnology. 6: 3–16.

Ghai, S.K., Rao, D.L.N. and Batra, L. 1988. Nitrogen contribution to wetland rice by green manuring with *Sesbania* species in an alkaline soil. Biology Fertility Soils. 6: 22-25.

Ghassemi F., Jakeman A.J. and Nix H.A. 1995. Salinization of land and water resources: Human causes, extent, management and case studies. UNSW Press, Sydney, Australia, and CAB International, Wallingford, UK.

Gill, H. S. and Abrol, I. P. 1985. Tree Plantation in Alkali Soils. Better Farming Series Salt-affected Soils. CSSRI, Karnal. 8: 11p.

Gill, H. S. and Abrol, I. P. 1986. Effect of site preparation and composition of filling mixture on the performance of *Eucalyptus* and *Acacia* in a highly sodic soil. Annual Report, CSSRI, Karnal. 25-27.

Grewal, S. S. and Abrol, I. P. 1983. Studies on rain water conservation for the establishment of an agro-forestry system in a highly sodic soil. Annual Report, CSSRI, Karnal. 31-33.

Gupta I.C. 1972. Note on relationship between pH of saturated soil paste and 1:2 soil water suspension. Indian J. Agricultural Science. 42 : 962-963.

Gupta, I.C. 1979. Note on the effect of leaching and gypsum on the detoxification of boron in saline-sodic soils. Current Agriculture. 4: 51-55.

Gupta, I.C. 1980. Effect of irrigation with high sodium waters on soil properties and growth of wheat. International Symposium Salt Affected Soils, Karnal, 287-292.

Gupta, I.C. 1984. Reassessment of irrigation water quality - criteria and standards. Current Agriculture. 8: 109-121.

Gupta, I.C. 1990. Use of Saline Water in Agriculture. Oxford and IBH Publishing Co. Pvt. Ltd. New Delhi, 308 p.

Gupta I.C. and Abichandani C.T. 1970. Seasonal variations in salt composition of some saline water irrigated soils of western Rajasthan I. Effect of rainfall. J. Indian Society Soil. Science. 18: 428-35.

Gupta I.C. and Bandyopadhya A.K. 1975. Effect of heating in reclamation of a salt affected soil. Proc. Int. Symp. New Developments in the Field of Salt Affected soils, Cairo, 1972. 525-528.

Gupta I.C., Bohra R.K. and Abichandani C.T. 1969. Relationships between certain indices of saline water irrigated soils of western Rajasthan. Annals Arid Zone. 8 : 27-32.

Gupta, I.C. and Chandra, H. 1972. Effect of gypsum in reducing boron hazard of saline irrigation waters and soils. Annals of Arid Zone. 11: 228-230.

Gupta, I.C. and Gupta, S.K. 2003. Use of Saline Water in Agriculture: A Study of Arid and Semi-arid Zones of India. Scientific Publishers (India), Jodhpur. 302p.

Gupta, I.C., Gupta, S.K. and Sharma, D.P. 2007. Alkali Wastelands Environment and Reclamation. MD Publications Pvt. Ltd., New Delhi. 411 p.

Gupta, I.C. and Yadav, J.S.P. 1986. Crop tolerance to saline irrigation waters. J. Indian Society Soil Science. 34: 279-284.

Gupta, I.C., Yaduvanshi, N.P.S. and Gupta, S.K. 2012. Standard Methods for Analysis of Soil Plant and Water. Scientific Publishers (India), Jodhpur. 224 p.

Gupta, R. D., Mahajan, G. and Goyal, B.R. 1995. Availability and quality of groundwater in Punjab state. Proceedings Symposium Water Management-Need for Public Awareness. PAU, Ludhiana. 18-35.

Gupta, R.K. and Abrol, I.P. 1990. Salt affected soils: their reclamation and management for crop production. In: Lal R and Stewart BA (Eds). Advances in Soil Science: III- Soil Degradatior Springer-Verlag, New York. 223-288.

Gupta, R.K. and Abrol, I.P. 1999. Salinity build-up and changes in the rice-wheat systems of the Indo-Gangetic Plains. Experimental Agriculture. 36: 273-284.

Gupta, R.K., Bhumbla, D.R. and Abrol, I.P. 1984. Effect of sodicity, pH, organic matter and calcium carbonate on the dispersion behaviour of soils. Soil Science. 137: 245-251.

Gupta, R.K., Chhabra, R. and Abrol, I.P. 1981. The relationship between pH and exchangeable sodium in a sodic soil. Soil Science. 131: 215-219.

Gupta, R.K., Chabbra, R. and Abrol, I.P. 1982. Fluorine adsorption behaviour in alkali soils: relative roles of pH and sodicity. Soil Science. 133: 364-368.

Gupta, R.K. and Gupta, R.K. 1989. Influence of mineral assemblage on the ion exchange equilibria and physico-chemical properties of salt affected soils. Annual Rep. (1989-90), CSSRI. Karnal, India. 25-29.

Gupta, R.K., Singh, N.T. and Sethi, M. 1994. Groundwater Quality for Irrigation in India. (Map 1:6 Million scale). Tech. Bull. 19, CSSRI, Karnal. 16p.

Gupta R.N. and Narain B. 1971. Investigations on some physical properties of alluvial soils of Uttar Pradesh related to conservation and management. J. Indian Society Soil Science. 19: 1-9.

Gupta, S.K. 2001. Extent and distribution of salt affected soils in India. Lecture delivered at the National Course on Agricultural Water Management. CSSRI, Karnal.

Gupta, S. K. (Ed.) 2014. Agricultural Land Drainage in India. Agrotech Publishing Academy, Udaipur. 308 p.

Gupta, S.K. 2015. Reclamation and management of waterlogged saline soils. Agricultural Research Journal. 52: 104-115.

Gupta, S.K. 2015. Assessing the hazards of high SAR and alkali water: a critical review. J. Soil Salinity and Water Quality. 7: 1-11.

Gupta, S.K. 2016. Integrated drainage solutions for waterlogged saline lands. In: Innovative Saline Agriculture (Dagar *et al.* Eds.). Springer (In Press)

Gupta, S.K., Bansal, K.N., and Verma, G.P. 1970. Distribution of nutrient cations in saline and sodic soils of Chambal commanded areas of Madhya Pradesh. J. Indian Society Soil Science. 18:77-81.

Gupta, S.K. and Gupta, I.C. 1987. Management of Saline Soils and Water. Oxford and IBH Publication Co., New Delhi, India. 399 p.

Gupta, S.K. and Gupta, I.C. 2014a. Salt Affected Soils: Reclamation and Management. Scientific Publishers (India), Jodhpur. 316 p.

Gupta, S.K. and Gupta, I.C. 2014b. Management of Saline and Waste Water in Agriculture. Scientific Publishers, Jodhpur. 315p.

Gupta, S.K., Sharma, D.P. and Swarup, A. 2004. Relative tolerance of crops to surface stagnation. J. Agricultural Engineering. 41: 44-48.

Gupta, S.K. and Sharma, S.K. 1990. Response of crops to high exchangeable sodium percentage. Irrigation Science. 11: 173-179.

Hardy, L.A. and Eugster, H.P. 1970. The Evolution of Closed Basin Brines. Mineral. Soc. Am. Special Publication. 3: 273-290.

Haroon, M.A.R. and Bose, S.C.M. 2004. Use of distillery spent wash for sodic soil reclamation, treated distillery effluent for ferti-irrigation of crops. Indian Farming. 53: 48- 51.

Harris, F. S. 1920. Soil Alkali: Its Origin, Nature, and Treatment. John Wiley & Sons, Inc., New York. 257 p.

Hasanuzzaman, M., Nahar, K., Alam, M.M., Bhowmik, P.C., Hossain, M.A., Rahman, M.M., Vara Prasad, M.N, Ozturk, M. and Fujita, M. 2014. Potential use of halophytes to remediate saline soils. BioMed Research International. 2014: 1-12.

Hermsmeier, L.F. 1966. Hydraulic Conductivity and Other Physical Characteristics of Some Wet Soils in Southwest Minnesota. USDA ARS, Washington DC. 41-127. 17p.

Hermsmeier, L.F. 1978. Drainage practices in Imperial valley. Transaction ASAE. 21: 105-109.

Hira, G.S. and Singh , N.T. 1978. Irrigation water requirement for dissolution of gypsum in sodic soil. Soil Science Society America Journal. 44: 930-933.

Hoon, R. C. 1955. Land Reclamation. Central Board of Irrigation and Power. Publication 2: 160-178.

Hoon, R. C. and Ahluwalia, G. S. 1953. The physico-chemical changes in deteriorated soils as a result of soil reclamation operation. II. J. Central Board Irrigation Power. 10: 38-42.

Hoon , R. C. and Dewan, R. 1949. The physico-chemical changes in deteriorated soils as a result of soil reclamation operations. II. J. Central Board Irrigation and Power. 6: 265-272.

Hoon, R. C., Jain, L.C. and Bhatnagar, B. B. 1950. The physico-chemical changes in deteriorated soils as a result of soil reclamation operation. IV. The occurrence of gypsum in soils. J. Central Board Irrigation and Power. 7: 60-68.

Hoon, R. C. and Mehta, M. L. 1937. A Study of the Soil Profile of the Punjab Plains with Reference to their Natural Flora. Research Publication Punjab Irrigation Research Institute. 3: 3 p.

Horton, R.E., 1940. An approach to the physical interpretation of infiltration capacity. Proceedings Soil. Science Society America. 5: 399-417.

HSMITC. 1984. Studies for the Use of Saline Water in the Command Areas of Irrigation Projects in Haryana, India – Appraisal Document (Summary).115 p.

http://www.agriculture.gov.sk.ca/Default.aspx?DN=2d20bb89-4290-4eea-b265-dfd3a155cc51. The Nature and Management of Salt-Affected Land in Saskatchewan (Opened on 20.04.2015)

http://www.rrcap.ait.asia/male/manual/national/12chapter12.pdf (opened on 05.05.2014) Chapter on Electrical Conductivity

Hydrology Project. 1999. How to Measure Electrical Conductivity. Training Module: WQ – 09 World Bank & Government of The Netherlands funded Hydrology Project, New Delhi. 21+4p.

Inglis, C. C. and Gokhale, V. K. 1928. Report on Drainage Experiments at Baramati, Bombay. Tech. Report. Public Works Department, Bombay. No. 14.

Isaac, R.K., Sharma, D.P. and Swaroop, N. 2010. New approaches in reclamation of degraded soils with special reference to sodic soil: an Indian experience. In: Degradation and Desertification: Assessment, Mitigation and Remediation. Springer, UK. 253-266

Jaggi T.N. 1978. Use of Amjhore pyrites as sodic soil amendment. Proceedings Seminar Use of Sedimentary Pyrites in Reclamation of Sodic Soils, Lucknow. 37-43.

Joshi, N. and Sharma, S. 2010. Physico-chemical characterization of sulphitation press mud, composted press mud and vermi-composted press mud. Report and Opinion. 2: 79-82.

Joshi, Y.C., Qadir, Ali, Bal A.R. and Rana, R.S. 1980. Sodium/potassium index of wheat seedling in relation to sodicity tolerance. International Symposium Salt Affected Soils, Karnal. 457-460.

Kaledhonkar, M. J., Tyagi, N. K. and Van Der Zee SEATM. 2001. Solute transport modeling in soils for irrigation field experiments with alkali water. Agricultural Water Management. 51: 153-171.

Kanwar, J. S. 1961. Reclaiming deteriorated soils. Fertilizer News. 6: 19-21.

Kanwar, J. S. 1962. Reclamation of saline-alkali soils. Proceedings Symposium Salinity and Alkalinity, IARI, New Delhi.

Kanwar, J. S. and Bhumbla, D. R. 1960. Reclamation of alkaline and saline soils in the Punjab. Rep. ICAR, New Delhi.

Kanwar, J.S. and Singh, S.S. 1961. Boron in normal and saline soils of the irrigated areas of the Punjab. Soil Science. 92: 207-211.

Katyal, J.C. and Sharma, B.D. 1980. A new technique of plant analysis to resolve iron chlorosis. Plant and Soil. 55: 105-119.

Keisling, T.C., Rao, P.S.C. and Jessup, R.E. 1978. Pertinent criteria for describing the dissolution of gypsum beds in flowing water. Soil Science Society America Journal. 42: 234-236.

Kelley, W.P. 1934. The formation, evolution, reclamation and the absorbed bases of alkali soils. The Journal of Agricultural Science. 24: 72-92

Kelley, W.P. 1951. Alkali Soils. Reinhold Publ. Corp., New York.

Kemper, W.D., Olsen, J. and De Mooy, C.J. 1975. Dissolution rate of gypsum in flowing water. Soil Science Society America Proceedings. 39: 458-463.

Khan, A. D. 1950. *Usar* soils: their identification and reclamation. Agric. Animal Husbandry. (Uttar Pradesh, India). I: 29-30.

Khan, W. A. M. and Yadav, J. S. P. 1962. Characteristics and afforestation problems of saline alkali soils. Indian Farming. 83: 259-261.

Khandkar, U.R., Tiwari, S.C., Verma, S.K., Sharma, R.K., Gupta, S.K. and Ambast, S.K. 2013. Raised and Sunken Bed Technology for Reclamation of Alkali Vertisols under Rainfed Conditions. Rajmata Vijayaraje Scindia Krishi Vishwa Vidyalaya, Indore. 16 p.

Khomo, L., Rogers, K., Hartshorn, A. and Chadwick, O. 2006. Geochemical Origin of Sodic Soil along the Phugwane. University of California, Santa Barbara and University of the Witwatersrand, Johannesburg. http://www.sanparks.org/parks/kruger/conservation/ scientific/

Khosla, B. K., Dargan, K. S., Abrol, I. P. and Bhumbla, D. R. 1979. Effect of depth of mixing gypsum on soil properties and yield of barley, rice and wheat on a saline sodic soil. Indian J. Agricultural Science. 43: 1024-1031.

Kittrick; J.A. 1971. Soil solution composition and stability of clay minerals. Soil Science Society America Proceedings 35: 450-454.

Kovda, V. 1954. Geochemistry of the Arid Zone in USSR, Publ. House of the Academy of Sciences, Moskva-Leningrad, U.S.S.R.

Kovda, V.A. 1965. Alkaline soda-saline soils. Agrokemia es Talajtan. Suppl. 14: 15-48.

Kovda V.A., Berg C.V.D. and Hagan R.M. 1973. Irrigation, Drainage and Salinity: An International Source Book. FAO/UNESCO, Hutchison and Co., London.

Krishna, G. C., Desai, A.D. and Krishnamurthy, C. 1960. Deep black cotton soils of Raichur. Bulletin Indian Council of Agricultural Research. 83: 53 p.

Krishnamurthy, S.L., Pundir, P., Singh, Y.P., Sharma, S.K., Sharma, P.C. and Sharma, D.K. 2015. Yield stability of rice lines for salt tolerance using additive main effects and multiplicative interaction analysis – AMMI. Journal of Soil Salinity and Water Quality. 7: 98-106.

Kudesia, K.R., Ran, M.I. and Pal, A. 2009. Reclaiming degraded land in India through the cultivation of medicinal plants. Botany Research International. 2: 174-181.

Kumar, A. and Abrol, I. P. 1983. Effect of gypsum on five tropical grasses grown in normal and extremely sodic soils. Experimental Agriculture. 19: 169-177.

Kumar, A. and Abrol, I.P. 1986. Grasses in Alkali Soils. Bulletin No. 11. Central Soil Salinity Research Institute, Karnal. 95 p.

Kumar, A., Abrol, I. P. and Dargan, K. S. 1980. Effect of exchangeable sodium on the growth and yield of five forage grasses. Forage Research. 6: 123-127.

Kumar, A., Sharma, P.C and Batra, L. 2006. Comparative performance of sole and intercrop of oat (*Avena sativa*) and Persian clover (*Trifolium resupinatum*) under alkali water irrigation. Indian Journal of Agronomy. 51: 119-122.

Kumar, P., Yadav, R.K. and Meena, R.L. 2015. Prospects of growing medicinal and aromatic plants with poor quality water. Lecture delivered at CSSRI for AARDO Training. CSSRI, Karnal.

Kumar, R. 1992. Genesis of alkali soils at different geomorphic locations in Haryana. Ph.D. Thesis, HAU, Hisar (Haryana).

Kumar, R., Ahuja, R.L., Singh, N.T. and Ghabru, S.K. 1994. Characteristics and classification of alkali soils on siwalik hills of Satluj-Yamuna divide, north-west India. Agropedology. 5: 29-35.

Kumar, R., Ghabru, S.K., Ahuja, R.L. and Singh, N.T. 1993. Elemental composition and mineralogy of alkali soils on the Siwalik Hills. Agropedology. 4: 59-68.

Kumar, R., Ghabru, S.K., Ahuja, R.L., Singh, N.T. and Jassal, H.S. 1993. Clay minerals in the alkali soils of Ghaggar river basin of Satluj-Ymuna divide of NW India. Clay Research. 12: 33-42.

Kumar, R., Ghabru, S.K., Singh, N.T, Ahuja, R.L. and Sharma, B.D. 2009. Origin of salinity in the Indus plains of the Indian Sub-continent. J. Soil Salinity and Water Quality.

Lado, M., Paz, A. and Ben-Hur, M. 2004. Organic matter and aggregate-size interactions in saturated hydraulic conductivity. Soil Science Society of America J. 68: 234-242.

Langeliar, W. F. 1936. The analytical control of anti-corrosion water treatment. J. American Water Works Society. 28: 1500-1521.

Leather J.W. 1897. Reclamation of *reh* and *usar* land. Agric. Ledger. 4: 195-253.

Leather, J.W. 1914. Investigation on *Usar* Land in the United Provinces. Govt. Press, UP. 87 p.

Lin, Z., Myhre, D.L. and Martin, H.W. 1988. Effects of lime and phospho-gypsum on fibrous citrus-root growth and properties of sodic horizon soil. Proceedings of the Soil and Crop Science Society of Florida. 47: 67–72.

Lindsay, W.L. 1979. Chemical Equillibria in Soils. John Wiley & Sons, New York.

Linsley, R.K. Jr., Kohler, M.A. and Pauthus, J.L.H. 1949. Applied Hydrology. Mc-Graw-Hill Book Co., New York. 689 p.

Lohani, D. N. 1979. Preliminary trials with high yielding varieties of *Luecaena leucocephola*. Indian Forester. 105: 199-202.

Low, P.F. 1980. The swelling of clay-II. Montmorillonites. Soil Science Society America Journal. 44: 667-676.

Maas, E.V. and Hoffman, G.J. 1977. Crop tolerance assessment. J. Irrigation and Drainage Div. Proceedings American Society Civil Engineering. 103: 115-134.

Maharudrappa, K., Baligar, U.C., Prabhakar, A.S. and Patil, S.V. 1972. Effect of gypsum application with and without FYM to Karl on the yield of jayadhar cotton. Proceedings Seminar Drought, Bangalore, 1968. 131-136.

Malik, K.A., Aslam, Z. and Naqvi, M. 1986. Kallar grass: a plant for saline land. Nuclear Institute for Agriculture and Biology, Faislabad, Pakistan. 93 p.

Manchanda, H. R., Gupta, I.C. and Jain, B.L. 1989a. Use of poor quality waters. In: Review of Research on Sandy Soils in India. International Symposium on Managing Sandy Soils. CAZRI, Jodhpur. Feb 6-10, 1989. 362-383.

Manchanda, H.R., Khosla, B.K., Sharma, D.R., Minhas, P.S., Chawla, K.L., Gupta, R.K. and Singh, N.T. 1989b. Irrigation and Crop Management with Brackish Waters. Bulletin Better Farming Series No. 16, Central Soil Salinity Research Institute, Karnal. 14p.

Mann, H.H. and Tamhane, V.A. 1910. The salt land of Nira Valley. Bull. Department Agriculture, Bombay. 35 p.

Massoud, F.I. 1974. Salinity and alkalinity as soil degradation hazard. FAO/UNDP Expert consultation on Soil Degradation (AGL:50/74/10), FAO, Rome.

Massoud, F.I. 1977. Basic principles for prognosis and monitoring of salinity and sodicity. In: Proc. International Conference on Managing Saline Water for Irrigation. Texas Tech. University, Lubbock, Texas. 16-20 August 1976. 432-454.

Mays, D.A., and Mortvedt, J.J. 1986. Crop response to soil applications of phospho-gypsum. Journal of Environmental Quality. 15: 78–81.

McNeal, B.L. 1968. Prediction of the effect of mixed salt solutions on soil hydraulic conductivity. Soil Science Society of America Journal. 31: 190-193.

McNeal, B.L., Pearson, G.A., Hatcher, J.T. and Hower C.A. 1966. Effect of rice culture on the reclamation of sodic soils. Agronomy Journal. 58: 238-240.

Mehta K.K. and Yadav J.S.P. 1977. Phospho-gypsum for reclamation of sodic soils. Indian Farming. 27: 6-7.

Mehta, M.L. 1940. The Formation and the Reclamation of *Thur* Lands in Punjab. Punjab Irrigation Research Institute. 3: 64p.

Mehta, M.L. 1951. Land Reclamation. Central Board Irrigation Power, New Delhi. 117 p.

Miller, J. J. and Brierley, J.A. 2011. Solonetzic soils of Canada: Genesis, distribution, and classification. Canadian Journal Soil Science. 91: 889-902.

Minhas, P.S. and Gupta, R.K. 1992. Quality of Irrigation Water – Assessment and Management. Indian Council of Agricultural Research, New Delhi. 123p.

Minhas, P.S. and Sharma, D.R. 2006. Predicability of existing indices and an alternative coefficient for estimating sodicity build-up using adj R_{Na} and permissible limits for crops grown on soils irrigated with waters having residual alkalinity. J. Indian Society Soil Science. 54 : 331-338.

Minhas, P.S., Sharma, D.R. and Sharma, D.K. 1996. Perspective of sodic water management for paddy-wheat cropping system. J. Indian Water Resources Society. 2 : 57-61.

Minhas, P.S., Sharma, O.P. and Patil, S.G. 1998. 25 years of Research on Management of Salt-affected Soils and Use of Saline Water in Agriculture. Central Soil Salinity Research Institute, Karnal. 220p.

Ministry of Agriculture. 1994. Draft Report on Status of Land Degradation in India. Ministry of Agriculture, New Delhi. 102 p.

Misra, B. 1996. Highlights of Research on Crops and Varieties for Salt Affected Soils. Research Bulletin. Central Soil Salinity Research Institute, Karnal. 28 p.

Mitchell, J.K. 1976. Fundamentals of Soil Behaviour. John Wiley and Sons Inc., New York.

Miyamoto, S., Prather, R.J. and Stroehlein, J.L. 1975. Sulfuric acid and leaching requirements for reclaiming sodium-affected calcareous soils. Plant and Soil. 43: 573-585.

Mondal, A.K., Obi Reddy, G.P. and Ravisankar, T. 2011. Digital database of salt affected soils in India using Geographic Information System. Journal of Soil Salinity and Water Quality. 3:16-29.

Muhr, G.R., Datta, N.P., Sankarasubramoney, H., Dever, R.F., Leley, V.K. and Donahue, R.L. 1963. Soil Testing in India. United States Agency for International Development Mission of India: Delhi.

Nangre, D.D. and Singh, K.G. 2010. Effect of blending of fresh and saline irrigation water on yield and quality of tomato. Abstracts-44th Annual Convention of ISAE, IARI, New Delhi. 3.35 p.

Narayana, V.V. Dhruva. 1980. Reclaiming Alkali Soils-Engineering Aspects. Bulletin No. 6. Central Soil Salinity Research Institute, Karnal. 61 p.

Narayana, V.V. Dhruva, Kalra, V.D. and Singh, O.P. 1980. Design of a farm layout for irrigation with limited discharge. Agricultural Water Management. 3: 457-464.

Nasir, N. M. 2006. Better Management Practices for Cotton and Sugarcane. WWF - Pakistan, Ferozepur Road, Lahore - 54600, Pakistan.

National Commission on Agriculture (NCA). 1976. Reports of the Commission. Ministry of Agriculture and Irrigation, Government of India, New Delhi.

Nayak, A.K., Mishra, V.K., Sharma, D.K., Kale, P. and Srivastava, M. 2008. Annual Report 2007-08. Central Soil Salinity Research Institute, Karnal. 73-74.

Nayak ,A.K., Sharma, D.K., Mishra, V.K., Minhas, P.S. and Verma, C.L. 2008. Reclamation of saline sodic soil under a rice wheat system by horizontal surface flushing. Soil Use and Management. 24: 237-243.

Nayak, A.K., Sharma, D.K., Mishra, V.K., Jha, S.K., Singh, C.S., Shahabuddin, Md., Srivastava, M., Kale, S.P. and Singh, G.B. 2011. Phospho-gypsum: An Alternate Chemical Amendment for Sodic Lands. Tech. Bull.1/2011. CSSRI, RRS, Lucknow. 42 p.

NBSS & LUP 1986. Benchmark Soils of India (Murthy R.S. et al., Ed.). NBSS&LUP, Nagpur.

NBSS & LUP. 2002. Soils of India. Pub. No. 94, National Bureau of Soil Survey and Land use Planning, Nagpur, 130 p.

NEDCO 1968. *Bhal* Reclamation Scheme (Govt. of the Netherlands and State of Gujarat, India). Delft Hydraulics Laboratory, Delft, The Netherlands. Vol. 1-6.

Neelakanthan S. and Pahwa, M.R. 1977. Note on the distribution of micro-organisms in normal, saline and alkali soils. Current Agriculture. 1: 77-78.

Northcote, K.H. and Skene, J.K.M. 1972. Australian Soils with Saline and Sodic Properties. CSIRO Australia, Division of Soils Soil Publication No. 27.

NRSA. 1996. Mapping Salt-affected Soils of India on 1:250,000 scale. NRSA, Hyderabad, India.

NRSA and DoLR. 2005. Wastelands Atlas of India. Department of Land Resources (DoLR), Ministry of Rural Development, New Delhi.

Oldeman, L.R. 1991-92. Global extent of soil degradation. ISRIC, Bi-annual report, Wageningen, The Netherlands. 19-39.

Oldeman, L.R., Van Englen, V.W.P. and Pulles, J.H.M. 1991. The extent of human-induced soil degradation. In: Oldeman, *et al.*: World Map of Status of Human-Induced Soil Degradation: An explanatory Note. Wageningen: International Soil Reference and Information Centre (ISRIC). 27-33.

Overstreet R., Martin J.C. and King H.M. 1951. Gypsum, sulphur and sulphuric acid for reclaiming a sodic soil of the Fresno series. Hilgardia. 21: 5.

Paikray, S.P. and Singh, K. 2012. Microbial reclamation of saline and sodic soil. WO 2012046098 http://www.google.com/patents/WO2012046098A1.

Pal, D.K., Dasog, G.S., Vadivelu, S., Ahuja, R.L. and Bhattacharyya, T. 2000. Secondary calcium carbonate in soils of arid and semi-arid regions of India. in: Global Change and Pedogenic Carbonates (Lal *et al.*, Eds.). Lewis Publishers, Boca Raton, Florida, USA. 149-185

Parsad R. 1989. Technology for the reclamation of sodic soils. Indian Farming. 39: 26-27

Parsad, R. and Singh, P. 1980. Some socio-economic constraints in the adoption of alkali soil reclamation technology by the farmers. Annals of Arid Zone. 19: 351-358.

Pasricha, N.S. and Randhawa, N.S. 1971. Available molybdenum status of some recently reclaimed saline sodic soils and its effect on concentration of molybdenum, copper,-sulphur and nitrogen in *berseem* (*Trifolium alexandrinum*) grown on these soils. In: International Symposium on Soil Fertility Evaluation. 1971. 1017-1026.

Patel, K.P., Singh, B. and Chhiba, I.M. 1990. Crop yields and properties of a sodic soil as influenced by S containing amendments under draining and non-draining conditions. Arid Soil Research & Rehabilitation. 4:173-180.

Patra, D. D. and Singh, D. V. 1998. Medicinal and aromatic crops. In: Salinity Management in India (Eds. Tyagi, N. K. and Minhas, P. S.). Central Soil Salinity Research Institute, Karnal. 499-506.

Pavan, M.A., Bingham, F.T. and Peryea, F.J. 1987. Influence of calcium and magnesium salts on acid soil chemistry and calcium nutrition of apple. Soil Science Society of America Journal. 51: 1526–1530.

Pawluk, S. 1982. Salinization and solonetz formation. Proc. Alberta Soil Science Workshop, Edmonton, 23-24 February 1982. 1-23.

Pearson, G.A. 1960. Tolerance of Crops to Exchangeable Sodium. US Dept. Agriculture Information Bull. 216: 4 p.

Pearson, K.E. 2009. The Basics of Salinity and Sodicity Effects on Soil Physical Properties Information Highlight for the General Public. MSU Extension Water Quality Program, Montana State University, Bozeman, MT (USA).

Phogat, V. Sharma, S. K., Kumar, S., Satyavan and Gupta, S. K. 2010. Vegetable Cultivation with Poor Quality Water. Bulletin, CCS HAU, Hisar. 72p

Planning Commission, India. 1963. Study of Wastelands including Saline Alkali and Waterlogged Lands and their Reclamation Measures. Research Committee of Natural Resources, New Delhi. 237 p.

Planning Commission. 2013. Report of High Level Expert Group on Water Logging in Punjab. Planning Commission, New Delhi. 95 p.

Ponnamperuma, F.N. 1977. Screening for Tolerance to Mineral Stresses. IRRI Res. Pap. Ser. No. 6. International Rice Research Institute. Los Banos, Philippines. 21 p.

Ponnamperuma, F.N. Martinez, E. and Loy, T. 1966. Influence of redox potential and partial pressure of carbon dioxide on pH values and the suspension effect of flooded soils. Soil Sci. 101: 421-431.

Prather, R.J., Goertzon, J.O., Rhoades, J.D. and Frenkel, H. 1978. Efficient amendment use in sodic soil reclamation. Soil Science Society America J. 42: 782-786.

Puri, A.N. 1930. A new method for estimating total carbonates in soils. Imperial Agricultural Research Institute, Pusa Bull. No. 206. 7 p.

Puri, A. N. 1933. The Relation between Exchangeable Sodium and Crop Yield in Punjab Soils and a New Method to Characterize Alkali Soils. Research Publication, Punjab Irrigation Research Institute. 5: 4 p.

Puri, A. N., Taylor, E. M., Mckenzie and Ashgar, A. G. 1937. Soil Deterioration in the Canal Irrigated Areas of Punjab. III. Formation and Characteristics of the Soil Profiles in the Alkaline Alluvium of the Punjab. Research Publication, Punjab Irrigation Research Institute. 4: 7 p.

Puri, G.S. 1960. Indian Forest Ecology. Oxford Book and Stationary Co., New Delhi.

Qadir, M., Oster, J.D., Schubert, S., Noble, A.D. and Sahrawat, K.L. 20107. Phytoremediation of sodic and saline-sodic soils. Advances in Agronomy. 96: 197–247.

Qadir, M. and Oster, J.D. 2004. Crop and irrigation management strategies for saline-sodic soils and waters aimed at environmentally sustainable agriculture. Science of the Total Environment. 323: 1-19

Quershi, R.H. and Barret-Lennard, E.G. 1998. Saline Agriculture for Irrigated Lands in Pakistan: A Handbook. Australian Centre for International Agricultural Research, Canberra, Australia. Monograph 50. 142 p.

Ranade, D.H. and Gupta, R.K. 1987. Salt effect on hydraulic conductivity of a Vertisol. Catena. 14: 45-54.

Raychaudhury, S. P. and Shankaran, A. 1952. Studies on the saline soils of Delhi state. I. Some typical profiles of the Jamuna *khadar* area (north). Indian J. Agricultural Science. 22: 209-222.

Raychaudhury, S. P. and Sinha, N. S. 1957. Studies on the saline and alkaline soils of Delhi state villages. IV. Indian J. Agricultural Science. 27: 347-361.

Redly M., Darab K. and Csillag J. 1980. Solubility and ameliorative effect of gypsum in sodic soils. International Symposium Salt Affected Soils, Karnal. 313-321.

Rengasamy, P. 2006. World salinization with emphasis on Australia. Journal of Experimental Botany. 57: 1017-1023.

Rengasamy, P. 2010. Soil processes affecting crop production in salt-affected soils. Functional Plant Biology. 37: 613–620

Rengasamy, P. and Olsson, K.A. 1991. Sodicity and soil structure. Australian Journal of Soil Research. 29: 935–952.

Rhoades, J.D. 1968. Mineral weathering correction for estimating sodium hazard of irrigation waters. Proceedings Soil Science Society America 32: 648-652.

Rhoades, J.D. 1977. Potential for using saline drainage waters for irrigation. Proc. Water Management for Irrigation and Drainage. American Society of Civil Engineering, Reno, Nevada. 85-116.

Rhoades, J.D. and Maas E.V. 1974. Response of crops to sodicity and calcium concentration. Ann. Rep. U.S. Salinity Laboratory, California.

Richards, L.A. (Ed.). 1954. Diagnosis and Improvement of Saline and Alkali Soils. USDA Handbook 60. U. S. Salinity Laboratory, Riverside, USA. 160 p.

Sakai, Y., Ma, Y., Xu Chi, Wu, H., Zhu, W. and Yang, J. 2012. Phytodesalination of a salt-affected soil with four halophytes in China. Journal of Arid Land Studies. 22: 17 -20.

Sandhu, S. S., Acharya, C. L. and Abrol, I. P. 1980. The effect of exchangeable sodium on the water retention, transmission and modulus of rupture of a soil. Proceedings International Symposium on Salt Affected Soils, Karnal. 185-194.

Satyanarayana, K.V.S. 1961. Studies on the soils of Rann of Kutch I: Description of some profiles and distribution of salts and gypsum in the profiles. Bull. Indian Society Soil Science. 6: 125-133.

Satyanarayana, M. 1953. Crop and land improvement under the Krishna East Bank Canal (Madras state). Madras Agricultural J. 40:293-294.

Saxena, P.N. 1960. Algal flora of *usar* lands in Uttar Pradesh. Science Culture. 25:542-543.

Schofield, R.V and Kirkby, M.J. 2003. Application of salinization indicators and initial development of potential global soil salinization scenario under climatic change. Global Biogeochemical Cycles. 17: 4.1- 4.13.

Schoonover, W.R. 1952. Examination of soils for alkali. Univ. of California. Extension Service, Berkley, California.

Sehgal, J. and Abrol, I.P. 1994a: Soil degradation in India—status and impact. Transactions. 15th International Society of Soil Science. International Society of Soil Sciences, Acapulco, Mexico, Vol. 7a: 212 p.

Sehgal, J.L. and Stoop, G. 1972. Pedogenic calcite accumulation in arid and semi-arid regions of the Indo-Gangetic plains of Punjab. Geoderma. 8: 59.

Sekhon, G.S. and Bajwa, M.S. 1993. Effect of incorporation of materials and gypsum in controlling sodic irrigation effects on soil properties under rice – wheat maize system. Agricultural Water Management. 24: 15 25.

Shah R.K. 1975. A note on quick and cheap reclamation method for saline-sodic soils using dairy waste sulphuric acid. Agricultural Research Communication Centres, Newsletter. 33: 4-6.

Shah, R. K. and Trivedi, A. M. 1954. An investigation of the alkali soils in Gujarat, preliminary report. J. Soil Water Conservation. 3: 28-31.

Shah, R. K., Vora, J. C. and Trivedi, A. M. 1958. Possibilities of reclamation of alkali soils of the Little Rann of Kutch. Indian J. Applied Chemistry 30: 96-99.

Sharma, D.P. 1986. Effect of gypsum application on long-term changes in soil properties and crop growth in sodic soils under field conditions. J. Agronomy Crop Science. 156: 166 – 172.

Sharma, D.P. and Mehta, K.K. 1981. Effect of different cropping patterns on alkali soil properties during reclamation on farmers' fields. Annual Report, 1981, CSSRI, Karnal. 155-156.

Sharma, D.P., Mehta, K.K. and Yadav, J.S.P. 1981. Effect of reclamation practices on soil properties and crop growth on farmer's fields – a case study. J Indian Society Soil Science 29 : 356-60.

Sharma, D.K., Kumar, Ashok and Singh, K.N. 1990. Effect of irrigation scheduling on growth yield and evapotranspiration of wheat in sodic soils. Agricultural Water Management. 18: 267-276.

Sharma, D.K. and Singh, K.N. 1992. Effect of irrigation on growth, yield and evapotranspiration of mustard (*Brassica juncea*) in partially reclaimed sodic soils. Agricultural Water Management. 23: 225-232.

Sharma, D.K., Singh, K.N. and Chillar, R.K. 1985. Effect of time of first irrigation on yield of wheat in sodic soil. Indian Journal Agronomy. 30 : 459-463.

Sharma, D.K., Thimmppa K., Chinchmalatpure, A.R., Mandal, A.K., Yadav, R.K., Chaudhary, S.K., Kumar, S. and Sikka, A.K. 2015. Assessment of Production and Monetary Losses from Salt-affected Soils in India. Tech. Bull. ICAR0CSSRI/Karnal/2015/5. Central Soil Salinity Research Institute, Karnal. 97 p.

Sharma, O. P. and Gupta, R. K. 1986. Comparative performance of gypsum and pyrites in sodic Vertisols. Indian J. Agricultural Science. 56: 423-429.

Sharma, P. and Swarup, A. 1990. Microbial oxidation of pyrites in relation to its efficiency for reclamation of sodic soils. Ann. Rep. 1990-91. Central Soil Salinity Research Institute, Karnal.

Sharma, P.K., Sehgal, J.L. and Sharma, K.R. 1976. Mapping salt-affected soils in Sangrur district (Punjab) using aerial photo-interpretation technique. Journal of the Indian Society of Photo-Interpretation. 4: 43-54.

Sharma, R.C., Abrol, I.P. and Bhumbla, D.R. 1975a. Quality of Groundwaters of Kaithal Tehsil, Haryana. Report No. 3, Central Soil Salinity Research Institute, Karnal. 63p.

Sharma, R.C., Abrol, I.P. and Bhumbla, D.R. 1975b. Quality of Groundwaters of Jind District, Haryana. Bull. No. 3, CSSRI, Karnal. 63p.

Sharma, R. C. and Bhargava, G. P. 1978. Soil survey of the operational research project area for the reclamation of alkali soils. CSSRI Rep 8. 72 p.

Sharma, R.C. and Bhargava, G.P. 1981. Morphogenic changes in an alkali soil pedon during amelioration through gypsum application. J. Indian Society Soil Science. 29: 274-75.

Singh, D. and Nijhawan, S. D. 1932. A study of the physico-chemical changes accompanying the process of reclamation in alkali soils. Indian J. Agricultural Science. 2:1-18.

Singh, G. B. 1987. Agronomic investigations on production of mesquite in highly alkali soils. Annual Report, CSSRI, Karnal. 120-128.

Singh, G. B. and Abrol, I. P. 1986. Agronomic investigations on production of mesquite in highly alkali soil condition. Annual Report, CSSRI, Karnal. 98-107.

Singh, G. B., Abrol, I. P. and Cheema, S. S. 1988. Agroforestry in alkali soils: Effect of planting methods and amendments on initial growth, biomass accumulation and chemical composition of mesquite with inter-space planted with and without Karnal grass. Agroforestry Systems. 7: 135-160.

Singh, G. B. and Dagar, J. C. 1998. Agroforestry in salt- affected soils. In. Agricultural Salinity Management in India (Tyagi, N. K. and Minhas, P. S. Eds.). Central Soil Salinity Research Institute, Karnal. 473-487.

Singh, G. B. and Dagar, J. C. 2005. Greening Sodic Lands: Bichhian Model. Tech. Bull. 2/2005. CSSRI, Karnal. 51 p.

Singh, G. B., Dagar, J. C. and Singh, N.T. 1997. Growing fruit trees in a highly sodic soil- a case study. Land Degradation and Development. 8: 257-268.

Singh, G.B. and Sharma, D.P. 1984. Nitrogen management for wheat in sodic soils. Indian Farming. 34: 37-39.

Singh, G. B. and Singh, N. T. 1993. Mesquite for the Revegetation of Salt Lands. Bull. 18. CSSRI, Karnal. 24 p.

Singh, G. B., Singh, N. T. and Abrol, I. P. 1994. Agro-forestry techniques for the rehabilitation of salt land. Land Degradation and Rehabilitation. 5: 223-242.

Singh, G. B., Singh, N. T. and Tomar, O. S. 1993. Agro-forestry in Salt-affected Soils. Bull. 17. CSSRI, Karnal. 65 p.

Singh, G. B., Singh, N. T., Singh, H., Dagar, J. C. and Sharma, V. P. 1997. Evaluation of agriculture, forestry and agro-forestry practices in a moderately alkali soil in north-western India. Agro-forestry Systems. 37: 279-295.

Singh, I. S. 1996. Agro-forestry for Salt Affected Lands. Bulletin, N. D. University of Agriculture and Technology, Faizabad, UP. 23 p.

Singh, I. S., Singh, R. K. and Khanna, S. S. 1994. Management of Sodic Soils Through Planation. Bulletin, N. D. University of Agriculture and Technology, Faizabad, UP. 51p.

Singh, K., Yadav, J.S.P. and Singh, V. 1992. Tolerance of trees to soil sodicity. Journal of Indian Society of Soil Science. 40: 173-179.

Singh, M. and Garg, M. 2002. Production of beneficiated phopho-gypsum for cement manufacture. J. Scientific and Industrial Research. 61: 533-537.

Singh, M. and Garg, M. 2005. Study on anhydrite plaster from waste phospho-gypsum for use in polymerized flooring composition. Construction and Building Materials. 9: 25–29.

Singh, M., Poonia, S.R. and Pal, R. 1986. Improvement of irrigation water by gypsum beds. Agricultural Water Management. 11: 293-301.

Singh, M.V. and Abrol, I.P. 1986. Split application of Zn better than in a single dose. Indian Farming. 36:35-37.

Singh, M.V., Chhabra, R. and Abrol, I.P. 1982. Effect of zinc levels on the growth and yield of rice and wheat in a sodic soil. CSSRI Annual Report, 1982.

Singh, N. T. 1974. Physico-chemical changes in sodic soils incubated at saturation. Plant Soil. 40: 303-311.

Singh, N.T. 1992. Salt affected soils in India. In: Land and Soils (Khoshoo, T.N. and Deekshatulu, B.L. Eds.) Proceedings Indian National Science Academy, New Delhi. 65-100.

Singh, N.T. 1998. Reclamation and management of alkali soils. In: Agricultural Salinity Management in India (Tyagi and Minhas Eds.). CSSRI, Karnal. 261-278.

Singh, N.T., Hira, G.S. and Bajwa, M.S. 1981. Use of amendments in reclamation of sodic soils in India. Agro. Es. Talaj. 30: 158-177.

Singh N.T., Hira G.S. and Singh R. 1978. Comparative effect of pyrites and gypsum on physico-chemical properties of a sodic soil. Proceedings Seminar Use of Sedimentary Pyrites in Reclamation of Sodic soils, Lucknow. 69-80.

Singh, O.P. 1980. Analysis of the hydrologic changes taking place in alkali soil watersheds during reclamation. Ph.D. Thesis, JNTU, Hyderabad.

Singh, O.P., Narayana, V.V.D. and Tiwari, A.K. 1979. Analysis of the hydrologic changes taking place in alkali soil watersheds during reclamation. Canadian Agricultural Engineering. 21: 147-150.

Singh, P. 1980. An appraisal of the economics of reclamation of alkali soils. Proceedings International Symposium Salt-Affectd Soils, Karnal. 555-561.

Singh, P. and Gupta, J. N. 1979. Income and employment generation in reclamation of alkali soils on marginal farms. Current Agriculture. 3: 109-114.

Singh, R.N. 1950. Reclamation of *usar* lands in India through blue green algae. Nature (London) 165: 325-326.

Singh, R., Sharma, D.K., Joshi, P.K., Tripathi, R.S. and Kumar, S. 2015. Strategies of resource conservation and mini-sprinkler on crop productivity under rice-wheat cropping system in semi-reclaimed sodic soil. Annual Report 2014-15. CSSRI, Karnal. 25-27.

Singh, S. and Singh, B. 1966. Trace element studies on some alkali and adjoining soils of Uttar Pradesh II- Profile distribution of cobalt. J. Indian Society, Soil Science. 14: 177-181.

Singh, S.B., Chhabra, R. and Abrol, I.P. 1979. Effect of exchangeable sodium on the yield and chemical composition of *Raya* (*Brassica juncea*). Agronomy Journal. 71: 767-770.

Singh, S.B., Chhabra, R. and Abrol. I.P. 1980. Effect of soil sodicity on the yield and chemical composition of cowpea (*Vigna sinensis* L.) grown for fodder. Indian J. Agricultural Science. 50: 852-856.

Singh, Y., Khind, C.S. and Singh, V. 1991. Efficient management of leguminous green manures in wetland rice. Advances in Agronomy. 45: 135-139.

Singh,Y.P., Sharma, D.K., Singh, Gurbachan, Nayak, A.K., Mishra, V.K. and Singh, R. 2008. Alternate Land Use Management for Sodic Soils. Tech. Bull. No. 2/2008. CSSRI, RRS, Lukhnow. 16 p.

Singhla, S.K., Bhargava, G.P. and Gupta, I.C. 1974. Relationships between certain diagnostic indices of salt affected soils of Indo-Gangetic alluvial region. Ann. Arid Zone. 13: 218-23.

Slade, P.G., and Quirk, J.P. 1991. The limited crystalline swelling of smectite in $CaCl_2$, $MgCl_2$ and $LaCl_3$ solutions. 1. Colloid Science 144: 18-26.

Smeck, N.E., Runge, E.C.A. and Mackintosh, E.E. 1983. Dynamics and genetic modelling of soil system. in: Pedogenesis and Soil Taxonomy. 1. Concepts and Interactions (Wilding *et al.*, Eds.). Elsevier, New York. 51-81.

Smedema, L.K. and Rycroft, D.W. 1983. Land Drainage: Planning and Design of Agricultural Drainage Systems. Cornell University Press, Ithaca, New York. 375 p.

Srinivasamurthy, C.A., Bhaskar, S. and Balakrishna, A.N. 2009. Effect of distillery spent wash and bio-compost on soil properties, growth and yield of crops. Platinum Jubilee symposium of ISSS "Soil Science in Meeting the Challenges to Food Security and Environmental Quality (Abstracts)", 22-25 Dec. 2009, New Delhi. 298.

Srivastava, P., Bhattacharyya, T. and Pal, D.K. 2002. Significance of the formation of calcium carbonate minerals in the pedogenesis and management of cracking clay soils (Vertisols) of India. Clays and Clay Minerals. 50: 111-126.

Soil Classification Working Group. 1998. The Canadian system of soil classification. Research Branch, Agriculture and Agri- Food Canada, NRC Press, Ottawa, ON.

Soil Science Society of America 1974. Glossary of Soil Science Terms. Soil Science Society of America Madison. USA.

Soil Survey Staff. 1975. Soil Taxonomy. USDA Handbook No. 436. Washington, DC.

Soil Survey Staff. 1998. Keys to Soil Taxonomy. USDA, Natural Resource Conservation Service, Washington, D C.

Srivastava, P., Bhattacharyya, T. and Pal, D.K. 2002. Significance of the formation of calcium carbonate minerals in the pedogenesis and management of cracking clay soils (Vertisols) of India. Clays and Clay Minerals. 50: 111–126.

Suarez, D.L. 1981. Relationship between pH_c and SAR and an alternate method for estimating SAR of soil or drainage water. Soil Science Society America J. 45:469-475.

Sukhmal Chand, Singh, D., Arun Prasad and Patra, D.D. 2005. Influence of soil amendments on nutrient concentration, chemical properties of soil and productivity of German chamomile (*Chamomilla recutita*) in sodic soil. Journal of Medicinal and Aromatic Plant Science. 27: 740-745.

Sumner, M.E. 1990. Gypsum as an Amendment for the Subsoil Acidity Syndrome. Florida Institute of Phosphate Research, Final Report, Project 83–01–024R.

Sunil Kumar, Agrawal R.K., Dixit, A.K., Rai, A. K, Singh, J.B. and Rai, S.K. 2012. Forage Production Technology for Arable Lands. Bulletin Indian Grassland and Fodder Research Institute, Jhansi-284003. 50 p.

Swarup, A. 1980. Effect of submergence and farmyard manure application on the yield and nutrition of rice and sodic soil reclamation. J. Indian Society Soil Science. 28: 532-534.

Swarup, A. 1981. Effect of flooding on physico-chemical changes in sodic soils. Z. Pflanzen Bodenk. 144 : 136-42.

Swarup, A. 1988. Effect of *Sesbania bispinosa* decomposition and sodicity on yield of rice and N contribution. International Rice Research Institute Notes. 13: 28-29.

Swarup, A. 2000. Nutrient management for crop production in alkali lands. In. Salinity Management in Agriculture (Gupta *et al.* Eds.). CSSRI, Karnal. 99-108.

Swarup, A., and Yaduvanshi, N.P.S. 2004. Response of rice and wheat to organic and inorganic fertilizers and soil amendment under sodic water irrigated conditions. IRRI Newsletter. 29.1: 49-51.

Szabolcs, I. 1974. Salt Affected Soils in Europe. Martinus Nijhoff, The Hague. 63 p.

Szabolcs, I. 1979. Review of research on salt affected soils. Natural Resources Research XV, Unesco, Paris.

Szabolcs I.1989a. Salt-affected Soils. Boca Raton, Fl: CRC Press.

Szabolcs, I. 1989b. Saline and alkali soil-commonalities and differences. In: International Symp. Salt Affected Soils, Karnal, India. Feb, 20-23, 1980. 1-6.

Szabolcs, I. 1998. Salt build-up as a factor of soil degradation. In: Methods for Assessment of Soil Degradation (Lal *et al.* Eds.). CRC Press, Boca Raton, FL, USA. 253-264.

Talati, R. P. 1941. Damaged lands in the Daccan and their classification. Indian J. Agricultural Science. 11: 959-977.

Talati, R. P. 1947. Field experiments on the reclamation of salt land in Baramati of Bombay Deccan. Indian J. Agricultural Science. 17: 153-174.

Taylor, E. M., Malhotra, J. K. and Mehta, M. L. 1933. An Investigation of the Rise of Water Table in the Upper Chenab Canal Area, Punjab. Res. Pub. Punjab Irrigation Res. Inst. 1, 39 p.

Taylor, E. M. and Mehta, M. L. 1941. Some irrigation problems in the Punjab. Indian J. Agricultural Science. 11: 137-169.

Taylor, E. M., Puri, A. N. and Asghar, A. G. 1934. Soil deterioration in the canal-irrigated area of the Punjab. I. Equilibrium between Ca and Na ions in base exchange reactions. Indian J. Agricultural Science. 4: 8.

Thimmppa, K., Singh, Y.P., Raju, R., Kumar, S., Tripathi, R.S., Pal, G. and Reddy, A.A. 2015. Reducing farm income losses through land reclamation: a case study from Indo-Gangetic plains. Journal of Soil Salinity and Water Quality. 7: 68-76.

Tiwari, A.K., Pandey, R.N. and Gupta, S.K. 1981. Transient behaviour of infiltration rate in alkali lands under reclamation. J. Agricultural Enggineering. 18 : 21-8.

Tiwari, K. N., Sharma, D. N. and Tripathi, S. K. 1989. Salt-Affected Soils of U.P.: Their Reclamation and Management. C.S. Azad University of Agriculture and Technology, Kanpur.

Tripathi, R.S. 2007. Economic analysis of subsurface drainage projects. In: Agricultural Land Drainage: Reclamation of Waterlogged Saline Lands (Eds.: Ambast, Gupta, Gurbachan (Eds.). Central Soil Salinity Research Institute, Karnal, India. 222-231.

Tripathi, R.S. 2013. Return to investment made for alkali land reclamation technology in India. J. Soil Salinity and Water Quality. 5: 65-72.

Tripathi, R.S. and Sharma, D.K. 2011. *Kshariya Bhumi Sudhar evam uske samajik-aarthik Labh*. Tech. Bull. CSSRI/Karnal/2009/02. Central Soil Salinity Research Institute, Karnal. 12 p.

Tyagi, N.K. 1978. Some desgn aspects of ponded borders with cut back stream in alkali soils. Current Agriculture. 2: 13-18.

Tyagi, N.K., Gupta, S.K. and Datta, K.K. 1996. Rural Land Development through Alkali Land Reclamation: The Gurusikaran Model. CSSRI, Karnal and AFPRO, New Delhi. 33p.

Tyagi, N.K. and Singh, O.P. 1979. Investigation on the effect of soil moisture on performance of land levelling implements. J. Agricultural Enggineering. 16: 67-72.

US Water Research Council. 1983. Economic and Environmental Principles and Guidelines for Water and Related Land Resources Implementation Studies. US Water Research Council, Washington. 137 p.

Vadjunina A.F. and El-Swaby El-kouni 1975. Chemical and electrical methods for reclaiming sodic-saline soils. International Symposium New Developments in the Field of Salt Affected Soils. Cairo. 505-512.

van Beekom, C.W.C., van den Berg, C., de Boer, Th.A., van der Molen, W.H., Verhoeven, B., Westerhof, J.J. and Zuur, A.J., 1953: Reclaiming land flooded with salt water. Netherlands Journal of Agricultural Science. 1: 153-163.

Varsha Rani. 2007. Effect of Sodic Water on Nutritional and Sensory Characteristics of Vegetables and Their Products. Ph.D. Thesis, Department of Food and Nutrition, CCS Haryana Agricultural University, Hisar, Haryana.

Venkat Krishnan S. 1980. Mineralization of green manure (*Sesbania aculeata*) nitrogen in sodic and reclaimed soils under flooded conditions. Plant Soil. 54: 149-152.

Verma, S.K., Tomar, S.S., Khandkar, U.R., Tiwari, S.C., Bangar, K.S. and Sharma, R.K. 2006. Black Alkali Soils: Survey, Characterization and Management. JNKVV, College of Agriculture, Indore. 132 p.

Voelcker, J. A. 1897. Report on the Improvement of Indian Agriculture.

Wadia, D.N. 1979. Geology of India. English Language Book Soc. MacMillan & Co. Ltd. London, 536 p.

Wang, M., Liang, Z., Yang, F., Ma, H., Huang, L and Liu, M. Effects of number of seedlings per hill on rice biomass partitioning and yield in a saline-sodic soil. 2010. Journal of Food, Agriculture & Environment. 8: 628-633.

Whitcombe, E. 1971. Agrarian Conditions in Northern India. Vol. 1. United Provinces Under British Rule. 1860-1900. University of California Press, California-USA.

Whitcombe, E. 1972. Effect of British raj on irrigation and agriculture in India. Proceedings Symposium Soil and Water Management, Hisar. 1969. 313.

Whitting, L.D. and Janitzky, P. 1963. Mechanism of formation of sodium carbonate in soil. J. Soil Science. 14: 322-333.

Wilcox, L. V. 1955. Classification and Use of Irrigation Waters, USDA, circular 969, Washington DC, USA.

Wilcox, L.V. 1948. The Quality of Water for Irrigation Use. US Department of Agriculture Technology, Bulletin 962, Washington DC.

Working Group 1991. Problem Identification in Irrigated Areas with Suggested Remedial Measures, 1991. Ministry of Water Resources, Government of India, New Delhi.

Yaalon, D.H. 1965. Downward movement and distribution of anions in soil profiles with limited wetting. In: Experimental Pedology (Hallsworthy and Crawford, Eds.). Butterworths, London.

Yadav, J.S.P. 1973. Annual Report. All India Coordinated Project on Water Management and Soil Salinity. Central Soil Salinity Research Institute, Karnal, India.

Yadav, J. S. P. 1978. Studies on reclamation and utilization of salt-affected soils at Central Soil Salinity Research Institute. Fertilizer News. 23: 14-21.

Yadav, J. S. P. 1980. Salt-affected soils and their afforestation. Indian Forester. 106: 259-272.

Yadav, J.S.P., Bhumbla, D.R. and Sharma, O.P. 1972. Performance of certain forest species on a saline sodic soil. Proceedings Symposium New Development in the Field of Salt-Affected Soils. Cairo. 683-90.

Yadav, J.S.P. and Girdhar, I.K. 1980. Effect of varying Mg/Ca ratios and electrolyte concentrations in the irrigation water on the soil properties and wheat crop. Plant Soil. 56 : 413-428.

Yadav, J.S.P. and Girdhar, I.K. 1981. The effect of different magnesium: calcium ratios and sodium adsorption ratio values of leaching water on the properties of calcareous versus non-calcareous soils. Soil Science. 131: 144-158.

Yaduvanshi, N.P.S. 2001. Ammonia volatilization losses from integrated nutrient management in rice fields of alkali soils. J. Indian Society Soil Science 49, 276-280.

Yaduvanshi, N.P.S. 2001. Effect of five years of rice-wheat cropping and NPK fertilizer use with and without organic and green manures on soil properties and crop yields in a reclamation sodic soil. J. Indian Society Soil Science 49: 714-719.

Yaduvanshi, N.P.S. 2003. Substitution of inorganic fertilizers by organic manures and the effect on soil fertility in a rice-wheat rotation on reclaimed sodic soil in India. Journal of Agriculture Science (Cambridge). 140: 161-169.

Yaduvanshi, N.P.S. and Sharma, D.R. 2008. Tillage and residual organic manures/chemical amendments: Effect on soil organic matter and yield of wheat under sodic water irrigation. Soil and Tillage Research. 98: 11-16.

Yaduvanshi, N. P. S. and Yadav, D.V. 1990. Effect of sulphitation press mud and nitrogen fertilizer on biomass and nitrogen economy, plant composition in sugarcane and on soil chemical properties. J. Agricultural Science. 114: 259- 263.

Zhang, J.Y., Wang, Z.Y., Jenks, M.A., Hasegawa, P.M. and Jain, S.M. 2007. Recent advances in molecular breeding of forage crops for improved drought and salt stress tolerance. In: Advances in Molecular Breeding towards Salinity and Drought Tolerance. Springer. 797-817.

SUBJECT INDEX

S

T